Advance Praise for
WASTE WARS

"*Waste Wars* is perhaps the most comprehensive indictment of consumer capitalism since Rachel Carson's *Silent Spring*. Fearless as he travels to some of the least appealing places on earth, Alexander Clapp lifts the heavy stones of greenwashing to reveal the literal and moral filth that Western societies have been dumping on their poorer cousins for decades. Always engagingly written, with jaw-dropping anthropological detail, this book introduces us to courageous tragic characters compelled to clean up the mess of Western material avarice, from the bizarre electronic slums of Ghana to the death yards breaking up ships in Turkey, Pakistan, India, and Bangladesh. If you wish to know how the world really works, read this book."

— Misha Glenny, author of *McMafia*

"In the seconds it has taken you to scan the few lines of this blurb, tens of thousands of plastic bottles have been discarded. Pause to consider that awesome fact for a moment, and those bottles are followed by tens of thousands more. One million per minute, every minute, every hour of every day. We are burying our planet in trash that, thanks to the plastic revolution, will outlive us by thousands if not hundreds of thousands of years. Where does all this waste go? The fact that this shocking disaster is largely invisible to the rich world is no accident. As Clapp shows, it is the result of a ghastly form of globalization that dumps the garbage of the rich on the poor. A mind-altering and unforgettable read, *Waste Wars* is an essential and deeply disturbing book."

— Adam Tooze, author of *Crashed*

"Superb reporting that definitively answers the question we really never ask: Where on earth does all that stuff go when we're done with it? This majestic account will transform the way you look at trash — and hopefully it will spur some real change at the highest levels."

— Bill McKibben, author of *The End of Nature*

"An infuriating, eye-opening, and spellbinding account of the globally uneven and unjust politics of trash. Clapp shows how the rubbish the affluent people of rich countries produce travels to poorer countries for processing, creating mountains of toxic waste in the Global South or whirlpools of plastic in our oceans. A must-read for those concerned with the health and hygiene not only of the planet but also of the people who populate it!"
— Laleh Khalili, author of *Sinews of War and Trade*

"*Waste Wars* cracks open standard recycling rhetoric to expose the toxic truths within. No study of global inequality is complete without the information in this excellent book."
— Malcolm Harris, author of *Palo Alto*

"Since its inception centuries ago, the global economy has consisted of complex commodity chains: extracting, transporting, manufacturing, distributing, profiting. Only very recently in historical times have we added the 'downstream' chains, dumping the resultant consumer waste—somewhere, somehow. In rich countries, we pay high municipal fees for the removal of our refuse. We may even feel good about sorting to recycle it. Does this do any good? If the news is really so bad, better that someone as sober and courageous as Clapp delivers it to us."
—Georgi M. Derluguian, author of *Bourdieu's Secret Admirer in the Caucasus*

"*Waste Wars* is the *Star Wars* of trash, a witty and brave account of Clapp's journey into the underbelly of modern life. You'll meet garbage-spotting drones and journalists who register pet fish as waste brokers, and go on a hunt for the El Dorado of poison. As Clapp explains, we live in a world where our ability to create garbage has surpassed Earth's ability to generate life. The consequences are terrifying, but this great book somehow leaves you awe-inspired by the sheer outrageousness of the human ingenuity that has created this toxic mess."
— Jeff Goodell, author of *The Heat Will Kill You First*

WASTE WARS

The WILD AFTERLIFE of YOUR TRASH

ALEXANDER CLAPP

LITTLE, BROWN AND COMPANY

NEW YORK BOSTON LONDON

Copyright © 2025 by Alexander Clapp

Hachette Book Group supports the right to free expression and the value of copyright. The purpose of copyright is to encourage writers and artists to produce the creative works that enrich our culture.

The scanning, uploading, and distribution of this book without permission is a theft of the author's intellectual property. If you would like permission to use material from the book (other than for review purposes), please contact permissions@hbgusa.com. Thank you for your support of the author's rights.

Little, Brown and Company
Hachette Book Group
1290 Avenue of the Americas, New York, NY 10104
littlebrown.com

First Edition: February 2025

Little, Brown and Company is a division of Hachette Book Group, Inc. The Little, Brown name and logo are trademarks of Hachette Book Group, Inc.

The publisher is not responsible for websites (or their content) that are not owned by the publisher.

The Hachette Speakers Bureau provides a wide range of authors for speaking events. To find out more, go to hachettespeakersbureau.com or email hachettespeakers@hbgusa.com.

Little, Brown and Company books may be purchased in bulk for business, educational, or promotional use. For information, please contact your local bookseller or the Hachette Book Group Special Markets Department at special.markets@hbgusa.com.

Maps by Le Cartographiste. Map data from OpenStreetMap, openstreetmap.org/copyright.

Excerpts: p. 29 from *Selected Poems of Pablo Neruda*, copyright © Pablo Neruda and Fundación Pablo Neruda. Translation copyright © 1961 by Ben Belitt. Used by permission of Grove/Atlantic, Inc. Any third-party use of this material outside of this publication is prohibited; p. 83 from "Toxiques." Words and music by Youssou N'Dour and Habib Faye. Copyright © 1990 BMG Rights Management (France) SARL. All rights administered by BMG Rights Management (US) LLC. All rights reserved. Used by permission. Reprinted by permission of Hal Leonard LLC; p. 198 from Nikos Kavadias, "Mal du départ," inheritor of rights of Nikos Kavadias.

Book interior design by Marie Mundaca

ISBN 9780316459020
LCCN 2024941719

Printing 1, 2024

LSC-C

Printed in the United States of America

For my dad

CONTENTS

Introduction: Mayhem in Mesopotamia 3

Part One: Toxic Tropics

   1. Banana Republic 29
   2. The Chemical Century 36
   3. Cash for Trash 42
   4. Debt and Development 47
   5. Merchants of Disease 58
   6. Guns and Germs 64
   7. Trash Ash Odyssey 76
   8. Rising Up 83
   9. American Exceptionalism 91
  10. The Waste Trade Strikes Back 94

Part Two: E-Waste on the Odaw

 11. State and Slum 103
 12. To the Quays of Tema 112
 13. Treasure 116
 14. Logging On 125
 15. Technological Tinkering 137
 16. The Flexible Mine 143
 17. Start-Up Cesspools 150

CONTENTS

    18. A New Agbogbloshie?...154
    19. Going Fishing...159
    20. Magical Things...167

Part Three: Aegean Abomination
    21. Global Junk Heap...177
    22. Shipping Out..183
    23. Into the Heart of Anatolia.....................................190
    24. Deadly Business...198
    25. Scrap Shepherds..207
    26. Scrap Nation..216
    27. At Europe's Edge...226
    28. Greeks Bearing Gifts...237
    29. Coming Home..241

Part Four: Pacific Plastic
    30. A Long Journey...245
    31. Plastification..249
    32. The Greatest Miracle Yet.....................................261
    33. One-Man Multinational......................................268
    34. Plastic China..274
    35. Mad Scramble...283
    36. A Trash Chief..291
    37. A Trash Scion...313
    38. Back to the Pacific...322

Conclusion: Whither Waste?..*327*
Acknowledgments.. *341*
Notes.. *343*
Index..*375*

WASTE WARS

INTRODUCTION

Mayhem in Mesopotamia

A Strange Arrival

>He who possesses many things is constantly on guard.
>— ancient Sumerian proverb[1]

ON A CHILLY evening in late 2016, a few miles from the Turkish city of Adana, a Kurdish farmer named İzzettin Akman was sitting on the second-floor balcony of his concrete ranch house when a white construction truck backed up to the edge of his citrus groves, paused, then dumped a great load of trash along the roadside. Before he pulled away, the truck's driver set a paper bag on fire and tossed it atop the garbage, triggering an outpouring of flames blacker than the night sky into which they ascended. *"Ji dil?"* Akman leapt up, put on his sandals, and sprinted out along his dirt driveway toward the crackling trash pile fuming several hundred feet in the distance. "Seriously?"

The trash, by the time İzzettin Akman got to it, was a hissing mass of fire; plastic is less flammable than wood or paper but burns more intensely, with a higher heat of combustion, and is at least as capable of

INTRODUCTION

getting swept up in a gust of wind and setting some fifty acres and six-thousand-odd orange and lemon trees alight. *"Kurê qahpê!"* Akman wheeled around, ran back home, located a bucket, then rushed back to the conflagration, which he began dousing with water lifted out of a stream running along the edge of the road. "Son of a bitch!"

Akman kept pouring. After about an hour, the flames started to dampen, then die, revealing with their retreat a bed of thousands of half-incinerated fragments of garbage, not unlike the way the topsoil of nearby Adana gets brushed away by archaeologists to reveal underlying tiles of ancient mosaics. Akman knelt down — disaster averted! — to examine the strange, unsolicited new arrival to his farm, turning over slices of candy wrappers and makeup containers with his fingers, before being struck by something peculiar. The writing on the packaging wasn't Kurdish. It wasn't Turkish either. Akman kept clawing through the still-scalding plastic, now looking for price labels to be sure. He found several. They were not in Turkish liras. They were in euros and pounds.

For decades, İzzettin Akman — a slim, middle-aged man with a face peppered by scraggly red-brown stubble — had, like generations of Akmans before him, made his living harvesting oranges and lemons and exporting them to Europe. Now Europe appeared to be sending its trash in the opposite direction, to the very cusp of his groves, where Akman couldn't help but be bemused by the occasional charred carton of juice jutting out of the pile.

"That might have been made with my oranges," he told me as we walked the edge of his farm, where, six years after its unceremonious dumping, the heap of garbage — a lumpy mound of ash-cum-plastic that resembled an unmarked grave coated with filthy confetti — still abutted acres of redolent trees.[2]

A month or so after the trash was dumped beside Akman's property, something more bizarre happened. For the first time in more than thirty years of farming, the leaves of scores of his citrus trees started turning yellow. Then their oranges and lemons began dropping to the

INTRODUCTION

ground. A year later, by which time Akman's losses had brought his family to the brink of serious financial trouble, the trees bore no fruit at all. It turned out that a truckload of garbage set alight along the side of a citrus farm, even if it burns for just an hour, can be the catalyst of much longer-term damage, the environmental equivalent of a delayed-fuse bomb. The smoke that continued to waft from the trash pile after its extinguishing hadn't just streaked across the sky for an evening; it had killed off parts of the bee population that helps citrus trees pollinate. And the innumerable pieces of half-melted plastic that had washed into the creek that provides water for Akman's irrigation system hadn't merely floated away to some distant place downstream; they had broken down into billions of microplastics and contaminants that circulated toward his groves, before eventually getting sucked up into the trees themselves, crowding their roots like particles of fat in human arteries.

İzzettin Akman's farmhouse sits just west of Adana, on the outskirts of the bare agricultural village of Küçükçıldırım, two hours'

drive from the Syrian border, in a lush plain across which snowmelt from the Taurus Mountains to the north trickles out toward the Mediterranean Sea shimmering to the south. It's a stunning landscape; the roads really do smell of fresh oranges, the rocky outcrops are ringed with medieval monasteries and ancient fortresses, and the fertility has been legendary since — literally — as long as anyone can remember. For good reason, it was here, of all possible places on Earth, that humanity likely first made the shift from so many tens of thousands of years of a wandering, hunting existence to a settled, agricultural one. A hundred miles east of Akman's farm sits the mysterious Neolithic mound of Göbekli Tepe, perhaps a sacred monument to that transition. And to the south of it unfurls the riverine lands that once constituted Mesopotamia, where writing was first invented, stars were first mapped, mathematics was first attempted, and civilization itself was first endeavored.

By the time I met Akman, his orange and lemon trees had begun to recover. But the land around Adana had not. The several tons of trash that had been dumped along the edge of his farm, it turned out, was no one-off. It was the vanguard of something larger, more organized, and more insane to come.

In the summer of 2017, Turkey's First Lady emerged on a stage in the capital of Ankara and announced a grand new plan for Akman's nation. Over the next fifteen years, Emine Erdoğan proclaimed, Turkey would be turning itself into a "zero waste" country. Sure, other countries began their pivots to a green future by slashing fuel emissions or constructing wind farms or taxing carbon outputs. But Turkey's transition, Emine Erdoğan explained, would begin elsewhere. It would begin within the homes of eighty-five million Turkish citizens.

Turks would be eliminating their trash.

True enough, their country's recent track record of discarding garbage had been dreadful. Over the previous generation, Turkey had become as addicted to plastic as any other place on Earth. Its network of public fountains — a tradition dating back half a millennium to the

INTRODUCTION

Ottoman sultans who aspired to adorn every community of their domain with marble *sebils*, "kiosks" of free-flowing water — had stood no chance against the unrelenting convenience of a water bottle made of polyethylene terephthalate, or PET, introduced to Turkey in 1984 and which, by the early 2000s, Turks were purchasing in the tens of millions every day. Street bazaars that sold fruits and nuts to shoppers bearing cotton sacks had given way to supermarkets that inserted every conceivable purchase into a low-density polyethylene bag — those plastic bags that are so flimsy you can see through them — which by 2010 Turks were discarding at a thirty-five billion annual clip.[3] More than 90 percent of all of this plastic was ending up in landfills, the countryside, or the sea, a travesty captured in real time in Fatih Akin's *Garbage in the Garden of Eden*, in which the acclaimed Turkish German filmmaker, returning after a long absence to his grandparents' picturesque tea-growing village in the mountains above the Black Sea, chronicles a plan to convert its outskirts into an open-air dumpsite. No one in the village wanted the landfill; the authorities schemed behind their backs and zoned it anyway; the result is the entirely foreseeable problem of plastic sloshing down into town, leading Akin to a grim — albeit self-evident — conclusion: "Trash is the global excrement of our society."[4]

That Turkey, assured First Lady Erdoğan, would soon be just another sour memory in the long history of Anatolian tragedies. Her campaign would effect a "clean Turkey" through a state-sponsored campaign that would "prevent uncontrolled waste" by collecting plastic efficiently and recycling it, resulting in a "livable world for future generations."[5]

A clean Turkey! A livable world! In the years to come, the Zero Waste Project would garner Emine Erdoğan accolades — "Zero Waste Project is not just campaign[,] it is an emotion," gushed one Istanbul daily — and award after award courtesy of global institutions ranging from the United Nations to the World Bank.[6] She would write a book on her initiative, *The World Is Our Common Home*, and read it aloud to Turkish children herded into the garden of Ankara's Presidential

Complex, her husband's 1,150-room palace, whose construction had recently razed an ancient forest. The Zero Waste Project would even be deployed as an instrument of foreign policy, espoused by Turkey's 257 diplomatic missions around the world to underscore its standalone commitment in the environmental badlands of the Middle East to combating the climate crisis. "As members of a religion where waste is forbidden and a civilization that kisses bread on the ground and puts it on their forehead, we have assumed a leading role against this threat," vowed Turkey's minister of foreign affairs, Mevlüt Çavuşoğlu.[7]

Only there was one small problem with Turkey's self-coronation as a "zero waste" nation worthy of such international emulation. No sooner had First Lady Emine Erdoğan announced her initiative than Turkey emerged as one of the biggest recipients—and one of the biggest dumpsites—of plastic waste anywhere on the planet.

Global Waste Mismanagement

> There is a gap between what the citizens know about their waste and what actually happens to their waste.
> — Yeo Bee Yin, former minister of Malaysia, 2018[8]

Just a few months after a truckload of Western garbage was set alight next to İzzettin Akman's citrus trees, and only weeks after Emine Erdoğan pronounced Turkey a "zero waste" nation, the Chinese Communist Party informed the world that it, too, was recalibrating its relationship with trash.

It would no longer be accepting it.

Since the early 1990s, when your discarded plastic Coke bottle first emerged as a major object of global commerce, China had been the recipient of *half* the plastic placed into a recycling bin *anywhere* on Earth. If you're reading this now, consider for a moment that hundreds and hundreds of pounds of trash that you've discarded over the course of your life and probably never thought about again went on to live a strange, hot-potato second existence. Dusty bags of cereal, crumpled soda fountain straws, squished Styrofoam egg cartons — for years all these things you deemed so worthless you were willing to freely dispense with them became the objects of arduous, globe-spanning, carbon-spewing journeys, getting trucked tens, perhaps hundreds, of miles from your house to a nearby materials recovery facility and thereafter to a port, then shipped thousands of miles beyond that to any number of hundreds of Chinese villages that specialized in processing the contents of your recycling bin.

From the United States, much of it was transported aboard cargo container ships that had first crossed the Pacific loaded with cheap consumer goods — dog toys, key chains, selfie sticks, you name it — before returning to China packed with (what else?) the plastic and paper in which those goods had been packaged.

By the early 2000s, America's biggest export to China was the stuff Americans tossed away. At least as much plastic was getting jettisoned out of the European Union, from self-congratulating environmental stewards like Germany, whose state recycling quotas were often reliant on a filthy secret: Much of the plastic that Germans claimed was getting "recycled" was in fact getting shipped to the far side of the world, where its true fate was far from clear.

In 2017, China may have informed the world that it would no longer be accepting its plastic waste.[9] But this hardly stopped rich countries from angling to get it all as far away as possible. Many just located desperate new buyers — or unguarded borders — and continued to insist that it was getting recycled. Within months, Greek garbage started surfacing in Liberia. Italian trash wrecked the beaches of Tunisia. Dutch plastic overwhelmed Thailand. Poland would be forced to charter a special police unit to patrol for waste getting trucked in from Germany, while French cops who had once busied themselves with checking the fenders of cars arriving from neighboring Belgium for heroin became tasked instead with inspecting trunks for bags of garbage.[10] Trash exports from Europe to Africa quadrupled, Malaysia became the world's greatest recipient of US plastic waste, and the Philippines threatened Canada with war for dispatching containers of dirty diapers to the capital of Manila.

And within less than a year of Mrs. Erdoğan's launch of the Zero Waste Project, more than 200,000 tons of plastic waste that would have headed to southeastern China at any point in the previous thirty years made its way instead to... southeastern Turkey.

At its most innocuous, the global waste trade shifts garbage from the world's richest countries to those places that can least afford to handle it. At its most nefarious, the global waste trade is an outright criminal enterprise.

Turkey was to prove a showcase in both. Most of its imported plastic was arriving from the United Kingdom, whose waste brokers — the businesses that function as intermediaries between the (often) publicly funded collection of your trash and the (often) privatized business of what becomes of it — had narrowed in on an egregious incentive for exporting garbage. They received paychecks from a state that, in the wake of Brexit, struggled to find truck drivers and port workers, resulting in surging transport costs and massive delays and mounting piles of refuse.[11] Just when China had stopped taking the world's plastic, the United Kingdom threw up its hands and offloaded the task of waste management onto anyone willing to take a stab at it. In exchange for claiming to have collected one ton of household plastic for "recycling," a British waste broker could receive up to £70. More than 250,000 waste brokers in the United Kingdom would eventually be found to be operating without legal permits, garbage parvenus looking to make quick cash off the UK's desperate attempt to appear like a global paragon of environmentalism — and its even more desperate need to turn its plastic waste into someone else's problem. So absurd was the situation, one journalist would endeavor to register his long-deceased pet fish as a professional waste broker. Within four minutes, Algernon the Goldfish had received his very own license to start transporting British trash.[12]

INTRODUCTION

The best part? No one seemed very concerned about what became of all that waste. Soon, half the plastic garbage the United Kingdom insisted was being "recycled" was being shipped abroad, approximately half of it to Turkey.[13]

And that was just year one. Within three years of Mrs. Erdoğan's announcement of the Zero Waste Project, more than 750,000 tons of old plastic was shipped to Anatolia from across Europe, turning the allegedly wasteless nation of Turkey into what was in fact the single greatest recipient of plastic waste on the planet. The equivalent of one dump truck full of foreign garbage was entering the country every six minutes.

To be fair, some of the plastic waste that got shipped to southeastern Turkey really would be put to use. Its fate, however, was almost never to cycle *back* into its earlier form, not to become a new candy wrapper or a new makeup container but to get turned into shoddy home goods. Through an astonishingly energy-intensive and toxins-unleashing process, Western plastic was cleaned, shredded into flakes, chemically reduced, then converted into polyester, which in recent years had begun to replace world-renowned Turkish cotton as the principal feedstock of the country's garment industry. If it wasn't turned into carpet padding or dish towels, some of the plastic was burned in any number of Turkey's cement factories, providing cheap — or even free — fuel for a construction industry that profited from erecting battalions of drab apartment buildings across Anatolia (not a few of which would crumble to smithereens in February 2023 after a large earthquake cracked the region).

But a lot of the plastic that headed to southeastern Turkey was too worthless or dirty to convert into a bath mat or incinerate as fuel. Its fate would be that of the garbage İzzettin Akman observed getting set alight on the edge of his farm: to get covertly dumped somewhere in the countryside and spend the next tens of thousands of years breaking down into millions of minuscule plastic pieces that would enter the sea, devastate croplands, and sprinkle hillsides.

INTRODUCTION

Beginning in 2021, activists and journalists around Europe struck upon the idea of inserting GPS chips into empty bottles of laundry detergent or dishwasher soap, depositing them in local recycling bins, then tracking their movements thousands of miles to the east, to the most distant edge of Turkey, occasionally via wild odysseys that beggared belief in the dizzying amount of effort expended on moving material of such — apparently negligible — value. In one instance, journalists observed as a plastic bag dropped off at a storefront recycling bin outside a London franchise of Tesco, a British supermarket chain that liked to publicize its commitment to sustainability, got routed eighty miles from London to the port town of Harwich, from there to the Netherlands by ship, then to Poland by truck, before finally getting sent two thousand miles south to the outskirts of Adana, where it was found in an industrial yard layered with European garbage.[14]

A single used plastic bag! Transported three thousand miles by ship and truck! A little flickering light cast into the wilderness of globalized waste trafficking that had turned the Kurdish lands of Turkey into its latest — and perhaps most unsuspecting — victim. Suffice it to say, by 2022 so much foreign trash was getting dumped under cover of night around Adana, across valleys or along rivers or, indeed, on the edge of farms, the only way for local environmentalists to track its arrival was to monitor the region from several thousand feet in the air, with drones.

"About once a month we find a big new pile of garbage," Sedat Gündoğdu, a marine biologist at Adana's Çukurova University, told me.[15]

I said goodbye to İzzettin Akman after a few pleasant days in Adana in which spring seemed to elbow its way out of winter almost overnight, turning the city's legion orange trees into glorious shocks of white blooms. It was only after leaving the bursting Levantine landscape behind me, while scrolling through my phone somewhere on the thirteen-hour bus ride back to Istanbul, that I stumbled upon a news

INTRODUCTION

article about one more Turkish government plan aimed at achieving "a significant reduction in the carbon footprint" of the country.

It was a plan focused on, of all places, the one I had just left — a slice of sun-shellacked Mediterranean coast exactly due south of Akman's farm. In October 2021, President Erdoğan flew to Adana to lay the foundation stone for a propane dehydrogenation plant that would come to occupy a beachhead stretching the length of two thousand football fields. The Turkish Wealth Fund, which was fronting $10 billion for the "Ceyhan Mega Petrochemical Industry Zone," insisted on its environmental bona fides: turning southeastern Turkey into a "global hub of petrochemicals" would ultimately reduce the country's reliance on imported polyethylene, thereby freeing up Turkish capital to combat climate change in the longer term — logic that sounded almost like a mocking parody of the arguments made by advocates of the green energy transition, that accelerating carbon output in the next few years is worth the eventual decoupling from hydrocarbons it may guarantee for the rest of time.[16]

Adana was no longer just going to take in trash, in other words. And Turkey was no longer going to continue to feign any commitment to a "zero waste" future. Instead, they were going to throw themselves into the madhouse business of manufacturing plastic — three billion pounds of it a year, the equivalent of a hundred billion plastic water bottles. You would no longer need drones to track it. It would be right there in front of you, in full view, getting cooked into existence.

And in all this — the transformation of the Fertile Crescent, the place from which human civilization had first spread out across the globe, into one of the largest recipients of plastic on the planet, a landscape so liable to being despoiled that it required surveilling by remote-control flying robots, and a place that seemingly had no option but to open itself to the production of the very material trashing its hills and rivers and farms — it was hard not to detect a certain unnerving symbol of our age, as well as a dire warning for our future.

INTRODUCTION

Trash Empire

> [We] saw products as garbage even when they sat gleaming on store shelves, yet unbought. We didn't say, What kind of casserole will that make? We said, What kind of garbage will that make?
>
> — Don DeLillo, *Underworld*, 1997

In December 2020, *Nature* published a report detailing a cataclysmic shift in humanity's relationship with Earth. The total mass of the world's human-made objects, its authors explained, had come to equal the entire biomass of the planet itself. That is to say, the weight of everything created by our hands — skyscrapers, automobiles, iPads, plastic straws — was on the verge of exceeding that of all trees and all plants, all animals and all humans, indeed the mass of all living things put together.[17]

Let's put this another way: You are currently living in a world in which the human ability to create garbage — or eventual garbage — has surpassed Earth's ability to generate life.

Reading *Nature*'s findings, my first thoughts went to Vance Packard. The son of Pennsylvania dairy farmers, Packard foresaw — maybe earlier than anyone — a world in which our obsession with pumping out new products would surpass our care for the resources that go into them. Packard was something of a seismographer of the tectonics underlying the American dream, a journalist who liked to pull the curtain back on the story of economic mobility that purported to liberate Americans from old-world hardships and identify the charlatans and chancers quietly cashing in. He remains best known for his 1957 book, *The Hidden Persuaders*, a cutting account of the myriad ways in which US consumers were being marionetted by a rising class of advertising executives into purchasing things they did not need — and it never occurred to them they wanted. It's a variation on a theme — who has an interest in offering you the illusion of fulfillment? — that he would

INTRODUCTION

revisit seven years later in *The Naked Society,* an examination of how an addiction to spending was increasingly turning Americans into bundles of personal data ripe for corporate exploitation.

But in many respects Vance Packard's most prescient work would prove to be a slim but pugnacious exposé released halfway between the publication of those other two books. In *The Waste Makers,* Packard proposed his own what's-the-real-story-here interpretation of the United States' mesmerizing economic growth in the years following the Second World War.

Mid-century America, contended Packard, was quite unlike anything else that had ever existed. Earlier societies — and one could go back hundreds, even thousands of years — had produced materials and objects that were meant to be durable, cherished, and useful to successive generations. Hunter-gatherers passed down stone tools. The Greeks and Romans reworked marble statuary as their ruling dynasties — or religions — changed. Medieval manuscripts were used, scrubbed down, reused. As Packard saw it, in the great sweep of history it largely had been the case that producing something like an article of clothing or a cup for drinking took a considerable amount of natural resources and human effort, a commitment generally worth undertaking only if that object was intended to last as long as possible.

Of course, there was no shortage of examples of objects that never did last long. But beginning in the 1920s, and increasingly after 1945, something very basic had changed. Out of the carnage of the Second World War charged an American "hyperthyroid economy" that functioned like a mangled inversion of earlier systems of production and consumption.

Postwar America, insisted Packard, had little interest in limiting waste. On the contrary, it was determined to create as much of it as possible. Packard attributed this phenomenon to "growthmanship," an obsession with registering more and more prosperity with each new year, and measuring that prosperity in increases in economic output. This had always been a thread within the American story, a frontier people confronted with a vastness of land and an imposing scale of resources. But in the years following the Second World War this obsession was compounded by geopolitical necessity: The United States needed to demonstrate material supremacy over Soviet Communism.

"Growthmanship" had created a strange society. To sell more and more products that were being produced ever more efficiently, corporate America had resorted to something Packard called "forced consumption." It amounted to tricking customers into buying stuff they did not want or need. Postwar industry, Packard maintained, had become as much about producing objects as conjuring up a desire for them, a psychological exercise in "artificially stimulating" Americans into new spending habits. The profits from those products derived not from crafting objects of any long-term value — this would be detrimental to the corporate bottom line — but in churning them out in increasingly instantaneous succession. "Our enormously productive economy demands that we make consumption our way of life, that we convert the buying and use of goods into rituals, that we seek our spiritual satisfaction and our ego satisfaction in consumption," Victor Lebow, an American economist, explained in a 1955 issue of the *Journal of Retailing*. "We need things consumed, burned up, worn out, replaced, and discarded at an ever increasing pace."[18]

In the decade following the Second World War, the national output of goods in the United States doubled. Private consumption came to account for two-thirds of the gross national product. And whether it was a poncho or a Pontiac, the result was much the same. Increasing prosperity required increasing outputs of waste. "Two aspects of American civilization strike almost everyone," the essayist John Atlee Kouwenhoven would claim. "The abundance it enjoys, and the waste it permits."[19] There had never been anything quite like this, insisted Packard: a "throwaway society" in which the fate of products as trash *presupposed* their apparent uses as garments, containers, utensils. And for the first time in distant memory, perhaps the first time in all of history, more waste didn't equate to a lost economic opportunity, an inefficiency gap between the resources that went into an object and those that ultimately had to be discarded. No: *More* waste meant *more* profit.

And herein may have been the most contentious argument within *The Waste Makers*.

Waste was, and often remains, characterized as a problem of consumption. This was backward, according to Vance Packard. Waste was a problem of production — of *over*production, to be exact.

Though Packard may have made American industry the object of his ire, within a generation, the absurdity he described — a society so hell-bent on registering annual upticks in economic growth that it thought little of resigning stupendous quantities of its own productive output to the waste bin — had become commonplace across much of the globe, emanating outward from the United States like a tidal swell. As historian Victoria de Grazia has chronicled in *Irresistible Empire*, her account of twentieth-century consumption, the most consequential victory of that century may not have been the triumph of liberal democracy over fascism or, four decades later, over Soviet Communism: The *real* victory was the commercial one, in which American mass consumer culture — armed to the teeth with the supermarket, brand-name goods, and advertising billboards — vanquished the creaking, centuries-old guild manufacturing societies of bourgeois Europe. The Cold War had turned Packard's "throwaway society" into an imperial project. Funded by the Marshall Plan, consumer culture would synchronize American hegemony over the globe with a US-based overhaul in production itself, a revolution in which naturally occurring materials — wood, metal, cotton — would give way to unnatural, petroleum-based, synthetic ones of rubber, plastic, and polyester. The world wasn't just going to be led by the US. It was going to produce like the US. It was going to consume like the US. And, sure enough, it was going to discard like the US.

The global balance sheet of trash today is astronomical. Humans currently manufacture their own weight in new stuff every week, only about one percent of which has been estimated throughout the world to be in use six months after its purchase.[20] Our consumption patterns — resulting from the production and use of those goods and services — now stand responsible for more than half of all carbon emissions.[21] Every day, the world discards 1.5 billion plastic cups, 250 million pounds of clothes, 220 million aluminum cans, 3 million tires.[22] For every human being alive right now, there exists slightly more than one ton of discarded plastic out there somewhere, scattered on land or layered in the ground or adrift at sea; there is little question that most of it will outlive our own planetary presence by thousands, possibly hundreds of thousands, of years. In the ocean alone, per every human, there exist 21,000 pieces of plastic, a net mass of shopping bags and six-pack rings and bottle caps that by 2050 will exceed the weight of all fish put together and is expected to double every six years for the foreseeable future.[23] Meanwhile, in just the minute it took you to read this paragraph, another million plastic bottles have been discarded and another garbage truck full of plastic has entered the seas.[24]

That's the wide-angle view. It only gets bleaker when you zoom in:

To Guatemala, a country that has had to resort to blockading the delta of its Río Motagua — a body of water that contains three percent of the world's plastic pollution — with two-story wire barricades to prevent a thousand tons of trash from entering the Caribbean Sea each month.[25] To India, where a garbage mountain in a landfill east of New Delhi gains thirty trash feet in elevation a year and now tops out higher than the Taj Mahal.[26] To Norway, where in 2017 the digestive system of a beached goose-beaked whale was found to be stuffed with more than thirty plastic bags and other forms of garbage, including chicken containers from Ukraine, an ice-cream wrapper from Denmark, and an empty package of potato chips from Great Britain.[27] To the Galápagos Islands — one of the few ecosystems in the world that has been spared major human intrusion — where researchers from a UK-based environmental watchdog recently concluded that you can now never be more than seventeen inches from a particle of plastic.[28] To the Maldives and the island of Thilafushi, which was once not an island at all but a collection of sandbar formations — until 1991, when the country's environmental ministry designated Thilafushi a "landfill" and began layering bales of trash across its dunes at a clip of 330 tons per day, resulting in what is now a mass of aggregated garbage the size of twenty-five aircraft carriers, replete with methane bottling plants, cement factories, and an incrementally lengthening perimeter of plastic.[29] And finally to the planetary exosphere, where the preponderance of satellite and rocket debris has now become so dense as to have come perilously close to reaching something known as the Kessler syndrome — the threshold at which it will no longer be possible for rockets to enter space owing to the overaccumulation of human-produced debris in Earth's low orbit.[30]

But at the same time, just as trash was piling up around the globe in terrifying quantities, something else started happening. For it was not just new notions of consumer culture that began sweeping across borders in the second half of the twentieth century.

Trash itself started to move.

Toxic Terrorism

> Garbage. All I've been thinking about all week is garbage....
> I mean, we've got so much of it. You know, I mean, we have
> to run out of places to put this stuff, eventually.
> — *Sex, Lies, and Videotape,* 1989

The story of İzzettin Akman's farm prompts a simple question: How did we get to a world in which a plastic bag placed in a recycling bin in the United Kingdom becomes the bane of a Kurdish farmer three thousand miles away in Turkey?

The problem Vance Packard unpicked was one in which waste, however great in output, and however deliberately produced, nevertheless stayed in the country in which it was discarded. Not anymore. Over the last forty years, trash disposal has morphed from a local problem into a global one. Sure, there's a long history of waste being bought and sold.

INTRODUCTION

And there's a twentieth-century story to be told about the growing geographic and psychological displacement between waste generation and waste disposal. But in the 1980s something peculiar started happening. Instead of going to your local landfill, some of your trash started traveling across national borders and even oceans. It went from being your detritus—something you dropped into the nearest bin and never thought about again—to your national export. And not waste per se but waste *movement* became incomprehensibly profitable.

This wasn't like other globalization stories that burst forth from the dusk of the Cold War. It wasn't like networks of organized crime, which emanated out across Eastern Europe and beyond with the Soviet Union's fall. It wasn't like the black-market caviar trade that sprung forth from the mountains of the war-torn Caucasus, or the hijacked-automobile and sex-trafficking businesses that surged with the dismemberment of Yugoslavia. This was a globalization story hiding in plain sight in which the engine of profit became *you*. Over the last forty years, great quantities of your garbage has been quietly relocated, at a profit, to the poorer countries of the world, often ending up in states that not so long ago released themselves from Western imperialism, only to find they have been turned into receptacles of Northern consumerism.

Let's play a game. You come from a poor country. I come from a rich one. The place where you were born has little money and less opportunity. But it has a bunch of *real* wealth. There are forests, mines, cotton fields, oil palms, millions of acres of fertile soil. Your only problem is that it's difficult to make use of it all. Long before you or I can remember, my ancestors acquired much of it. And somehow, generations later, even after my people have agreed to leave, your economy is still predicated on shipping your resources to me. It's an injustice, I concede, so let me offer something back to you. I can't let you have your trees, no, but what if I were to sell you...all this old junk mail that may have been made from them? I can't let you have a gold industry of your own. But what if I sold you...these used cell phones that possess minuscule traces of gold that could have come from your land? I can't offer you your cotton fields. But how about these... secondhand clothes?

Fair deal, right?

For hundreds of years, European empires enriched themselves by treating the Southern Hemisphere as a place from which they could take what they needed. This hasn't drastically changed, but US growthmanship has instilled it with a mischievous new dimension. In the 1980s, the so-called Global South increasingly became a place from which you not only took things but *put* them, too, because these things were either too expensive, too dangerous, too unsightly, or too great in quantity to discard within your own borders. Poor countries no longer just propped up your living standard; they also cleaned up your environment. Today's waste trade occasionally operates by covertly spewing loads of dumpster trash across the coasts and hills of African and Asian and Latin American nations. But most of the time it works through some version of what I previously described, a chameleon attempt to convince the poorer countries of the world that the things you throw away — and which are often made of materials that have been appropriated from their lands — are invaluable economic opportunities they should *pay for*.

Almost immediately this was recognized as wrong. "The words 'toxic terrorism' and 'garbage imperialism,' I think, are words that will be increasingly heard in foreign countries if we don't do something about it," a US congressman insisted to his colleagues three months before the fall of the Berlin Wall in 1989.[31] And yet over the next decade, as a new world began to open up, a world of free-ranging capital and untrammeled commerce, something odd happened. Outrage over trash's export continued, yes. Efforts to ban its movement were signed by dozens of nations, yes. Yet the waste trade kept getting exponentially more massive. What had begun in the 1980s as the sporadic trafficking of dangerous but rather obscure forms of industrial residue — old asbestos, expired pesticides, spent airplane fluids — had, by the 1990s, transmogrified into the hourly movement of almost everything you

INTRODUCTION

can possibly imagine. Trash didn't merely globalize. It became a globalization pillar. Slimy plastic spoons, broken TV remotes, raggedy clothes — to this day, every day, thousands of cargo containers of it get dispatched thousands of miles around the world.

One result of all of this? Offshoring our consumption footprint has only encouraged waste's proliferation. The more trash rich countries have sent away over the last forty years, the more trash they have continued to produce. And yet few citizens of the Netherlands or Canada would necessarily know to see it like this. For the waste that travels across the globe and often inflicts irreversible environmental damage is *not* the trash that — to so much chagrin — goes into the garbage bin and then the local landfill. It's the stuff you place in the recycling bin in the conviction that doing so is helping the planet. Recyclability is no lie, but in the 1990s you were encouraged to "recycle" all sorts of objects — electronics, batteries, Styrofoam, Tetra Pak, plastic everything — when their ability to be effectively or infinitely resurrected was difficult or downright impossible. Systemic overproduction was reframed as a narrow matter of personal ethics: As long as you chose the recycling bin, nothing was wrong. Yet much of what you have been led to believe was getting "recycled" over the last generation has never been helping the planet. It has the opposite effect of what you have imagined. It functions as a Trojan horse of pollution, redistributing material *packed* with toxins onto poorer corners of the world, all while allowing wealthier nations to engage in what has essentially become a morality performance — replete with blue bins and chasing arrows and pithy slogans — that absolves consumerist guilt and dissuades self-examination about why we insist on producing as much as we do.

Another result? Across much of the Global South, enormous and complex trash economies now dominate. Like the arrival of the potato in Europe, or the reintroduction of the horse into the Americas, the impact of sending millions upon millions of pounds of foreign synthetics — new and used — into lands with no history of or capacity for handling them is a story we might only fully come to understand

INTRODUCTION

decades from now. But already it's no exaggeration to say that across much of the equator today, waste—gathering it, sorting it, burning it—has come to replace thousands of years of farming as the default occupation of humanity. Urbanizing often entails entering slums and eking out a sub-existence by scaling scrap networks; of the world's fifty largest dumpsites, all but two are to be found in developing nations, informally employing many millions.[32] And, again, this occurs not because most of these countries are necessarily poor. On the contrary, it occurs because many are rich, in minerals and metals and cropland and fresh water. That those very places full of such desirable resources came to be repositories for *your unwanted garbage* is not some happenstance irony. They are interconnected stories that can be traced back to the later years of the Cold War.

Many books have been written about trash. It's an important topic, one that often stands as proxy for value systems and how societies writ large function. How often is your trash collected? Is it managed by the state or outsourced? To whom is it outsourced?

If the handling of its trash points to certain deeper truths about a society, what does the globalized movement of waste have to tell us about our world today?

This is a book about why your trash started to travel between continents and how the seemingly pedestrian act of throwing something away spawned an earth-spanning exchange that functions like a funhouse-mirror inversion of the globalized economies of extraction, production, and consumption. Forty years after it first began crossing oceans, it's hard to imagine a more incisive barometer of international hierarchies and inequalities than the question of who must accept waste and why. And, to be clear, the question is not one of mere aesthetics. By this century's end, the stuff we discard will have emerged as the greatest unchecked contributor to climate change. Sending shiploads of it—millions of tons of sequestered carbon injected with countless toxins—to poorer nations amounts to a pitiful evasion of responsibility by those countries that have the greatest obligation to rein in their material outputs and confront a climate crisis overwhelmingly of their own making.

But before I continue, a confession. I began my investigation into the geopolitics of garbage probably not unlike the person reading these words: unclear as to why the waste trade exists and how it quite concerns me, beyond a vague premonition that trash from a rich country probably shouldn't be mashed into a twenty-foot cargo container and dispatched to an impoverished country. I still cannot claim any credentials as an environmentalist. I have little expertise on the complexities of how the chemical byproducts of natural gas get cracked into polyethylene terephthalate and then proceed to produce a see-through bottle of Gatorade. I still eat meat but avoid plastic when I can, even as I remain skeptical about what role — if any — individual morality plays in addressing the world's plastic pandemic.

But after innumerable flights, nineteen bus rides, fourteen train journeys, six ferry voyages, countless taxi trips, and the odd rickshaw jaunt; after a night in a one-star Guatemalan hotel that featured squadrons of cockroaches scrambling along my floor in concert with several geckos slithering across my walls; after dispensing six thousand shillings to a pair of police officers in Nairobi who pulled me out of my taxi and into a frightening cell for not having my passport on my person ("We'll purchase something nice for our wives!"); after learning to never wear shorts to an interview in the balmy highlands of Java and to always sip your tea and nod amiably when being hosted in the frosty innards of Turkey; and after two years spent roaming five continents loaded with jaw-dropping jungles and waterfalls and volcanoes and beaches and deserts and archaeological ruins, precisely none of which I saw, living instead out of a backpack and frequenting only the most hideous and putrefying landfills and ports and slums all those places had to offer, here's what I *do* know:

Globalized garbage is a bizarre, illogical industry. The companies that generate the bulk of what you consume — Coca-Cola, Nestlé, Apple, Samsung, Procter & Gamble, Unilever — are household names, crossword-puzzle clues, publicly traded entities. You know them. You may own stock in them. You may know someone who works for them. You may work for one of them yourself.

The companies that reach down into your recycling bin and rinse out the last rancid drops of profit? It's not just that you haven't heard of many of them. It would be almost impossible to do so. They are postboxes in Anaheim or Hong Kong. They are kinship networks of Nigerians and Indians and Lebanese and Chinese. They work out of slapped-together warehouses in Port Klang or Dar es Salaam. They change names from one month to another. They don't have websites. They have WhatsApp numbers, Google Translate, a cousin in Newark or Croydon. Sure, when it comes to most major commodities, such as steel or oil, the market is lorded over by legendary trading houses such as Cargill or Trafigura. But when it comes to the biggest commodity of all—everything humanity tosses away—you tend to be dealing with grifters and hustlers. They learn market prices by scrolling through internet forums. They can examine a container of dirt-caked plastic or battered copper wiring and have a good sense of where it came from—and where it should be sent next.

"In the world of scrap, all you really need is a smartphone," Nathan Fruchter, a former Glencore scrap steel trader based in the post-Soviet republics, told me.[33] "The most important thing is knowing who to know," I was told by Patty Moore, one of California's biggest plastic traders.[34] "When you think of the trash trade, think of the drug trade," Teodor Niță, a Romanian state prosecutor who tracks waste shipments from Western Europe, explained to me. "Only trash moves from rich places to poor."[35]

They owe their existence to globalization. But in many ways they operate in defiance of it, wedging open legalistic loopholes in international trade clauses, exploiting the murky definitional differences separating "waste" from "scrap" from "resources," and prospering off the fact that, sure enough, of the forty thousand cargo containers that will get loaded or unloaded in the ports of Shenzhen or Rotterdam today, a fraction are opened and fewer still are inspected. In a world in which we seem to possess a preponderance of data about what goes where, even the broad-brush parameters of the waste trade remain anyone's best guess.

The European Anti-Fraud Office estimates *illegal* waste trafficking to be more profitable than human trafficking, while the United Nations recently came to the conclusion that over the last thirty years, the global trade in plastics has been 40 percent higher than previously believed — meaning that for every two pieces of plastic thought to have been transported around the world since 1992, the first year such statistics got tracked, there have in fact been three pieces, a revelation that turns what was once a trillion-dollar annual business into something more valuable than the global weapons, timber, and wheat trades — combined.[36] As for Interpol, in 2017 it devoted thirty days and the considerable heft of its resources and expertise to attempting to understand where exactly your garbage goes. More than forty countries participated. Dozens of ports were monitored. Highways were surveilled. Countrysides were patrolled. The greatest revelation? Not the hundreds of instances of waste caught illegally entering West Africa and Southeast Asia or the tens of millions of dollars in undeclared profits. No: It was the fact that almost no waste trader anywhere in the world appeared compelled to disguise or hide what they were sending.

"With the illegal lumber or ivory trades, countries are losing commodities," Joseph Poux, who coordinated the Interpol investigation, told me. "When waste exits their borders, they tend to be losing a liability. The incentive is to just let it go. We are talking about an unbelievably massive business."[37]

Such was the strange, evasive, unbelievably massive business of globalized garbage I set out to try to understand. And to do so, I thought it best to attack in depth. A different country, a different form of waste, a different story of how and why it got there — and what it seemed to be doing to our planet.

PART ONE

Toxic Tropics

1

BANANA REPUBLIC

> When the trumpet sounded
> everything was prepared on earth,
> and Jehovah gave the world
> to Coca-Cola Inc., Anaconda,
> Ford Motors, and other corporations.
> The United Fruit Company
> reserved for itself the most juicy
> piece, the central coast of my world,
> the delicate waist of America.
> — Pablo Neruda, "United Fruit Co.," 1950

PUERTO BARRIOS MAY be Guatemala's biggest port, but it is an unassuming and squalid place, a hot concrete town comprised of a thousand or so low-slung homes crowding a gridiron of unpaved roads. From Guatemala City, you reach it by a seven-hour bus ride through the Sierra de las Minas, the final stretch of which features mile upon mile of Dole and Chiquita cargo trucks backed up along a one-lane highway that coils its way from the haciendas of the interior to the Caribbean coast. Each shipping container bears a refrigeration system and is emblazoned with a mattress-sized replica of that small sticker found on the bananas at your local supermarket; some may also contain cocaine, or so you're informed by many Guatemalans you meet. Where the highway meets the sea, at Puerto Barrios' waterfront, no more than a mile or two long, you encounter port workers whiling away the last of the sweaty afternoon sipping on bottles of Gallo beer. Iguanas sun themselves on wooden docks extending out over a porcelain-blue sea splotched with brown runoff that might be sewage. Mangy stray dogs sniff outsiders and chase them for sport.

Puerto Barrios exists because of the banana. Or, better put, Puerto Barrios exists because just over a hundred years ago Americans became obsessed with eating bananas. As late as the 1880s, few Americans had ever seen the fruit. Yet by century's end, tariffs on its importation from Central America and the Caribbean had been slashed and the fruit had turned into a national craze. A chaotic scramble began to ship it north. In 1901, a Boston-based firm called United Fruit Company homed in on the opportunity offered by Guatemala. Previously, US merchant ships had mostly sailed to Central America and purchased bananas by the crate from dockside merchants. United Fruit Company struck upon a different idea. It decided to set up its own plantations in Guatemala and raise the crop itself.

Guatemala possessed almost none of the infrastructure needed to support United Fruit Company's novel scheme, though. And so, in exchange for sprawling land concessions from Guatemala's ruling class, the company agreed to finance and construct most of that infrastructure itself. Over the next decade, with the help of President Theodore Roosevelt's Army Corps of Engineers, United Fruit Company connected Guatemala City to the shipping lanes of the Atlantic with a 180-mile-long railway. It built suspension bridges. It dug mines. It founded Guatemala's first postal service and set up its first major radio and telegraph services.

Before long, United Fruit amounted to, as one Guatemalan observer noted, a "country within a country."[1] The biggest landowner in Central America's largest nation, it employed more workers than almost any other company in the Western Hemisphere. And as for Puerto Barrios, by 1930 it had blossomed into the biggest port in Guatemala, though it would be hard to say it was Guatemalan in any true sense. A company headquartered three thousand miles away owned its wharves. It owned its warehouses. It owned its lone hotel, where scenes of *The New Adventures of Tarzan* would be filmed in 1934. And United Fruit of course owned the fleets of refrigerated steamers that chugged out of Puerto Barrios' harbor bearing more than five million bunches of bananas to the United States every year.

Puerto Barrios is a set piece in how a certain type of colonialism worked in the twentieth century. Even in their apparent willingness to help bestow the tools of development and progress upon poorer countries, countries like the United States were typically only ever rigging up a system of exploitation, extraction, and exportation whose beneficiaries proved to be their own financiers, industrialists, and consumers.

Go to Puerto Barrios today and it's hard not to get the sense that suspiciously little has changed. There is still no train in Guatemala that can transport anything other than bananas or coffee beans from the country's interior to its coast; for Guatemalans, there is the bus. And a town still lacking any reliable sewage system, consistent electricity supply, or clean drinking water continues to huddle in the shadow of the monstrous scaffolding of six-story industrial cranes that spend all hours of the day and night plucking up cargo containers of bananas — now the most consumed fruit in the United States — and stacking them atop bulk carrier ships bound for Delaware and New Jersey. The banana bosses, needless to say, still reign. For United Fruit Company never really ceased to exist. In 1984, it just renamed itself: Chiquita Brands International.

Puerto Barrios may be one example of colonialism at work. But I endured the seven-hour bus ride to the port to behold a bewildering, latter-day attempt at another.

In 1992, ninety years after United Fruit Company first arrived in Guatemala, the country's newly democratic government made plans to construct a separate port twenty miles farther up the Caribbean coast. It was designed to do the exact opposite of what United Fruit had undertaken at the beginning of the century in Puerto Barrios. The port, which was to be zoned at a section of seaside jungle known as Cocolí, would be built to *receive* something from the United States. And that something was toxic sewage sludge.

Guatemalan bananas were heading north. Now American shit was to head south.

By 1992, Guatemala could already boast a bleak history as the serial target of toxic waste dumping by US cities and corporations. "Our heads were spinning with all the incoming allegations and rumors," Erwin Garzona, who tracked hazardous waste shipments into Central America for Greenpeace in the early 1990s, told me.[2] In 1985, a firm out of Miami had attempted to offload wastewater on a section of coastline not far from Puerto Barrios; it was to be dried on concrete platforms near the port and turned into powder, then used as crop fertilizer. A year later, a decrepit Liberia-flagged vessel called the *Khian Sea* approached Guatemala with unsolicited plans to dump along its beaches more than 30,000 pounds of Philadelphia's incinerated garbage—the sooty remnants of "cigarette butts, old shoes, dead batteries," according to the Associated Press.[3] In 1987, the city of Los Angeles propositioned the opposite, western coast of the country to take in 125,000 tons of its sewage sludge, to be strewn across Pacific swamplands that US taxpayers would "rent" in exchange for $14 million a year.[4] "We have reviewed the proposal and have no objections," the US Embassy in Guatemala City claimed at the time. "One can imagine that shipping sewage sludge...will incite some unfavorable press in Guatemala. With this caveat, we have no problem."[5] Even by the early 1990s it was not uncommon for well-to-do Guatemalan families with large coastal estates to be propositioned by US waste management firms with contracts to import trash and bury it on their land.[6]

Cocolí was to be something else. The sludge entrepôt-to-be was to take the name of a beautiful local waterfall, Las Escobas ("The

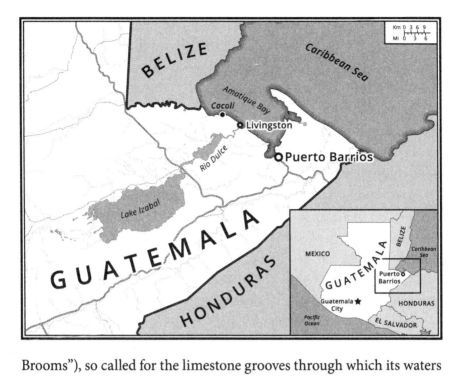

Brooms"), so called for the limestone grooves through which its waters tumble. It was to receive shipments of waste not sporadically but regularly, with the sewage originating from a handful of US cities, including Miami and Galveston, which had arranged to pay Guatemala's government handsomely to accept it. From Las Escobas, the sewage was to be pumped inland, parallel to the Rio Dulce, before getting deposited along the shores of Lake Izabal, a preserve for endangered manatees.

Already by the spring of 1992, according to *Prensa Libre*, Guatemala's largest-circulating newspaper at the time, plans were afoot. Five *caballerías*, or about a thousand acres, had been cordoned off for Las Escobas' construction. Two hundred Garifuna families — members of an Afro-Caribbean minority — who could trace their roots in Cocolí back "more than sixty years," had been asked to relocate from four surrounding villages, "losing their homes and the farms they need to survive," and were herded to Puerto Barrios to start new lives; as compensation for the loss of their property, the families had been promised monthly public-sector salaries in cash, which had already begun arriving at the port in state helicopters.[7] All the same, reported the investigative Guatemalan magazine *Crónica,* "the peasants claim to have been subjected to harassment and request an investigation."[8]

There is no road to Cocolí; it can only be reached by sea. On a steamy May morning I hired a boat from one of Puerto Barrios' fishermen and motored north toward the border of Belize. You would be hard put to locate a more gorgeous place on Earth. The shoreline of Izabal Department, the swampy region of Guatemala that abuts the Atlantic, leaves an impression on all who see it. You could be on the set of one of those Corona beer commercials. *Find Your Beach.* Howler monkeys skip across the tops of palm trees that form a curtain-like backdrop to a miles-long ribbon of sand so pristine it looks like it must have been imported from a golf course. After about an hour, I nosed my boat into the small triangle of jungly headland known as Cocolí. At the shore, several shirtless Garifuna men were hacking away at brush with machetes. They agreed to let me disembark in exchange for a few quetzals, though claimed to know nothing about a curious plan dating back thirty years to convert the ground beneath their feet into an American septic tank. I spent the last of the morning walking around the coast, meandering through acres of palms, contemplating the land's near-miss fate.

That the port of Las Escobas was never constructed is largely the result of one improbable figure. In June 1992, the then-governor of Izabal Department, a former social worker named Lilian Vásquez de Guzmán, leaked the details of the waste importation scheme to the Guatemalan press. In a staggering litany of allegations, De Guzmán pointed her finger at the very top of Guatemala's new democracy. The mastermind and greatest potential profiteer of Las Escobas, she claimed, was Jorge Serrano Elías, who had been elected president of Guatemala eighteen months earlier on pledges to bring human rights to a country that had spent the greater part of the previous half century waging war on its own indigenous populations. Serrano had agreed to build the port at Cocolí because he stood to personally pocket tens of millions of dollars, or so claimed Governor De Guzmán. The allegations triggered months of reputational crossfire in the pages of Guatemala's newspapers, with Serrano in turn shooting accusations of corruption and backhanders back at De Guzmán. Eventually plans for Las Escobas' construction were cancelled — though not before the governor, fearing for her life after men began tailing her around Puerto Barrios, was forced to flee Guatemala. Three months after De Guzmán was running one of her country's most important provinces, she was working the afternoon shift at a pharmacy in New Jersey.

Thirty years later, by the time I met De Guzmán, she felt safe enough to be back in Guatemala, in no small part because Serrano himself now sat in exile in Panama City, barred from returning to the country he once presided over, owing to allegations of titanic corruption — unrelated to Las Escobas — while serving as president. Now seventy-six, De Guzmán was frail. At a modest house in the outskirts of Zacapa, a dusty agricultural town under the thumb of the cocaine cartels that sits a hundred miles inland from Puerto Barrios, small dogs jumped in and out of her lap as we chatted next to a swimming pool filled with stagnant green water. De Guzmán stood by her claims with undiminished alacrity. Las Escobas was a double crime, she insisted to me, shuffling yellowed newspaper clippings with wrinkled fingers. A small clique of Guatemala's richest men had planned on getting richer by poisoning their own land and people.

"Wouldn't you have tried to stop it?"[9]

2

THE CHEMICAL CENTURY

> Better Things for Better Living... Through Chemistry.
> — DuPont advertising slogan, 1935 to 1982

THE STORY OF the international waste trade starts thirty years before plans were concocted to dump US sewage sludge along the gorgeous coastlines of Guatemala. It starts far from scheming businessmen in Western cities and purchasable elites in the Global South.

It starts with a remarkable woman and the publication of her shocking book. And it starts with the most admirable of intentions: to create a cleaner world for future generations.

In 1960, journalist Vance Packard offered an analytical rebuke of the artificially stimulated consumption driving US postwar economic growth. But two years following the publication of *The Waste Makers* came a more disturbing critique of American prosperity. In 1962, a marine biologist named Rachel Carson peeled back the cover on the astonishing extent to which lethal toxins had infiltrated modern society. *Silent Spring* plunged deeper than *The Waste Makers*. It didn't criticize American consumption; it performed a chemical autopsy of it. In nearly three hundred pages of eloquent prose, Carson introduced the average American to the boggling array of substances that had been concocted in laboratories since the 1940s, then poured, sprayed, and dumped into their farms and gardens and rivers. There was dichloro-diphenyl-trichloroethane, or DDT, a carcinogenic insecticide that had been used during the Second World War to limit the spread of malaria among troops, only to be resold thereafter as a household mosquito repellent. There was 2,4-dichlorophenoxyacetic acid, or 2,4-D, a deadly herbicide routinely sprayed across farms, from which it then infiltrated the cell systems of grazing animals, before eventually making its way into humans. Some two hundred chemicals in all — the bulk of them invented and mass-produced with little or no understanding of their consequences on public health — had saturated American society, and Americans themselves.

Vance Packard's "throwaway society" was, it turned out, one in which very little actually went "away." Deadly chemicals invented and mass-produced in the space of less than twenty years — the "synthetic creations of man's inventive mind," as Carson put it, "hurled against the fabric of life" — were bound to invisibly course through our water systems and soils for the rest of time.[1]

The US environmental movement of the 1960s was a complex mass of disparate interests. But it would be difficult to overestimate what the publication of *Silent Spring* almost single-handedly brought about. Friends of the Earth and Greenpeace cite it as an inspiration behind their founding. Joni Mitchell and Glenn Frey sang songs about it. After reading an advance excerpt of the book, John F. Kennedy proceeded to appoint a presidential science advisory committee to probe its claims, which were promptly confirmed. "We need a Bill of Rights against the twentieth-century poisoners of the human race," declared Supreme Court Justice William O. Douglas after reading it for himself.[2]

In the 1950s, chemicals were widely deemed "miracles" of modern American life; in the 1960s, they were increasingly recognized as its quiet killers. By 1964, *Silent Spring* had sold more than a million copies, a milestone Rachel Carson just barely lived to see. In April of that year, after a long struggle with breast cancer, possibly contracted through the very toxins she had spent years investigating, she died.

During the fifteen years following the publication of *Silent Spring*, and in great part because of it, the United States would drastically overhaul its relationship with the environment. At the heart of the legislation was the matter of not just what chemicals should be allowed to be produced but also where those that *could* be produced should be allowed to end up. In other words, in recalibrating its relationship with the environment, the United States was just as directly addressing its relationship with waste.

For the purposes of the garbage trade, the most crucial piece of legislation passed in the direct wake of *Silent Spring* was one you might not suspect: the Federal Environmental Pesticide Control Act of 1972, which banned dozens of the chemicals, including the notorious DDT insecticide that had featured so prominently in Rachel Carson's account. Admirable in aspiration, the act's legacy was to prove problematic and, ultimately, undermined the exact problem it attempted to address. For the FEPCA turned the United States into a country that, legally speaking, was full of thousands of tons of toxic materials that could no longer be used — and that Congress had rendered increasingly difficult and expensive to get rid of.

To grasp what follows, it's important to understand that America's newfound commitment to environmentalism came with a little secret: It didn't extend to other countries. In the 1970s, an uneven geography of ecological regulation was placed atop a preexisting geography of unequal power and economic dependency. Nothing stood in the way of crafty American companies looking *beyond* the United States to sell stockpiles of chemicals whose use had been outlawed at home. The short-term legacy of the FEPCA? Americans were safer. The longer-term one? Residents of many poorer countries were not. In the decade that followed the act's passage, more than five hundred million dollars' worth of highly dangerous chemicals would be shipped south, occasionally by the US government itself, and sometimes to the enrichment of the unlikeliest figures.[3]

In 1973, two brothers from New Jersey began renting warehouses across the United States. Jack and Charles Colbert had briefly worked in the small-arms trade. But the environmental movement afforded them the chance to become, in their own words, "middlemen" for corporate America. Over the late 1970s, the Colbert brothers began amassing recently banned chemicals and other dangerous substances from signature US companies. The Colberts acquired material from Ford, Exxon, General Motors, DuPont; their biggest supplier was the government, with the Colberts sourcing huge quantities of pesticides, paint thinners, asbestos, and even plutonium from the State Department, Pentagon, and US Navy. From some, the Colberts offered to purchase the stuff for a dollar or less per ton. From others, they were handed it for free.

The Colberts weren't mere collectors. They were also salesmen. Shortly after amassing their stockpiles, the brothers began contacting government after government across the developing world, claiming to possess "good products." The Colberts approached countries in Africa, offering industrial quantities of toilet paper at bargain rates; what they were *really* selling were rolls of lead-tainted engraving paper they had picked up off the US Treasury Department's Bureau of Engraving and Printing. The brothers approached countries across Central America, offering cut-rate deals on DDT, recently outlawed by the FEPCA; it worked fine, they insisted to potential buyers. Indeed, to distant clients the Colberts didn't appear to be selling dangerous or banned chemicals at all. They seemed to be in possession of good products at great prices. By the early 1980s, the brothers had earned $180 million from buyers across more than a hundred nations.[4]

The Colbert brothers would probably remain unknown today were it not for a shipment they sent to a company in Zimbabwe in 1984. That spring the Colberts purchased several hundred drums of hazardous waste from a Cleveland-based industrial garbage collector. At 60¢ a gallon, the toxic chemicals cost the Colberts slightly more than $12,000.

Several weeks later, after spotting an advertisement for "dry cleaning and fluid solvents" in a US trade catalog, a company in Zimbabwe agreed to buy the material off the Colberts for $54,000.

Only it wasn't technically the company's own money that would be used to purchase the several hundred drums of solvents that left Ohio that summer. To finance the deal, the Zimbabwean firm had secured tens of thousands of dollars in funding from the United States' own Agency for International Development. The Colberts, in other words, hadn't just duped an unsuspecting dry-cleaning company in distant Zimbabwe; they'd managed to get US taxpayers to front a small fortune for the steel drums — each bearing a label that read POISON along with a sticker that read UNITED STATES OF AMERICA — which began corroding at their bottoms shortly after touching down at Harare International Airport.

What happened to the waste? No one knows. Yet complaints filed by the Zimbabwean company to Washington ultimately resulted in an audit of the Colberts' finances — and the jaw-dropping realization that over the previous fifteen years, thousands of tons of chemicals and toxins outlawed *within* the United States had been sold and shipped to the poorest and most desperate nations of the world by two criminal bottom-feeders from New Jersey. US suburbs might have emerged from the 1970s blighted by less and less toxic residue, but the transformation was owed in no small part to anyone canny enough to hawk hazardous materials to buyers overseas.

How commonplace were charlatans like the Colberts? Also unclear. The EPA kept dismal track of such shipments, which fell largely to recipient countries to monitor, and periodically destroyed its own records.[5] The US Agency for International Development is nevertheless known to have itself dispatched large stockpiles of DDT to India, Ethiopia, Nepal, Indonesia, and Haiti, while in 1973 the World Health Organization estimated that 250,000 residents of the Global South were poisoned every year by pesticides imported — by transparent means or otherwise — from countries like the United States.[6]

"We were, in a sense, innovators ahead of the times because what you had was a whole definition in the environmental area that isn't really defined yet," Charles Colbert would tell a reporter years later during an interview at a federal correctional institution in Otisville, New York.

"You are asking me a question: am I sorry that I sold chemicals to the Third World? No. Let me ask you a question, OK, all right, so now you have 2,000 tons of pesticide that's been produced in America, OK, still on sale in the rest of the world," added Jack Colbert. "Now what do you want it to do? Do you want it to be buried in America or do you want it to be sold in a Third World country, which would you prefer?"[7]

3
CASH FOR TRASH

> Just between you and me, shouldn't the World Bank be encouraging MORE migration of the dirty industries to the LDCs [least developed countries]?...I think the economic logic behind dumping a load of toxic waste in the lowest wage country is impeccable and we should face up to that.... I've always thought that under-populated countries in Africa are vastly UNDER-polluted[;] their air quality is probably vastly inefficiently low compared to Los Angeles.
>
> — Lawrence Summers, chief economist of the World Bank, 1991[1]

THE COLBERT BROTHERS may have been crooks. But the discrepancies they exploited between the elevating environmental standards within rich countries and the pitiful — or nonexistent — ones within poor countries were all too real. Indeed, in light of what was to come, the most striking thing about the Colberts is that much of their material actually *was* of value. The DDT pesticides they sold to Central American countries may have been poisonous, but they worked; so did the Kepone, predominantly an ant killer, they sent to African nations.

The toxic world Rachel Carson had put under a microscope was up for export. But soon came something worse. By the late 1980s, it would no longer be necessary to send developing countries dangerous material of even negligible value. A "middleman for corporate America" could make unfathomable sums of money just by shipping the wasted, worthless *residue* of legal chemicals. Old asbestos, cyanide-laced mining tailings, drums of polychlorinated biphenyls, hydraulic fluids from airports, infectious hospital waste, sewage sludge — it turned out that with enough incentive, you could locate willing takers for all these things too.[2]

The Colberts had found a way to make tens of millions of dollars off the Federal Environmental Pesticide Control Act. The entrepreneurs who came in their wake would owe their greater fortunes to the Resource Conservation and Recovery Act (RCRA), a piece of legislation that attempted to amend the chaotic, often ad hoc disposal habits of postwar Americans.[3] Passed into law in October 1976, it was the first federal legislation to designate hazardous waste as distinct from solid waste; its stipulations required that such material be tracked from "cradle to grave," with all facilities producing it being required to keep running manifests of what they discarded. In its attempt to curb excessive chemical production and curtail such waste, the RCRA also raised and standardized prices of disposal. The results would take almost a decade to materialize, but catastrophe struck when they did. In 1980, it cost $15 to bury a ton of asbestos or paint sludge in the United States. By 1988, when the RCRA had come to be nationally enforced, that price had risen to $250.[4]

Once again, US environmental "progress" was to come at grave cost to distant parts of the world. Because imagine for a second that you are an American company and that you could ship your hazardous waste to a far-off country where the cost of burying a ton of toxic material was not $250 but $100. For every ton of waste produced, $150 stood to be saved in disposal costs. Were you to ship, say, 1,000 tons of hazardous waste, you might even manage to recoup the losses incurred by shipping and transportation and insurance, which in the late 1980s were slim, amounting to $100 per ton or less.[5] And were you an

American company in the business of producing, say, 10,000 tons of industrial waste a year, enough money could eventually be saved to make it worth your while to contract a third-party company to handle waste export on your behalf.

The cost of burying a ton of hazardous waste in most countries of Africa in the late 1980s was not $100. It was less than $3.[6] And the average amount of money to be saved on shipping toxic material across the Atlantic instead of dumping it within the United States was not marginal. It was enormously cost saving, a lucrative enterprise in its own right, amounting to about $200 per every ton of hazardous material produced. For its part, legislation passed across Western Europe in the 1980s was to carve out a similar profit incentive, generating a situation resembling that in the United States: Rules that applied across most of the continent did not extend beyond it.

And, in time, it only got easier. Were you able to bribe local elites in developing nations — and you typically were — the terms of a big garbage deal could be sweetened. Countries, cities, and companies in the Global North would prove inventive with the prizes they dangled before Global South states and their elites in the late 1980s. They offered to build roads. They offered to construct hospitals. They offered to supply an African government struggling to put down a rebellion with all the necessary weapons, only to then water down the deal and court the rebel faction instead. They promised jobs and tourists. Liberia was offered $1 million in unnamed prescription drugs.[7] A California company offered the daughter of King Tāufaʻāhau Tupou IV of Tonga a 40 percent stake in its operations if her father would agree to import its waste to his Polynesian island realm.[8] A company in New York offered to provide a town in Peru with schools.[9] The South American nation of Guyana was offered "an agricultural system, an ice plant, and a slaughterhouse."[10] In June 1988, a Philadelphia waste management firm approached the government of Netherlands Antilles in the Caribbean with a proposal to construct a seventy-mile-long artificial reef out of the city's incinerated trash; in return for agreeing to the plan, the tiny volcanic island of Saba would receive one dollar per every ton of imported waste, in addition to "one ounce of fine gold, one 'yard fowl' and a basket of fruit and vegetables."[11]

Schools! Slaughterhouses! Baskets of fruits and vegetables! The second half of the 1980s proved to be a bonanza of such propositions, some ingenious, some preposterous, all resulting in one thing: Rich countries increasingly got all the benefits of industrialization while solving their "environmental problem" by jettisoning their most toxic consequences south. True, large amounts of industrial waste were transferred between industrialized nations themselves. The biggest recipient of US hazardous waste in the 1980s was Canada; France received huge quantities of industrial residue from West Germany. But those countries possessed proper toxic waste management facilities because they were in the business of producing industrial toxic waste themselves. Most Global South countries were not. "Like water running downhill, hazardous wastes inevitably will be disposed of along the path of least resistance and least expense," a congressman from New Jersey — a notorious departure point for hazardous waste to the developing world — concluded in 1989.[12]

Waste *diversion* exploded into an unbelievably profitable enterprise. You no longer needed to be like the Colberts and ship drums of banned chemicals to the developing world. You could make more money sending ships packed with the residue of *legal* chemicals. And, indeed, by the late 1980s, so much money was to be saved in exporting your hazardous residue abroad, rich countries could begin paying developing countries hard cash to take it.

Let's just call these arrangements what they were: bribes. The sums were frequently phantasmagorical. In 1989, Sudan was offered $300 million — $765 million today — to receive incinerated US garbage.[13] A company from Northern Europe proposed $84 million — more than $200 million today — to Zaire (Democratic Republic of the Congo) if it would agree to take its chemical refuse.[14] A waste disposal firm based in Germany presented a Mauritanian colonel named Brahim Ould

Cheibani Ould Cheikh Ahmed a personal gift of what would now be nearly $5 million if he could convince his government to accept 40,000 tons of hazardous material.[15]

"Thinking about making money? Hazardous toxic waste [is] a billion-dollar-a-year business. No experience necessary. No equipment needed. No educational requirements," ran an advertisement in the May 25, 1988, issue of the *International Herald Tribune*.[16]

4

DEBT AND DEVELOPMENT

> We cannot accept that at a time when industrialized nations refuse to buy our commodities at reasonable prices, these same countries are selling us death for ourselves and our children.
> — Economic Community of West African States, 1988[1]

IF THIS ALL sounds imperialist and exploitative, that's because it of course was. With the waste trade, one need not look very far to detect the embers-still-lit afterlife of colonialism. Sure, North American cities may have courted Latin American states, whose dictatorial regimes were often in hock to the US for the financial and military assistance needed to suppress plucky internal left-wing movements. But Western European countries were approaching countries whose ruling classes they had deemed their imperial subjects hardly a decade or two earlier.

Take the Republic of Benin, which in 1960 received its independence from France and, six military coups later, had fallen under the rule of Major Mathieu Kérékou, who proceeded to chart a Leninist course of development. By the mid-1980s, the tiny West African nation was in trouble. GDP per capita stood at half that of neighboring Nigeria, a third of that of Morocco. The government was struggling to pay its public sector, which comprised an astonishing 75 percent of its formally employed workforce. When striking schoolteachers began taking to Benin's streets, demanding four months of unpaid salaries, President Kérékou instructed his troops to shoot them. Fearing an overthrow, he'd ordered the entirety of his nation's ammunition stockpiles be stored beneath his two houses.[2]

Benin nevertheless had one advantage: Tucked into the Gulf of Guinea, and smaller than New England, Benin was a place not many international observers would likely countenance as a graveyard of the industrial output of the world's richest nations. In the mid-1980s, Kérékou made plans to raise much-needed cash for his nation by taking in toxic waste. "We in the Soviet Union firmly support the struggle waged by the African people against ecological imperialism, and together with them come out against any burial of nuclear or other industrial waste," declared Aleksey Litvinov, a Soviet radio commentator at the time.[3] Yet, according to a senior Benin government official forced into exile in the late 1980s, it was in fact the Soviet Union that was the first to dump "several tons" of radioactive waste in Benin — a nominal friend in the struggle against liberal capitalism — between 1984 and 1986.

Next came France, Benin's former ruler. In 1988, a deal was negotiated with Paris whereby Benin would receive a down payment of $1.6 million and thirty years of debt relief if it agreed to import French radioactive and industrial waste; President Kérékou made plans to bury the shipments in the outskirts of the ancestral village of Michel Aikpé, his erstwhile interior minister who'd been shot thirteen years earlier after he was caught sleeping with Kérékou's wife.[4] When that deal fell through owing to its leak to the French media, in came a Gibraltar-based British company, which arranged for Benin to receive at least fifty million tons of toxic residues over the next decade. It never happened. All the same, whether any of the $300 million intelligence officers would later find socked away in Kérékou's collection of Swiss bank accounts owed its existence to waste imports — how many schemes went undiscovered? — remains an open question.[5]

DEBT AND DEVELOPMENT

* * *

The uneven environmental regulations between the developing and developed worlds are one reason why the toxic waste trade got started. But it's not the only reason. The case of Benin points to another factor, one whose origins are worth pausing over to understand in detail.

One of the continued justifications of the waste trade today is that countries across the Global South are not *forced* to take garbage. They choose to import it. And this was true in the late 1980s too. Almost every country that received hazardous waste did so because it — or its ruling class — agreed to accept it. And it's worth asking why — why, that is, dozens of countries that had only recently unshackled themselves from decades or even centuries of colonial rule and had once held out such promise for their independent futures became so despondent in the late 1980s that they were willing to take in toxic pollution, often from their despised former colonial masters, and endanger the lives of their own newly liberated citizens.

Puerto Barrios, Guatemala's biggest port, is, as I mentioned, a set piece in how a certain type of colonialism worked in the twentieth century. Puerto Barrios may have had warehouses and quays and a railroad, but none of it belonged to Guatemala per se — and very little of the banana commerce it conducted enriched average Guatemalans. Across most of the Global South, however, even that fate would have been an altogether enviable one. Many states released from colonialism in the 1960s and '70s had a great deal less than Guatemala. At the time it was negotiating to take in radioactive waste from Paris, the Republic of Benin possessed just two networks of paved roads and a single harbor. Forced to specialize in the cultivation and exportation of a single raw commodity like cotton or cashews for the sake of consumers thousands of miles to their north, such nations entered statehood possessing pitifully little with which to build an economy that could raise living standards and employment rates and guarantee education and healthcare for their own dramatically increasing populations. And the problems, in the case of a place like Benin, were only poised to get worse; in 1980, half its population was under fifteen, even as its only major export, palm oil, was in decline owing in part to new competitors arising out of Southeast Asia.[6]

Developing countries needed to develop. They needed roads, factories, universities, hospitals. They needed to be able to invest in technology. They needed infrastructure, which they could use as collateral in order to borrow cash on the cheap. They needed balanced national economies that might be protected from shocks in commodity prices. And, at just the time when new environmental standards were overtaking their former colonial rulers, many of them dismissed fears surrounding pollutants and the environment: To nations that had existed for less than a generation and were thirsting for industrialization, talk of air quality or deforestation could smack of a luxury concern.

In 1973, however, something upended the Global South's attempt to break free from colonial inequalities. In October of that year, in partial response to the decision of the United States and other nations to rearm the Israeli military following the Yom Kippur War, the Organization of Arab Petroleum Exporting Countries spiked the price of oil, then began an "embargo" of Western states, which after the Second World War had become increasingly reliant on Middle Eastern hydrocarbons to run their cars, heat their homes, and power their factories.

Environmental historian Emily Brownell has argued that what became known as the oil crisis was to prove the all-important inflection point in the forthcoming history of waste as an object of globalized exchange. It spawned two important legacies. The first was financial. Within six months of the crisis's outbreak, by March 1974, the price of oil had risen from three to twelve dollars a barrel. Wealthy countries entered a near-decade-long period of economic stagnation, possessing less money to purchase agricultural goods or raw materials from developing nations, where costs of cultivation or extraction concurrently rose. As petrostates got rich off the boom in oil prices, they began divesting their supercharged revenues into Western banking systems — a phenomenon Henry Kissinger termed in 1974 "petrodollar recycling" — which in turn began lending that money out to developing nations. Previously reliant on public development banks such as the World Bank or the Inter-American Development Bank for the capital needed to industrialize, countries from Costa Rica to Cameroon now began borrowing cash from Wall Street. A self-reinforcing triangle snapped into place: Middle Eastern countries placed their revenue into Western banks, which in turn funded poor countries' ability to purchase ever more expensive oil from the Middle East. And, over time, a cruel, slow-motion trap was set for one country after another across the Global South: Owing to deteriorating returns on their commodity exports, many would prove incapable of paying back their loans. By the early 1980s, many were plunged into extraordinary debt.

Debt was the first legacy of the oil crisis. The second pertained to the question of what constituted a "resource." As global prices for raw commodities like aluminum and timber spiked, Brownell notes, there came a demand for cheaper, secondhand alternatives that could be pulped or melted down and cycled back into production streams.

By the late 1970s, certain types of waste — aluminum cans, newspapers, cardboard boxes — would get relabeled as "scrap" and rerouted from Western landfills to Western ports. The Global South had historically provided raw materials to the Global North. Now the Global North was exporting the contents of its garbage bins to the Global South, where they were purchased like raw materials, then used to manufacture new objects, much of it for Global North consumption.

As Brownell argues, these parallel aftermaths of the oil crisis — recycling petrocapital into banking systems and then issuing it to developing countries as "loans"; recycling aluminum cans and newspapers and then exporting them to developing countries as "raw materials" — amounted to variations of the same thing: "repackaging risk and profiting from its repurposing in the developing world."[7] The Global South, in other words, was no longer just a half of the world from which resources — bananas, palm oil, gold — could be appropriated and extracted. It was increasingly a repository where *over*accumulations could be *put*. As the case of the Colberts has already demonstrated, it was a two-tier system liable to abuse by bad actors. And as the post–Cold War explosion of the waste trade shows, it eventually resulted in the blurring of any meaningful distinction between "resource" and "waste." By the 1990s, if you didn't want something within your borders anymore, one recourse was to label it a "resource" or "recyclable material" and sell or even donate it to the Global South.

There was to be one additional twist of the screw. In the mid-1980s, developing countries increasingly turned to the International Monetary Fund to help them pay off the debt that had begun to accrue over the previous decade. The IMF had, for its part, been more than alive to the problem of the emerging toxins trade, asserting "deep concern... that the widespread emphasis on environmental issues in industrial countries is inconsistent with the frequent use of developing countries as 'pollution havens.'"[8]

And yet the IMF would ultimately play its own shortsighted and destructive role in helping alleviate poorer countries of that foreign pollution. For structural adjustment—the package of reforms and policy conditions a country must agree to in order to receive financial assistance from the IMF—often stripped developing nations of valuable resources around which they might build independent economies of their own and redress colonial-era imbalances. Gold in West Africa, copper in South America, forests in East Asia—all were subject to heavy-handed privatization in the 1980s, often to Western firms, leaving gaping craters where future economic lifelines had been imagined, and making the money that could be raised through toxic-waste importation a far more attractive option than it ever should have been to Global South nations. At the same time, structural adjustment tended to require developing countries to level barriers to international trade. In return for debt relief, what often followed was a deluge of cheap foreign products, undercutting local industry and steamrolling the capacities of existing—or scarcely existent—systems of waste management, whose budgets were simultaneously slashed to the bone. The very economic forces that put so much plastic in the Global South in the first place would also restrict its ability to manage it properly.[9]

Much of that was still to come. In the meantime, while many developing countries may not have had an abundance of factories or ports or roads, they did tend to have something else: big tracts of uninhabited land ideal for dumping toxic waste. In agreeing to take the asbestos or sewage sludge of the very countries to whose banks they were often indebted, Global South countries could receive hard currency with which they could attempt to pay off those debts. The indignity of the hazardous waste trade — and the iterations of the waste trade that were yet to come — stemmed not from the fact that developing countries agreed to accept trash. It's that they thought they had no other option. For many, the choice seemed to be poison or poverty.

"We are being asked to commit genocide for a few dollars more and so leave a legacy of contamination for our children," lamented Arthur Robinson, the prime minister of Trinidad and Tobago, in 1988. "You don't know, but you mortgage the entire history of your people," Omar Sey, the foreign minister of the Gambia, claimed that same year.[10]

And it wasn't just the Global South. When it came to garbage, the developed world's relationship with the Eastern Bloc was to prove nearly as exploitative. By 1987, the cost of disposing of a ton of hazardous waste in West Germany was ten times that of trucking it across the border and dumping it in the other Germany. Over the course of the late 1980s, millions of tons of domestic and hazardous waste bypassed the otherwise impenetrable Iron Curtain, traveling from West Germany to East Germany to be scattered in landfills that were off-limits to East Germans themselves. No greater destination existed than Schoenberg, "the beautiful mountain," which sat only five miles east of the West German city of Lübeck and would eventually become Cold War Europe's largest dump, taking in a million tons of West German, Belgian, Danish, and Dutch waste every year. In return, East Germany received annual installments of $50 million, a figure that by the Cold

War's end amounted to *more than 10 percent of its hard currency income*. "We need the foreign currency," Marianne Montkowski, an East German environmental official, explained.[11]

One of the legacies of the waste trade was to upend the future that developing countries had foreseen for themselves during the great decolonization movement a decade or two earlier. In the 1960s, the ability to generate waste had widely been considered the signpost of a healthy, prosperous society. "The colonist's sector is a sector built to last, all stone and steel," the Martinique-born political philosopher Frantz Fanon wrote in 1961. "It's a sector of lights and paved roads, where the trash cans constantly overflow with strange and wonderful garbage, undreamed-of leftovers."[12]

In the 1980s, waste was increasingly the indicator of a poor, exploited society. "The alarm expressed by enlightened citizens over the dumping of toxic waste in many African countries is yet another indication that our people still do not understand the nature of the relations between the West and the rest of us," observed the Jamaican political scientist Patrick Wilmot in 1988. "We are, in the immortal words of the rock and roll song, 'so near, yet so far away.'"[13] Developing countries became stuck with toxic pollution they themselves had not produced, while rarely getting any of the industrialization they had sought or, in many waste deals, been promised. "Africa already has enough problems of her own, without becoming the garbage bin of the wealthy northern nations," Robert Mugabe, the president of Zimbabwe, would tell the United Nations General Assembly in 1988, the year West Africa alone took in twenty-four million tons of hazardous waste from industrialized nations. "It is not fair that the poorest nations should be the ones to suffer the worst effects of a program in which they do not share."[14]

By 1988, as the historian Matthew Sohm has observed, something extraordinary had happened: The total value of toxic waste being shipped from the Global North to the Global South was larger in absolute dollar terms than *all* developmental aid flows between those same regions. Waste trade revenue hadn't just come to exceed the budgets allocated by countries like the United States to help poor countries build hospitals, fight diseases, educate their children. In certain respects, it had come to replace them.[15]

To consider how waste came to overtake foreign aid as a tool of development, and how cash for trash increasingly rivaled industrialization as an economic future for newly independent states, take a look at the truly bizarre case of the Marshall Islands.

In the late 1980s, California was struggling to handle its mounting piles of waste, much of it toxic. Landfill space was decreasing; disposal costs were rising. Joined by the states of Oregon and Washington, California looked west for a potential solution, five thousand miles away, to the Marshall Islands. The remote scattering of coral atolls had already suffered decades of environmental injustice as an epicenter of US nuclear weapons testing when, in 1989, ten years after its right to self-government was recognized by the United States, a board of Seattle businesspeople calling themselves the Admiralty Pacific Group — "having the philosophy of dealing with waste products (garbage) as a secondary resource," they claimed — approached the archipelago's first president, Amata Kabua, with an idea: taking their trash.[16]

California, Oregon, and Washington may have collectively constituted 333,000 square miles and the Marshall Islands a mere 72, but the three states also possessed 34 billion pounds of garbage they were willing to pay eye-popping sums of money to get rid of. And, owing to low shipping rates, it so happened that sending the West Coast's garbage five thousand miles away to the Marshall Islands cost twenty-two dollars per ton, or roughly a quarter of what it then cost California, Oregon, and Washington to truck their waste to their own landfills. As one member of the UN Environment Programme put it: "It's cheaper to barge it down there than to move it inland forty miles."[17] And, finally, while the Marshall Islands might have technically been independent of the United States by the late 1980s, that sovereignty was, at best, conditional. Sixty percent of the island chain's operating expenses were paid for by the US, which funded everything from its defense to its postal service.

In return for accepting two billion tons of US garbage a year, Admiralty Pacific offered the Marshall Islands a revenue stream of $139 million, which would more than cover whatever annual operating expenses were not paid for by the US. The company went on to assure President Kabua that certain types of US garbage, such as tires, could be put to practical use: boosting tourism by serving as construction stock for a new artificial causeway to be built atop coastal reefs linking the islands of Ebeye and Gugeegue.

"We realize there is no such thing as nontoxic, nonhazardous garbage," Dan Fleming, Admiralty Pacific's vice president, conceded, but even toxic trash had its uses: making sure Kabua's country didn't disappear, by shoring up its coastlines against rising sea levels with millions of six-foot-by-six-foot bales of compacted waste weighing one ton each, capable of increasing the Marshall Islands' landmass fivefold in twenty years if piled incrementally out from its beaches.[18] "It will murder us financially, but we'll probably have to dispose of that waste in the U.S.," Admiralty Pacific's executives lamented as President Kabua took his time weighing the merits of turning his sinking island nation into the West Coast's dumpster.

The problem, as critics of Admiralty Pacific pointed out at the time, was not just that a big, rich country had abundant financial and environmental incentive to unload its refuse upon a remote archipelago already vulnerable in the 1980s to the onset of the climate crisis. There was also the precedent that such a transaction could set.

Suppose an emerging nation like the Marshall Islands were to become dependent on the revenue stream that came with accepting the West Coast's garbage. What would happen should its annual expenses, say, increase? In the Marshall Islands this was all but guaranteed. Too remote to rely on tourism, too poor to industrialize, its future as a self-sustaining farming and fishing economy was already in doubt: More than a third of the Marshall Islands' population was below the age of fifteen in 1989, a time when swathes of its land were still unsuitable for agriculture owing to the sixty-seven nuclear bombs detonated on, in, and above its shores in the postwar decades. President Kabua acknowledged such concerns himself. "We are alarmed by the rate of our [population] growth," he answered critics of the Admiralty Pacific deal. "And I certainly want to leave this government with enough money to go on. I don't want to leave it broke."[19]

Furthermore, what was to stop the Marshall Islands from agreeing to take in more lucrative — and more dangerous — types of US garbage? The prospect was already being floated by Admiralty Pacific before Kabua could even reject its initial offer. In June 1989, one of the company's internal memos was leaked to a reporter at the *Marshall Islands Journal*, detailing Admiralty Pacific's "secret plan to haul nuclear waste to the Marshall Islands...under the name of another entity to be formed."[20] It was a scheme that left the residents of the Marshall Islands in the unusual position of pleading with the government of Japan — the nation that had formerly occupied them — to help save them from unwanted toxic incursions from the United States, the country that had purportedly liberated them. "We lost our homes when the U.S. conducted atomic bomb threats," ran a letter sent by the community of Bikini Atoll to Tokyo in late 1989. "Now they are planning to turn our islands into their toxic waste sites. To halt such insult to our people, we need your support."[21]

5

MERCHANTS OF DISEASE

> I consider [the waste trade] as important as the atomic bomb.
> It is very serious.
> — Thomas Kahota Kargbo, ambassador of Sierra Leone to
> the United Nations, 1988 [1]

NO GARBAGE EVER did travel to the Marshall Islands. And yet the very prospect of such an exchange — a Pacific island chain expanding fivefold in size owing to shiploads of toxic trash dispatched five thousand miles from Los Angeles and Seattle — underscores the absurd dimensions of the new economic order that waste diversion was generating.

We often like to think of the 1990s as the decade when globalization fully flattened and interconnected our world. But at the twilight of the Cold War, hazardous waste transfers were already knitting the most far-flung, improbable corners of the planet together in the strangest, most head-scratching of exchanges. Old toxins were sent to peripheries where little else had any reason to go. Waste shipments from Philadelphia and Rotterdam and Stuttgart were arranged for dispatch to the beaches, jungle ravines, and abandoned mines of Haiti, Congo, and Morocco. They specialized in going to war zones like Lebanon, wrecked states like Eritrea, and borderlands from Somaliland to the Sahel. The American country singer Kenny Rogers proposed building a waste incineration plant in the Bahamas. Cows in Nigeria were killed by fumes from polychlorinated biphenyls (PCBs) that had been used in the production of automobiles in Milan. Residents of Sierra Leone's capital of Freetown complained to their government of "choking and involuntary tears" after being exposed to ammonia shipped from Britain via a "Lebanese businessman" and dumped along the Rokel River.[2] Zulu villages along the banks of South Africa's Mngcweni River were poisoned with mercury that had been discarded in Louisiana and New Jersey.[3] Guinea and Nigeria resorted to taking hostages — the former a Norwegian honorary consul, the latter the crew of an Italian ship — in order to ensure foreign waste would be removed from their coastlines.[4]

"It was the Wild West. There were all manner of lawless people involved," Roland Straub, a Swiss lawyer, told me. Straub was approached by numerous Italian businesspeople over the course of the 1980s with plans to construct incineration plants in Africa that could burn Europe's hazardous waste. The Skeleton Coast of Northern Namibia — a stretch of desert so remote and unforgiving it is littered with the carcasses of wrecked ships dating back to the Portuguese explorers — was a location of recurring interest, he explained.

"Europe is too densely populated. You just can't build incineration plants with all the requirements that are needed. You would need to look into the tundra in Finland or something. Not even Germany has that kind of space. Building a plant in Africa, but especially in Namibia, was for many reasons quite ideal — good climate, possibly skilled labor, relatively good infrastructure, and relatively stable — whereas Angola next door only became more and more unstable over time. It made absolute sense."[5]

Among those who approached Straub was a Swiss Lebanese soldier of fortune by the name of Arnold Andreas Künzler. A veteran of the British forces of Kenya until their disbandment in 1964, later deploying in the Congo Crisis, in which he and a small contingent of European and South African mercenaries helped crush Simba rebels who had overtaken the breakaway province of Katanga, Künzler would eventually run in the inner circle of Idi Amin, the brutal dictator of Uganda. After serving in several of Africa's wars, Künzler later crisscrossed the continent supplying them — providing weaponry to dozens of conflicts and uprisings breaking out in the 1980s — before pivoting to the world of toxic waste, one he would then never leave.

"My heart beats for Black people," Künzler assured *Die Tageszeitung*, or *Taz*, the German newspaper.

Oil, diamonds, gold — in the 1980s the rest of the world was trying to extract all these things from Africa and ship them north. Künzler was bent on hustling garbage south and utilizing sub-Saharan deserts to burn and bury it. Post-colonial Africa, he contended, needed its own version of the Marshall Plan, a developmental project funded by the money that could come with taking in the detritus of Western industrialization. It was Künzler who first narrowed in on the opportunity of Namibia, a scheme in which he claimed to have received financial backing from figures ranging from Swiss telecom executives to Texas oilmen to the sultan of Brunei. He later shifted his plans to Angola, a country in its second decade of civil war and desperate for capital. In exchange for taking five million tons of chemical waste, to be burned in a pair of Swiss-constructed incinerators, Angola would receive two billion dollars in developmental aid, fifteen thousand jobs, and an unspecified number of hospitals.[6] "If it is good for the Swiss, it is good for the blacks!" he fired back at skeptics.[7] "Künzler was a colorful man," Straub told me; he turned down the chance to invest in Künzler's Namibia scheme.

Toxic waste export was not a crime in the 1980s. Yet it still tended to attract dubious characters. Künzler, by 2004, had relocated to Mexico, where, now in his seventh decade, he had hatched a new scheme to import truckloads of US waste and incinerate it in facilities constructed across the state of Tlaxcala.[8] Out of Detroit there was Robert Zeff, a personal-injury attorney who had made his name building casinos and jai alai frontons from Nevada to Connecticut, and owned a museum's worth of Salvador Dalí paintings; he also had brokered the divorce of Henry Ford II from his wife Cristina. Zeff arranged waste export deals to ship millions of tons of paint sludges and pesticide residues to West Africa. Out of Argentina there was Isidro Kicillof, a lawyer who struck garbage deals between Miami and Central America. Out of Turkey there was Mehmet Büyüktemiz, a lieutenant who worked for a contingent of the West German air force headquartered at a NATO radar station in Central Anatolia; Büyüktemiz had a reputation for strutting around with a holstered pistol, and used his connections to Bonn to arrange the shipment of industrial waste from the factories of Baden-Württemberg to the plains of western Turkey.[9]

Though toxic waste export was not yet a crime, it nevertheless helped that evidence of what had been done could disappear from sight, typically dumped in deserts or mines or the tarmac of any number of roads laid across developing countries. In El Salvador a highway leading out of the capital, San Salvador, was paved with a mixture of asphalt and toxic, probably American, waste.[10] In Somalia, at the height of the country's civil war, mysterious "roads to nowhere" were laid through the deserts of Puntland, leading Italian journalist Ilaria Alpi to travel to Mogadishu in early 1994 to investigate the remarkable allegations surrounding their origins: that the warlord Ali Mahdi Muhammad had brokered a deal with the 'Ndrangheta, an Italian organized crime group, to accept three thousand tons of toxic waste from Italy — Somalia's former colonial ruler — in exchange for weaponry with which to launch his bid for Somali leadership after the ousting of its longtime dictator in 1991. According to the United Nations and the Italian Parliament, the "roads to nowhere" may have been paved out of colossal quantities of tarmac manufactured to hide the toxic waste. The claims remain difficult to verify. That March, Alpi and her camera operator were assassinated in Mogadishu — because, or so confessed a member of the 'Ndrangheta put on trial in Italy fifteen years later, the two had witnessed the arrival of separate shipments of hazardous waste to the Somali port city of Bosaso some weeks earlier.[11]

The question of whether Somalia did receive hazardous waste from Western countries was to have a contentious afterlife, one that would circle back to some of those nations that may have been on the exporting end. In 2008, following that September's successful hijacking of the MV *Faina*, a Ukrainian freighter, a spokesperson for the Somali pirates defended their actions to the *New York Times* with a reference to what foreign toxic waste had done to the coastal waters they had once fished for a living and now no longer could. "We don't consider ourselves sea bandits," explained Sugule Ali. "We consider sea bandits those who illegally fish in our seas and dump waste in our seas and carry weapons in our seas."[12]

The most alarming aspect of the hazardous waste trade? It's not what we do know. It's what we don't. For there is a reason why a relatively large number of details exist on those schemes that targeted countries like the Marshall Islands or Angola: They were leaked to embassies, customs officials, or newspapers before the transactions could be completed. "What we were able to report on is certainly a fraction of what actually happened," Bill Lambrecht, who investigated toxic waste schemes targeting South Africa and the Caribbean for the *St. Louis Post-Dispatch* in the late 1980s, told me. "From a journalistic perspective, waste export fell between the cracks. The foreign bureaus and the environmental desks — such as they were in the 1980s — just weren't talking to one another about it."[13]

And the problems that could come from exposing a scheme might prove unending. "We jumped on the story right away," Sam Oduche told me. Oduche was studying architecture as an exchange student in Pisa, Italy, in 1987 when he and several other Nigerians read local press reports about thousands of tons of toxins that had been shipped from Italy to their native land, then proceeded to inform their national press by telephone. "And as a result of what we did, most of us have never been able to go back to Nigeria. You have no idea how dangerous this was. We were never treated like heroes in Nigeria. To our politicians and military, we became an enemy." [14]

Those deals that *did* happen occurred precisely because they never got exposed — and likely never will.

6

GUNS AND GERMS

On the fifth day of our march, the pioneers, who went in front with the guide, came to a great lake, looking like an arm of the sea.
— Hernán Cortés, *The Fifth Letter of Hernán Cortés to the Emperor Charles V*, 1526

WHAT HAD ORIGINALLY led me to Guatemala was not the aborted fate of the Las Escobas sewage sludge port in the summer of 1992. It was rumors of something that may have happened around that same time, or perhaps a few years earlier, far from the crystal teal waters of the Caribbean Sea, some hundreds of miles to the north, square in the middle of one of the world's biggest jungles.

It all began with a letter I tracked down from a lawyer named Martina Langley to James Nations, an anthropologist from Texas who had spent years living in the Maya villages of southern Mexico and Belize. The letter was composed on a typewriter. It was signed by hand. And it bore a date: July 21, 1992.

"Hazardous Waste" ran its subject line.

United Fruit Company may have gotten its start in Guatemala cultivating bananas and shipping them north from the quays of Puerto

Barrios. But by the 1940s it had grown obscenely powerful, overseen by a board of directors that had come to include the head of the CIA and his brother. So powerful had United Fruit become, in fact, that in 1954 the United States helped plot the overthrow of the newly elected president of Guatemala, Jacobo Árbenz, who had campaigned on a platform that called for the redistribution of the company's sprawling plantation systems. The coup was successful, a rural uprising mounted against it, and over the next thirty years the United States armed the Guatemalan state to the hilt to eliminate all resistance.

Martina Langley moved to Guatemala at the pinnacle of the fighting. Her extraordinary life was to bear not a few resemblances to that of Rachel Carson. A lieutenant in the US Women's Army Corps during the Second World War, Langley went on to earn a law degree and take it to Central America, where she began conducting fieldwork, often in villages in the Guatemalan highlands. Langley would spend years investigating the disaster that US chemicals exported to the rest of the world had inflicted on unsuspecting populations. And much of her time in Guatemala would go toward gathering evidence in support of a stunning claim.

In the fall of 1974, the United States government had begun offloading millions of contraceptive devices known as the Dalkon Shield onto the poorer countries of the world. The devices were not merely of bad quality; they were known to be extremely dangerous, capable of leading to pelvic infections and even death, for which reasons their sale had been banned in the US in the spring of 1974. Only several months later, stuck with millions of contraceptive devices it could no longer sell to Americans, the producer of the Dalkon Shield, A. H. Robins Inc., slashed its price in half and approached the US government with a plan to place "this fine product into population control programs and family planning clinics throughout the Third World."[1] It worked. Boosted by millions of dollars in US developmental aid, hundreds of thousands of unsterilized Dalkon Shield products were shipped south from the

United States, turning the women of countries like Guatemala into the unknowing recipients of a contraceptive deemed too dangerous for their counterparts in the US.

Langley's lifework was to help bring the executives of A.H. Robins to class-action court. Against considerable odds, the suit was successful. Even then, in the early 1990s, and by now in her late seventies, Langley wasn't done. She spent the next years flying back and forth between the United States and Guatemala, delivering suitcases of cash to the indigenous Maya women who had been the disproportionate victims of the program.

That was one devious import scheme that Langley heard about. But soon she started hearing about another. And here's where her letter to James Nations, the Texas-based anthropologist, comes in.

In the early 1990s, Langley began spending extended periods of time busing through the Petén, an expanse of jungle the size of Switzerland that juts out of the north of Guatemala like a dorsal fin. A land teeming with archaeologists and jaguars and narcos and cowboys and oilmen and mosquitoes the size of cockroaches, the Petén was where Langley first heard something so incredible it made the case of the Dalkon Shield contraceptives seem almost benign by comparison. Hundreds of miles from the nearest city, a thousand miles from the nearest industrialized nation, Maya villagers whom Langley met insisted that drums of foreign waste had secretly been smuggled onto their land. "They speak to me of dumps in the Petén and other remote places," Langley confided in her 1992 letter to Nations. "I believe the money involved in this field" — toxic waste importation — "will exceed that of drugs and may take its place as a fueling force in the country."

There was an opportunity here, Langley reflected, to investigate the villagers' disturbing claims — and perhaps shed light on yet one more instance of the United States turning Central America, and its desperate populations, into a receptacle of its own banned fertilizers, its flawed medicines, its overstock weaponry, its Cold War containment policies, and, maybe just maybe, its toxic waste. "It just seems to be a slot waiting for someone who does not have too much to lose," Langley concluded, lamenting that she had recently been diagnosed with leukemia and no longer possessed her former gumption to take up the investigation. She would spend her last days in Section 8 public housing along a highway in Central Texas, in a small apartment decorated with Guatemalan crafts, often dressing in a huipil, the traditional embroidered blouse of the Maya.²

Thirty years later, I took a propeller plane north from Guatemala City to the shores of Lake Petén Itzá, an emerald-blue dot in the middle of a shaggy carpet of olive green that stretches in every direction as far as the eye can see. The plane touched down outside the administrative outpost of Flores, where a Catholic church now squats atop the razed ruins of a Maya temple, where whiskey saloons abut machete shops, and where coach buses shuttle pasty tourists north toward the monumental ruins of Tikal.

The Petén is an exhilarating place. Landowners bear the very names of the conquistadors who sabered through its jungle lowlands half a

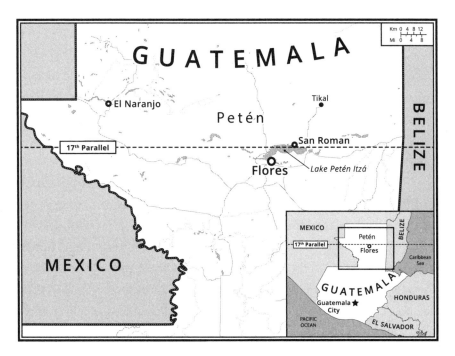

millennium ago, spreading disease and horse and musket. Ten-gallon hats are worn without irony. The pickup truck is driven almost without exception. There are workers who make meager wages excavating the tombs of ancient Maya aristocrats from whom they descend. There are Mennonites in Wranglers. There are criminals. There are conservationists with Glocks. There are state prosecutors trained in bombing the drug-trafficking airfields that scar the border with Mexico. There are dark memories of massacres and mass kidnappings, some quite recent, occasioned by uneasy glances around the room and related in nervous whispers.

On my first visit to the Petén, a sixty-six-year-old American birdwatcher had disappeared in the jungle some months earlier; a jaguar had claimed him, or so the Peteneros concluded, shrugging their shoulders. Latter-day hidalgos strut the grungy streets of San Benito and El Remate in cowboy boots, as do northern adventurers seeking a hideout — or a quick buck. Former "communist" guerrillas, whose quashing appeared to necessitate a Cold War, and the latest weaponry Washington could muster, can now be found in jungle communes, presiding over freshwater fish farms. Along the shores of Lake Petén Itzá, cattle ranches are so vast they could constitute their own zip code, though many are devoid of a single udder; some are owned by the narcos who require laundering fronts for their cash, others by the unbelievably corrupt politicians in Guatemala City who entered office bent on acquiring the best asset money can buy: a fat chunk of the manically developing jungle.

This is no ordinary place. If Guatemala has historically been an auditioning stage for US foreign policy, what the historian Greg Grandin has called "the empire's workshop," where "counterinsurgency" tactics and "nation-building" methods are tested and honed before they are brought to bear on more distant parts of the world, then the Petén is in many ways the ultimate laboratory.[3] For the Petén has historically been to Guatemala's own political and military elites what Guatemala itself has been to the United States: a frontier ripe for extraction, a land to be tamed, colonized, and exploited through a "civilizing mission."

And it is *remote*. In 1525, the Spanish conquistador Hernán Cortés managed to reach Lake Petén Itzá, where he left behind a wounded horse, but it took the Spanish a hundred years to find their way through the jungle back to its shores. It was a wandering pair of Catholic friars who finally managed to locate Itzá again, by which time — to the astonishment and consternation of the clergymen — its Maya community had begun worshipping idols dedicated to the limping four-legged creature Cortés had left their ancestors. The priests ordered the last Maya capital in the New World to be burned to the ground and the last Maya royal family to be imprisoned by a Spanish army; it would take nearly a hundred more years to do so.

Ever since, the Petén proved a place where Guatemala's problems seemed to exist in concentrated and exaggerated form. One-third of the country was home to one-twentieth of its population, indigenous communities scattered throughout an impenetrable wilderness that appeared immune to modernization and had become, by the 1960s, an incubator for anti-government activity. That decade, the United States helped its client elite in Guatemala City "civilize" their northern expanse through a policy of "whitening" its population. Surveyors from Washington, DC, mapped out a network of roads to be paved through the jungle. The World Bank funded the construction of an airport. Tourists and film crews were guided to ancient cities freshly excavated by US archaeologists. If the Petén wasn't going to join the rest of Guatemala, then the rest of Guatemala was going to push its way into the Petén. And as Norman B. Schwartz, the famous anthropologist of the Petén, has argued, the colonization of the Petén during Guatemala's civil war may outwardly have been an attempt to develop the region, but what it also offered was an opportunity for a rising echelon of military commanders to use all means imaginable — stealing land, smuggling timber, surveying for oil — to enrich themselves once its indigenous populations had been crushed or cleared out.[4]

"If security is finally established in the northern frontier — as we believe it will be — and development proceeds," reads a CIA memo on the Petén dating to 1982, "the region could help ameliorate the serious economic and social conditions plaguing Guatemala by providing land to many landless peasants, increasing the amount and variety of agricultural products for domestic consumption and export, and reducing balance-of-payments problems through forestry, tourism, and oil production."[5]

It's not hard to see why the Petén would have been as attractive a place as any to dump hazardous waste in the 1980s. It was a microcosm of the Global South. And with so many vehicles full of weapons and cement and timber and palm oil being directed in and out of a theater of conflict larger than the state of Maryland, who was going to ask any questions about the arrival of some trucks bearing heavy steel drums?

In the 1990s, with the end of military rule in Guatemala and the establishment of a putative democracy, it wouldn't be only Maya peasants telling foreigners like Martina Langley about the shipments of toxic waste they had witnessed getting hustled onto their lands.

In May 1992, at just the time when the plans to construct the Las Escobas port along the Caribbean coast were being made, the head of one of Guatemala's newly minted political parties came forward and informed the Guatemalan press that "thousands of metal barrels containing toxic and radioactive waste" had been discarded at the "edge of a little natural lagoon" located along the "17th parallel north latitude," which just so happens to cut across the northern shore of Lake Petén Itzá itself.[6] The man identified no year. He identified no exporting nation. He identified no importing party. But he was no stranger to the seedier aspects of the Petén's recent history: His father, a ruthless military dictator of Guatemala in the early 1970s, had played an enthusiastic role in clearing insurgents and indigenous peoples out of the Petén's jungles and handing its land over to ranchers, army pensioners, and oil speculators.

What, I wondered, was the story being related here?

Was the waste received by his father, the military dictator, during his reign in the early 1970s, then deposited in the Petén? If so, it would mean foreign toxic waste had been stashed in Guatemala years before it was dispatched to other parts of the Global South.

Or maybe, like most shipments of hazardous waste to the developing world, the Petén story dates to the late 1980s, and only became public in 1992. Was it an attempt by the son of a military dictator to dredge up an older story and pin it on the political class that had turned his father's army out of power in 1986 — and, by the early 1990s, was threatening to put members of that military on trial for war crimes?

Or say the claim dated to 1992, the year it was leaked to Guatemala's press. Would it really have been necessary for toxic waste to get trucked hundreds of miles into the jungle the same year that Guatemala was planning the construction of a massive sewage sludge import facility along its eastern coastline?

Or — just as possible a scenario as any — could it all have been made up?

Over the course of five weeks in Guatemala, I asked around about what exactly Roberto Arana — the man who had revealed the details about radioactive and toxic waste importation into the Petén and who had died in 2004 — had been referring to with a story that bore an uncanny resemblance to that of the hazardous "dumps in the Petén" that Maya peasants had told Martina Langley about in the early 1990s. What I found were dozens and dozens of Guatemalans and outsiders, from all sectors and classes of society and of all political bents, who, even at a time when the depravations of the last half century in Guatemala were finally coming out into the open, and even as Guatemala's judicial system was making an admirable effort to bring many of their perpetrators to justice, were reluctant to discuss whether such a toxic waste shipment even *might* have happened. It was a taboo, sinister, unapproachable subject.

Most people who agreed to speak to me were skeptical. "My father was governor of the Petén in the early 1990s," Francisco Asturias, director of the Petén's National Park Mirador–Río Azul, told me. "I used to hear about this toxic waste shipment all the time. But I have no idea if it's true."[7] Many wondered how such a story could have been kept quiet, even as they conceded that bloodier crimes that had occurred in wartime Guatemala—and particularly in the Petén— stood zero chance of becoming the object of public discourse. "Guatemala is full of rumors that keep getting regurgitated time after time. That said, anything is possible here," Eugenio Gobbato told me; he was an environmentalist who served as a consultant for Guatemala's social investment fund from 1986 to 1989.[8] Donald J. Planty told me, "Guatemala is a small place and it is very hard to keep a secret long in Guatemalan society." Planty served as US ambassador in Guatemala City from 1996 to 1999. "The Guatemalan media is notoriously unreliable and has a proclivity toward sensationalism."[9]

At the same time, some people believed it could have happened, and the consensus tended to be this: If it was true, if the shipment did occur, it was an army operation, and therefore a military secret whose last known caretakers were, in all likelihood, dead or imprisoned. "If Roberto Arana was so specific with this claim, it probably could have happened. If he was lying, he was probably trying to gain public exposure and a few minutes of fame," Jorge Serrano Elías told me by phone from Panama City; he was president of Guatemala at the time of the alleged shipment. "But the details in the claim seem very specific to me."[10] Magalí Rey Rosa, one of Guatemala's leading environmentalists and director of the Savia School of Ecological Thought in Guatemala City, was blunter: "Oh, of course it fucking happened."[11]

For weeks I drove round and round the Petén, searching for the waste, any sign of abnormality, any mound of mud that looked discolored, any parcel of jungle that appeared tainted or bleached. I brought a Geiger counter to measure radiation, held it out my car window, waiting for a quickening succession of beeps that never came. I asked lake fishermen, who stared at me like I was deranged. Doctors with decades of experience in jungle villages hadn't witnessed any inexplicable diseases.

The septuagenarian engineer from Catania, Sicily, who oversaw the construction of the Petén's airport from 1978 to 1981 knew nothing of any waste, though an industrialist I met in Guatemala City told me that he heard it was buried beneath its runway. An archaeologist who once heard the story of a local mayor contracting a near-death illness while illegally digging for Maya antiquities on the island of a Petén lake called Macanché located the former mayor on my behalf, only for me to learn that he had been exposed to no toxic fumes, no deadly sludge, no radioactive material, but a vicious parasite contracted via black jungle fly.

I called Roberto Arana's daughter, who works at the Guatemalan Embassy in Washington; she insisted her father was right, the waste shipment had happened. I tracked down Roberto Arana's cousin, who lives on a cattle ranch along the Mexican border only easily accessible by Cessna aircraft and who claimed to have no idea what I was talking about. I scoured Google Earth and conducted digital tours of the littoral of lagoon after lagoon after lagoon. Would the water need to be accessible by road? Almost certainly. Would the road need to be paved to accommodate a truck bearing heavy steel drums? Probably. While Roberto Arana had spoken of the waste being dumped at the "edge of a little natural lagoon" located along the "17th parallel north latitude," it was hardly a pinpoint location. The Petén, at that coordinate alone, is pocked with a dozen or so major lakes, hundreds of lesser ones, and thousands upon thousands of sinkholes; every year, scores of new water bodies are formed by the collapse of the jungle's limestone bedrock owing to accumulated rainwater, a phenomenon that sometimes releases a thunderous noise into the surrounding forest and leads locals to believe that a meteor has struck their land. "It's not impossible that whatever waste was dumped thirty years ago would still be there," Mark Brenner, a limnologist at the University of Florida, told me. He has spent decades studying the water bodies of the Petén. "But in the Petén, you're really hunting for a needle in a haystack."[12]

Finally, in the glum little village of San Roman along the shores of Lake Petén Itzá, I found someone who had seen something. Carlos

Salazar was in his late sixties. He was a short man, with strands of frizzy brown-gray hair running down from an oval head, whose large eyes seemed to glow as he spoke in a careful, syllable-by-syllable march of words. In May 1991, Salazar was working as a wildlife conservationist when he was sitting in a café in the Petén town of San Benito and witnessed something he's never been able to properly explain. Three orange pickup trucks passed through the community. Their beds were protected by meshed caging. Their license plates were Mexican. And, most curiously of all, across their doors was emblazoned the radiating trefoil symbol that is the international symbol of radioactive material. "I saw these trucks and said aloud to my wife, 'Oh, my fucking god! What is this?'" Salazar told me, brimming with excitement when I showed him the newspaper clippings detailing the claims Roberto Arana made in May 1992. Salazar dropped his eyes and asked which intelligence agency I worked for. "No one would believe me that it was hazardous waste. They thought that trucks like that must be coming from local oil wells. But I know what I saw. I know what I saw."[13] Salazar, who is suffering from an advanced form of cancer, said that in the weeks that followed, he witnessed similar trucks in several other towns of the Petén. He could never convince other environmentalists that their presence wasn't related to oil drilling.

Did it happen? Salazar's story bears a startling similarity to one additional claim I was able to dig up in the dust-frosted archives of the Biblioteca Nacional in Guatemala City. It dates to September 1990, eight months before Salazar said he saw the pickup trucks, and I found mention of it in a brief compilation of potential toxic waste shipments into Central America that had been cataloged by local Greenpeace researchers. It references the "arrival into the Petén of toxic and radioactive waste by land from Mexico," adding only that the barrels may have been dumped not in a body of water but "in a sanitary landfill," though no such thing exists in the entirety of the Petén. Greenpeace workers with experience in Guatemala in the early 1990s later told me they knew nothing of the story or where it had come from, though a

Mexican expert on radioactive waste, Bernardo Salas, considered the timing of the allegation eyebrow raising: Only two months earlier, Salas told me, Mexico's first nuclear plant, a facility outside the Caribbean port town of Laguna Verde, four hundred miles north of the Guatemalan border, had gone into commission. But in September 1990 the Laguna Verde Nuclear Power Station would not yet have produced radioactive waste requiring disposal — and besides, added Salas, there existed a preponderance of places within Mexico to, if necessary, stash such toxic trash.[14]

And so, once more, the trail went cold.

I left Guatemala convinced of only one thing: If it did happen, dumping toxic waste into a little natural lagoon somewhere in the middle of Central America's largest jungle really was the most perfect crime conceivable. Here was a multimillion-dollar exchange in which all evidence of wrongdoing buried itself — literally — beneath the subsoil and in the recesses of a gory past few wanted to remember and fewer wished to discuss. It's possible that an El Dorado of poison really does sit out there, somewhere in the vast, dank lowlands of the Petén. It's possible it doesn't. Here's what cannot be denied, and what seems to me to be the deeper tragedy: Guatemalans will be left forever wondering which it is.

7

TRASH ASH ODYSSEY

> It's enough that we have the contras. Now they want to dump more garbage on us.
> — Héctor Hernández Fuentes, president of the Unitary Federation of Honduran Workers, 1987 [1]

HERE'S A WASTE shipment we do know happened. It's arguably the most infamous, unbelievable, outrageous trash shipment of all time. And it emanated from what might seem to be an improbable place: the Roxborough neighborhood of Northwest Philadelphia.

In the early 1980s, Philadelphia was incinerating the majority of its garbage and transporting the resulting ash north to a nearby New Jersey landfill for burial. In 1985, after New Jersey declared that it would no longer be accepting it, much of which was deemed toxic, the city redirected its incineration ash south to a landfill outside Baltimore. Six months later, Maryland issued its own ban on Philadelphia's piles of — ever-mounting — waste ash.

And so Philadelphia's waste brokers began looking farther afield. They propositioned Ohio, then West Virginia, then South Carolina, then Florida — before finally, with no apparent US option forthcoming, they, like many Global North garbage handlers, began looking south. In early September 1986, an aging dark blue Liberia-flagged barge called the *Khian Sea* motored out of Delaware Bay bearing what its manifest certified to be *"ingrais pour du sol,"* or fertilizer; within its hull sat nearly thirty thousand pounds of sooty black trash ash that had been accumulating over the course of that summer in Northwest Philadelphia.

The *Khian Sea*'s destination? Ocean Cay, a human-made island in the Bahamas twenty miles south of Bimini. By late September the ship had moored in the territorial waters of the Bahamas — only to be denied entry to Ocean Cay by a customs official who had been sent a US EPA report certifying that the ship was carrying not fertilizer but burnt garbage.

The City of Brotherly Love was not to be rebuffed. "I would slash my wrists if I didn't think there is enough greed and avarice in the world to find somebody willing to take Philadelphia's trash," insisted its streets commissioner, Harry M. Perks. And so the captain of the *Khian Sea* went on searching for that willing taker. From Ocean Cay the ship re-charted its course a thousand miles north to the island of Bermuda. There, too, it was denied entry — as it was a month later back south in Puerto Rico, and in the Dominican Republic, and in Costa Rica, and in Guatemala, and in Honduras. It crossed the Atlantic, heading for the West African country of Guinea-Bissau, where it was turned away, then went back across the ocean, making for the Netherlands Antilles north of Venezuela, where it was refused once more, before unsuccessfully proposing to the government of Chile that it convert its defunct copper mines into landfills for Philadelphia's incineration ash.

As the *Khian Sea* changed course, so too did its identity. When its Liberia registration expired, Monrovia disavowed the vessel, which proceeded to navigate the seas answering to no nation's authority. And as its journey continued, the *Khian Sea* changed its "cargo" too. What had left Philadelphia as *"ingrais pour du sol"* would become, in turn, "general cargo," then "bulk construction material," then finally "top soil fertilizer."[2] In October 1987, as the ship left the coast of South America and approached its next destination, Haiti, it finally received some promising news: The country's commerce department would grant the *Khian Sea* a permit to unload its cargo along the Gulf of Gonâve. A company called Cultivators of the West—owned by brothers of Jean-Claude Paul, commander of the fearsome Dessalines Battalion of the Haitian military, which specialized in cocaine smuggling—agreed to accept the "fertilizer" in exchange for an undisclosed sum. That New Year's Eve, the *Khian Sea* motored into the port of Gonaïves, its hull "derelict," according to a US commercial attaché who came to observe the arrival, "completely rusted."[3] As Haitians took to the streets to protest the shipment, Colonel Jean-Claude Paul ordered them home, threatening "force of arms."

Three weeks later, approximately thirty Haitian farmers were ordered to shovel out the contents of the *Khian Sea*'s hull and layer it across a beach outside Gonaïves. The next morning, a representative of the shipping company that owned the *Khian Sea* took a bite out of its ash to demonstrate that reports of its dangers were sensationalized. "That's how worried I am of its toxicity," he boasted for TV news crews who had come to bear witness to the strange conclusion to Philadelphia's twisted garbage saga.[4] And yet, only days later, Haitians woke up to find that the *Khian Sea* had vanished. Still loaded with ten thousand tons of ash, it quickly left Haiti when the country's officials, finally bowing to public outcry, threatened to imprison the ship's crew.[5] "If they have so many acres of land in the United States, why should they dump their garbage on other people?" asked Michael King, executive director of the Caribbean Conservation Association, as the ship began sailing back north.[6]

In March 1988, the *Khian Sea* returned to Philadelphia, where something extraordinary happened—extraordinary even by *Khian Sea* standards. As the vessel reappeared in the city's harbor, the residents of Philadelphia were struck by a simple yet discomfiting realization. Why had so many different countries rejected their garbage? What was wrong with their garbage? If it was truly as toxic as all those other nations had insisted, why should the residents of Philadelphia have to take it?

TRASH ASH ODYSSEY

For the second time in as many years, the *Khian Sea* was sent away from Philadelphia. On and on it traveled. Over the next seven months the renegade vessel recrossed the Atlantic, steered through the Strait of Gibraltar, attempted to stash its ash on Tunisia's coast, skirted Sicily, swung up the Adriatic to Yugoslavia — where it underwent repairs and received a fresh coat of paint in what is now Montenegro — and acquired a new name, the *Felicia*, before attempting to dump its cargo on an uninhabited Greek island. It tacked back across the Mediterranean, headed south through the Suez Canal, entering the Indian Ocean and making for Sri Lanka, where authorities refused to let it dock, as did those in the Philippines and Indonesia.

And then, as mysteriously as it had all begun, the goose-chase voyage of the *Khian Sea* ended. In November 1988 the *Felicia*, by now rechristened the *Pelicano*, motored into Singapore's Jurong Port — one flag of convenience, one coat of paint, two names, and two years separated from its initial departure from Philadelphia — with nothing inside its hull. The ash had been unloaded, confessed the ship's captain, though he refused to say where.

For twenty-seven months the *Khian Sea* had zigged and zagged around three oceans, bearing a single season's worth of trash produced by a single neighborhood of a single US city. After thirty thousand miles, its crew had beheld five continents and at least twenty countries; several claimed to have not set foot on dry land once. The ash they deposited outside Gonaïves outlasted six Haitian governments and bore witness to two military coups. In 1989, a Haitian environmentalist organization, Les Amis de la Nature, began sending packets of it to the Philadelphia office of Mayor Wilson Goode and the Washington headquarters of William K. Reilly, then administrator of the EPA; thirteen years later, the island would successfully force Philadelphia to spend fifty thousand dollars for the ash to be shipped back to the United States, where, after an unsuccessful attempt to pay Oklahoma's Cherokee Nation to take it, the toxic ash can now be found lining a Pennsylvania landfill three hours west of the city that produced the world's most bewildering shipment of incinerated garbage.[7] The fate of the rest of the *Khian Sea*'s cargo — roughly ten thousand tons of poisonous soot — remains a mystery.[8]

The most shocking aspect of the *Khian Sea* isn't that it traveled for as long and as far as it did. It's that for more than two years it felt compelled to dump its waste in a foreign land rather than settle on the totally obvious solution of illegally unloading it somewhere in the middle of the ocean, as its crew almost certainly ended up doing. It demonstrates an important definitional line that was crossed in the late 1980s, one in which waste morphed into something that was not merely to be disposed of within nature but to be offloaded onto another society, an object of exchange rather than mere detritus.

And it was the ludicrous odyssey of the *Khian Sea* — in all its tawdry, what-on-earth details, dragged out over the course of more than two years in full view of the rest of the world and demonstrating American contempt for so much of it — that would ultimately spur the most earnest public debate about what exactly the United States was doing with its mounting piles of domestic waste, much of it toxic. The *Khian Sea* could not be written off as a sui generis case of a lone shipment of waste being dispatched by a sleazy waste broker to a distant land few Americans had heard of. Here was a US city — the birthplace of US democracy, no less — that had openly attempted to dump its toxic garbage on one country after the next. And every year, to be clear, Philadelphia incinerated enough trash to fill a ship the size of the *Khian Sea* many times over. Shipping it abroad was not a sustainable solution.

In the summer of 1988, four months before the vessel would dump its cargo somewhere in Southeast Asian waters, the US Congress began to examine what problems the *Khian Sea* presented and how they might be solved. That July, the House Committee on Government Operations convened what would be the first of four subcommittees on the topic of US waste export. "The latest status of the ship is that it's now off Cape Verde awaiting orders," Mike Synar, a Democratic congressman from

Oklahoma, declared as dozens of Congress members and environmental officials, waste management specialists and Philadelphia port authority employees, filed into Rayburn House in Washington, DC. Synar's opening requests were simple. He asked the committee to consider the two factors that had put a US waste management firm in Philadelphia in the position of attempting to send tens of thousands of pounds of burnt garbage to a tiny artificial island in the Caribbean Sea in the first place. There was, Synar said, "the lack of capital which drives many impoverished nations to consider accepting dangerous wastes which they may not have the capability to handle." And then there was "our own inability to minimize the amount of waste that we generate."[9]

Three things are striking about this first congressional waste export hearing. The first is the breathtaking ignorance concerning how exactly waste left the United States. For hours, members of Congress interviewed one official after another on the methods by which the EPA tracked toxic waste exports. And the picture that emerged was one of almost no monitoring at all. How much toxic waste left the United States, from which ports it departed the United States, the process by which countries could attempt to reject such shipments — no one could really say. "What we found basically was a program in shambles," a government auditor of the EPA admitted at one point.[10] It was enough to leave Democratic congressman John Conyers Jr. of Michigan proclaiming out of exasperation: "Have we examined whether or not we should just prohibit waste export from going out of [the] country?"[11]

Second, if one thing *was* clear to most members of the committee, it was that waste export was dramatically undermining US foreign policy. For the previous thirty years, the United States had been sending billions of dollars in aid to scores of developing or impoverished countries in an effort to thwart the influence of the Soviet Union. A single cargo container of US garbage offloaded onto any one of these places threatened to undo the entirety of that effort. "Hazardous waste exports from the United States to developing countries has cast the United States in a very unfavorable light as the 'ugly American' dumping garbage on other nations," Bonnie Ram, a fellow at the Federation of American Scientists, chided congressional members, feeling compelled to explain that the "U.S. national economy, environment, and health are inextricably linked to other nations."[12]

The third and most intriguing aspect of the 1988 testimony is the earnestness of the discussions around why so much waste existed in the United States in the first place. What was evident to most members of the testifying committee was that the problems stemming from waste exportation were not problems of waste per se. The problems owed to a US production sector gone rogue. After 1945, Representative Synar explained, "veterans of that war" returned to the United States and began turning civilian life into a material extension of the conflict they had just won. Dangerous chemicals that had been produced in the utmost desperation of battle had been redeployed back home and were now to be found everywhere. There was plastic, which possesses "five of the six" most toxic chemicals in existence. There were pesticides, dyes, flammable solvents, herbicides, arsenic agents.[13] And for the last forty years, US manufacturers have never been held liable for what happened to it all.

At last, after hours of testimony, Congressman Conyers vowed to reconvene the subcommittee the next summer, but not before issuing one last appeal:

"What bothers me is that great nations have great responsibilities. Great nations that generate great waste must have greater consciousness on how we are going to deal with it here with ourselves, and dadgumit, if it's not good enough for us, if we wouldn't want it in our neighborhood, then it is deplorable to think that we could ship it somewhere else."[14]

8
RISING UP

> Rich countries make toxic waste
> Why should they send it to me?
> Poor countries know toxic waste
> Why should they accept it?
> When I'm in bed
> I can't stop thinking about it
> When I'm awake
> I have to warn you
> We say it's true
> Many of the underdeveloped countries
> Are beginning to say No!
> — Senegalese singer Youssou N'Dour, "Toxiques," 1990

IN THE AUTUMN of 1987, on the other side of the Atlantic, a Belgian member of the European Parliament—François Marie Gabriel André Charles-Ferdinand Roelants du Vivier—was preparing dinner at home in Brussels when he received a document from an Africa-based researcher working for the United Nations. The scion of a storied line of Belgian politicians, dating back to a prime minister of the 1840s responsible for the construction of his country's rail system, Du Vivier was an early critic of toxic waste flows. And he had developed a strategy to begin tracking export schemes not weeks after they had already occurred but in real time, as they were happening. It was made possible with the refinement of a recent piece of technology. "It was something called the fax machine," Du Vivier told me. "Back then it took one day for the faxes to arrive. You sat around the machine and waited for it to print out. When this particular news came in from Guinea-Bissau, I could not believe what I was reading. I thought: *Someone must be playing with me. They want me to go to the press and give false information.*"[1]

But it was true. In October 1987, the small, West African coastal nation had signed a deal to accept fifteen million tons of toxic pharmaceutical waste—Western European and American in origin, to be shipped by a waste broker based in Italy—and dump it along the country's northern border with Senegal, across plots of swamp owned by President João Bernardo Vieira's brother. In return, Guinea-Bissau would receive $600 million. The figure exceeded its national debt twofold, its annual GDP fourfold, and amounted to thirty-five years' worth of yearly exports, which consisted largely of cashews and dried fish. "We need money," the country's trade and tourism minister had insisted to US ambassador John Blacken.[2]

Only after Du Vivier leaked the memo to Belgian newspapers—triggering disbelief and outrage not just in Europe but eventually from the citizens of Guinea-Bissau and the leaders of neighboring African nations—did Filinto Barros, Guinea-Bissau's minister of Natural Resources and Industry, agree to "regretfully suspend" the deal. But Du Vivier's relief was to prove short-lived. Just a year later he was faxed a new memo, this time contending that Guinea-Bissau had agreed to take in separate streams of toxic waste, only in even greater quantities; 10 percent of the European Communities' toxic exports were heading to Guinea-Bissau in 1988, or so the EU parliamentarian was informed by his sources in West Africa.

From the perspective of emerging new citizenries across the Global South, the hazardous waste trade was as insulting a sign as any of how little had changed since the heyday of colonialism. One people after another had thrown off the mantle of European imperialism only to find that they had been turned into receptacles of Northern industry. "You liberate your countries from colonialism and you talk about imperialism, apartheid. But what is more horrible than dumping nuclear and toxic waste?" Omar Sey, the foreign minister of the Gambia, protested in 1988.[3] Toxic waste export "re-echoes what Europe has always thought of Africa: a wasteland," observed Nigerian novelist Sam Omatseye. "And the people who live there, waste beings."[4] Oladele Osibanjo, a chemist who was sent by the Nigerian government to the coastal town of Koko to test a shipment of "fertilizer" shortly after it arrived in 1988, told me, "An Italian construction firm approached one of our illiterate farmers with a deal. 'We in Europe are too stupid to use this fertilizer. But you can use it,' they told him. And this farmer trusted the Europeans because they were rich and knew better.... He paid peanuts for two thousand drums of PCB. He dumped it across his farmland. Within a few days his son's hair turned gray as a result of inhaling the fumes. His son was only sixteen. Many others died as well."[5]

That Global South nations were ultimately able to unite and rise against the depravations of the hazardous waste trade is a little-discussed postscript to the decolonization movement of the second half of the twentieth century. And it's one in which it must be made clear just how much the developing world was up against when its people decided they had had enough of being cornered into accepting other countries' trash.

The late 1980s were not a time when erecting legal barriers against the flow of cash and goods was being very widely proposed. The Cold War was about to culminate in the triumph of the opposite worldview, one that elevated free-ranging capital, consumerist culture, and international markets above virtually all else. African and Caribbean leaders began plotting to halt toxic waste flows at a time when tens of thousands of East Germans were pushing their way into West Germany, when McDonald's was opening its doors in Pushkin Square in Moscow, when old state-owned mining facilities, oil refineries, and railroads in

Czechoslovakia and Poland were being snapped up by Western multinationals, when flashing Coca-Cola billboards were being raised atop battleship-gray socialist apartment blocks in Bucharest and Sofia, when economic activity — maybe more than at any other point in history — was being diffused across political borders.

Global South nations were going up against the General Agreement on Tariffs and Trade, GATT, enacted in 1948 to foster international commerce. They were going up against the World Bank, whose chief economist, Lawrence Summers, had made it all too clear that he deemed waste just one more commodity, something to be dispersed across national borders by market forces: "Under-populated countries in Africa are vastly UNDER-polluted." They were going up against the International Monetary Fund. They were going up against Global North industries and their consumers. They were going up against states that had, until not that long ago, lorded it over many of them.

And, perhaps most importantly, they were going up against a new kind of environmentalism. In the 1960s and '70s, much of the environmental movement had been suspicious of unregulated industry and endless economic growth. It feared population rise. It worried about the physical constraints of the planet. It wondered how much longer more and more people could keep consuming more and more resources. It spoke of an "age of waste." Two decades later, a new outlook had pushed many of those concerns aside. The "international community" that was finally in a position to pass globe-spanning legislation in the 1990s was convinced that the "economy versus environment" binary was a false one.[6] The future was in "sustainable development," a model of growth in which, through the proper organization of society and technology, the very market forces that had wreaked so much ecological devastation in the first place could also be enlisted to fix the planet. Multinationals could usher environmental standards into poor countries with no history of them. Free trade could spread more efficient technologies. And in terms of waste, the problem was not that we produced too much of it. The problem with waste was that we didn't utilize it.

Continuing to turn useful resources into useless waste would only cancel out future economic growth, argued proponents of sustainable development. Trash instead needed to be exploited for all that it could offer in the way of productivity. Think about how trees lose their leaves and then proceed to use them as fertilizer, argued David Pearce and R. Kerry Turner, two British economists who coined the phrase "circular economy" in an obscure academic text published three weeks after the fall of the Berlin Wall.[7] Humanity needed to operate more like the environmental systems it was trying to save, maximizing and assimilating everything around it and converting waste from an inert blight into an economic input. The best way to do this, proponents of sustainable development argued, was to commodify waste. Don't bribe other countries to take it; convince them of the value of your trash — and sell it to them.

As that new iteration of the waste trade got going — one in which poor countries would begin *paying* for containers full of things like plastic waste in the hopes that it might lead to economic opportunity — a small group of environmentalists were making every attempt to end the gargantuan iteration still festering in plain sight: the one in which poor countries were being bribed to accept drums of spent hazardous chemicals.

On a freezing late-December morning on the coast of Maine, I tracked down Jim Vallette, who played an indispensable role in helping bring about the Basel Convention on the Control of Transboundary Movements of Hazardous Wastes and Their Disposal, which would eventually emerge against the toxic waste trade. We met two hours north of the city of Portland, along a rocky promontory looking out on Sears Island, whose perimeter of beach appears at first glance to have been dusted by a light snowfall. Only I looked closer and realized that the ground wasn't covered in snowflakes at all. Thousands upon thousands of minuscule pieces of plastic speckled the island's shoreline.

One year before I met Vallette, a barge full of discarded plastic waste had arrived at nearby Searsport from Northern Ireland, purchased and imported as "fuel" to be incinerated in a local power plant;

upon reaching Maine, a two-ton bale of plastic consisting of, give or take, one million shards of old soda bottles and hot-sauce packets fell overboard and proceeded to break down and disperse, eviscerating segments of Maine's treasured fishing and lobster industries. "It's such an irony," Vallette told me as we surveyed the plastic-peppered sand. "The United States never ratified the Basel Convention. For years we've been shipping our garbage abroad. But it also means that other countries are allowed to ship all kinds of stuff to us."[8]

Vallette was in college when, in July 1985, French intelligence agents detonated the *Rainbow Warrior*, the flagship of the Greenpeace fleet that had been docked in New Zealand prior to a scheduled visit to the remote atoll of Moruroa, where it had planned to sail in protest of French nuclear weapons testing. Vallette found himself wanting to dismantle the "idea that if you're born in one place you're somehow superior to being born in another." And he had a penchant for organizing data too. In late 1985, Vallette began volunteering for Greenpeace's toxins campaign, which at the time was focused on the incineration of hazardous waste aboard ships anchored off the coast of New Jersey, just beyond jurisdiction of the US legal system. "We had our radar up for waste and its movement," Vallette told me. "In the course of tracking waste incineration, we started seeing these schemes for waste export. And I remember thinking, *This is going to be a nightmare.*"

Over the next three years, Vallette took up the task of forming an exhaustive inventory of every export waste scheme Greenpeace and its partner NGOs across the Global South could discover. The organization developed regional networks driven by data collection. It got access to port manifests and pored over subscriptions to PIERS, the Port Import/Export Reporting Service, an arm of the *Journal of Commerce*, which at the time had a monopoly on most of the world's trade data. The organization communicated with customs officials via telex and sent teams to investigate port warehouses and examine suspicious cargo, even securing fake IDs in Bangkok that certified their professions as waste traders.

"You would often only hear about an export shipment six weeks after the fact. And you usually only knew where it was getting imported, not where it was getting sent from. We had to start identifying the patterns of waste movement," Annie Leonard, one of Greenpeace's field operatives in the 1990s, told me. "Eventually I learned that the most valuable way to do this was to go to harbors across Asia and hang out at the bars and befriend port workers."[9] Greenpeace became a student of the internet, communicating via its own internal email service, Greenlink, as early as the late 1980s. It developed contacts in embassies. It swapped information with diplomats. It collaborated with journalists. "We had learned a lot from Greenpeace's earlier campaign to track the pesticide trade," Vallette told me. By 1989, more than a thousand waste export schemes out of the United States had been chronicled and exposed, hundreds more than were known to the Environmental Protection Agency, which at the time had two employees tasked with tracking the billion-dollar business of shipping US garbage abroad.[10]

Amid all the mysteries of the waste trade, one thing was soon obvious to Greenpeace's toxins trackers. Global North nations were never going to voluntarily come to the table and offer to stop exporting their own hazardous trash. Any attempt to end the waste trade would have to start from the opposite perspective. It would have to originate from developing countries themselves.

"We devised a Global South–driven campaign," Vallette told me. "These governments were anti-colonial, or at least most of them were. It would be hard for any of them to justify not signing a ban, because any government that didn't sign it would look suspicious to its neighbors."

Greenpeace first focused on the so-called kernels: tiny states like Tonga and Western Samoa and Costa Rica, which by 1986 had already begun the process of passing legislation against the import of waste. Gradually, like the disparate sections of a jigsaw puzzle coming together, the kernels banded into blocks. In 1987, CARICOM, a confederation of thirteen English-speaking Caribbean states, came out fully against the import of waste—"We do not want to become the outhouse for the United States," Cornelius Smith, a representative in the Bahamian House of Assembly, claimed—while a similar declaration followed from ECOWAS, the Economic Community of West African States.[11] Gradually, as one region after another issued one such pronouncement after the next, a global picture came into view.

The trickier countries, Vallette told me, were big developing states like India and Brazil, emerging industrial powers inclined to export waste to their own smaller neighbors. But by 1989 even most of these had reached moral clarity on the issue, as had the European Communities, precursor of the EU, which that May agreed to end waste exports to sixty-eight former colonies across Africa, the Caribbean, and the Pacific. "Waste traders are on the run," Greenpeace could pronounce by year's end, by which time the signature international legislation against the waste trade—the Basel Convention—had been signed by more than thirty nations.[12]

Over the next five years, the grainier details were hammered out at a succession of international gatherings. Could rich countries ship certain forms of toxic waste to one another? (Answer: yes.) Could a poor country accept toxic waste from a rich country *if it wanted the trash*? (Answer: initially yes, but only so long as it was warned in advance of the shipment, and in the unlikely event that the exporting country itself possessed no means of processing the waste.) By 1994, a more sweeping decision had been reached to entirely outlaw the movement of hazardous waste from developing countries to developed. A year later, by which time more than a hundred nations had signed the declaration, the so-called Ban Amendment was added, not just prohibiting all hazardous waste exports from rich countries to poor but also making it a criminal act to attempt such exports. It would nevertheless take until 2019 for it to come into full legal force following its ratification by three-quarters of the Basel Convention signatories.

The world's greatest waste producer, however, continued to be an outlier.

9
AMERICAN EXCEPTIONALISM

> What may be waste to us in the United States may be something that has a value, can be transformed into another product, lends itself to productive and profitable use.
> — William K. Reilly, administrator of the EPA, 1991[1]

IN THE EARLY 1990s, in one congressional subcommittee after another, US politicians continued to weigh the competing merits of signing or rejecting the Basel Convention. At the 1988 subcommittee prompted by the *Khian Sea,* there had been a caught-red-handed embarrassment over the fact that the United States was engaged in the business of exporting toxic garbage to poor countries. In the subsequent committees, though, that embarrassment hardened into irritation and eventually a kind of cynicism that US industry was being asked to apologize for its behavior. Concerns emerged. If US companies could no longer shift their waste abroad, wouldn't they lose ground to Canadian and Mexican rivals?[2] Wouldn't ending waste exports to poorer countries merely increase waste exports to poorer states? Why should the citizens of Haiti be subject to greater environmental protection than the residents of, say, Alabama? And, after all, in the grand new post–Cold War world, what was a national border, really?

"With the globalization of an economy, a border between a Canada and a Mexico becomes not that different [from] a border between a Texas and an Oklahoma, especially with the treaties as they have emerged and are emerging," Donald Ritter, a Republican congressman from Pennsylvania, would insist.[3]

As for the EPA, by the early 1990s it was sounding a lot like the Colbert brothers, arguing that there might not even be such a thing as waste export. For what was waste? "I think one of the most important insights that we discovered over the past 20 years since the environmental revolution began is the insight that a pollutant is a resource out of place," William K. Reilly — the head of the EPA who, at the time, was receiving packages of the *Khian Sea*'s toxic ash in his DC mailbox from infuriated Haitian activists — expounded to Congress.[4]

By 1991, Greenpeace had learned to stop calling its campaign the "International Waste Trade Project." It made it all sound so innocent.

"At first people were morally aghast by what we had found. There was such outrage. But what we then began hearing, and this happens all the time with industry, was 'You're right! This is terrible. We need to legislate it,'" Annie Leonard told me. "They wanted to take it from the realm of outrage to that of legitimate commercial activity. So we started calling it 'hazardous waste trafficking.' It had to be equated with weapons and drugs." As for Jim Vallette, his frustration with Congress had become clearer still. To the first three waste export subcommittees he had brought calculated arguments. To the fourth one, held in July 1991, he brought a jar of Philadelphia's burnt garbage that the *Khian Sea* had dumped in Haiti, samples of which Greenpeace had secured through the island's environmentalists.

"Now, is the ash per se poisonous?" protested the session's chairman as Vallette placed the jar on the table in front of him.

"It has high levels of lead and cadmium and dioxin, which is widely believed to be the most toxic industrial waste by far," Vallette replied.

For Jim Vallette, after years of cataloging waste exports and chronicling what they were doing to distant corners of the globe, there was one final point to be made about the waste trade. It was a warning about what a failure to ban the export of hazardous materials would ultimately do *to the United States itself*: Refusing to ratify Basel would result in a future whereby it would become more and more difficult to hold US companies accountable to *domestic* standards of environmental regulation and to keep Americans themselves safe because of the escape valve proffered by waste export.

"U.S. industry is growing reliant on the rest of the world as a dumping ground," Vallette explained to the committee. "Open borders for any hazardous waste exports will only encourage industry to churn out more and more deadly waste and avoid internal U.S. regulations designed to force industry to stop producing poison."[5] In other words, there was no point to the great environmental revolution of the early 1970s if waste could continue to cross international borders as it had in the late 1980s.

The United States would nevertheless refuse to ratify the Basel Convention. To this day it has yet to do so.

10

THE WASTE TRADE STRIKES BACK

Trash is our only growing resource.
— Hollis Dole, US undersecretary of the
Department of the Interior, 1969 [1]

THE ATTEMPTED CONSTRUCTION of the Las Escobas sewage sludge port in Guatemala is but one example of how, even after the Basel Convention was coming into form in the early 1990s, the toxic waste trade never truly ended. Waste brokers only continued to get more cunning and, crucially, learned to stop calling their business a "waste trade." They dealt in "construction materials," "scrap," "fuel," "recovered byproducts." And even when toxic waste was confined to flowing within and between rich countries, as opposed to beyond them, it continued to exacerbate North–South divisions. Within the European Union, a set of hills east of Naples in Italy became known as the "triangle of death" owing to their conversion into a chemical graveyard, with battery acids and hospital residues trucked in from as far away as Germany. "For us, rubbish is gold," a convicted mafioso would boast in 2008.[2]

In the United States, a new "third world" was located right in the heartland in the form of Native American reservations, which the government attempted to pay in the early 1990s to receive mounting stockpiles of nuclear waste. Native Americans, federal agents would insist in their discussion with tribal leaders, were in an advantageous position to receive a uniquely dangerous form of waste that required cosmic timescales to disintegrate. Indigenous societies were in greater touch with Earth's natural rhythms than white Americans. They had a more profound sense of moral purpose. And it was, after all, their land.[3]

And toxic waste shipments, when they did continue crossing international borders, would of course never stop being deadly. Take the West African nation of Ivory Coast. In 2006, a Panama-flagged vessel — owned by a Greek company based in Russia and chartered out to a Dutch firm — dumped five hundred tons of sulfide waste originating from Texas in the harbor of its capital, Abidjan. The calamity was instant. Though Ivory Coast authorities claimed that seventeen died as a result of the incident, at least one hundred thousand are known to have sought medical help. Abidjan la Belle (Abidjan the Beautiful), as it was known to its residents, became Abidjan la Poubelle (Abidjan the Trash Can).[4]

At the same time, just as the Basel Convention was getting ratified, opponents of the hazardous waste trade were confronted with a strange irony. By the mid-1990s, the Global North was less bent on diverting hazardous chemical waste to developing countries for the simple reason that it had begun producing much less of it. Industrialization itself was moving south. Factories that had once operated in places like Michigan or Northern Italy were dismantled, shipped out, and reassembled in countries like Honduras or Thailand, where they could be held to rock-bottom standards of environmental regulation and labor rights. Sure, this offshoring meant that *certain* developing countries finally got the industrialization they had so desperately wanted in the 1970s. But the price to be paid was the *entire* spectrum of pollution that got exported out of the Global North at a legal and financial discount.

And that was hardly the greatest paradox that recontoured the waste trade in the 1990s. In its understandable focus on the trafficking of known hazardous materials like asbestos or discarded paint solvents, the Basel Convention left the matter of innumerable other kinds of waste ambiguous or unaddressed. The result? Explicitly outlawing one form of waste export — toxic chemicals to poor countries — made the offloading of other kinds of waste appear legal and, indeed, unproblematic.

"The tragedy of the ban on hazardous waste is that it stopped with hazardous waste," Jim Vallette told me, explaining that to some degree the failure to build on the partial success of the Basel Convention in the early 1990s was due to Greenpeace's decision in the late 1990s to elevate the fight against climate change over that of environmental injustice.[5]

To crack open the floodgates still further, tucked within the terms of the Basel Convention was a stipulation that would be subject to quick and relentless manipulation by Western waste brokers. If an exporter declared that an object discarded in a rich country would be traveling to a poorer country for the purposes of *reuse,* not final disposal, then it was technically not considered waste at all. It was just another commodity. Historically always murky, the definitional differences between "waste" and "resource" became *deliberately* muddied in the 1990s, the purpose being very much that of the toxins trade of the 1980s: to divert to developing countries that which could not be safely or profitably discarded within developed ones.

The illusion that the waste trade had been subjected to a degree of international regulation in the 1990s resulted in the greatest of contradictions. That same decade, waste export *exploded.* Not the obscure industrial hazards offloaded by a factory in Pittsburgh or Düsseldorf but the things *you yourself* were throwing away within the confines of your home were circulating to the most distant ends of the globe at unprecedented rates. The highly lucrative world of post-industrial waste export gave way to the incomprehensibly lucrative world of post-consumer waste export. The plastic forks you discarded after a single use were going to villages in Vietnam. The broken TV you put out on your curb was going to slums in Nigeria. Your worn-out automobile tires were going to the interior of India. Your unwanted clothes were going to deserts in Chile. Your spent batteries were going to Mexico.

The remaining chapters of this book explain where some of these objects travel and why. But for the moment, it's important to underscore how this new waste trade differs from its 1980s predecessor. First, as mentioned, it rarely claims to be dealing in waste at all. The post–Cold War waste trade insists that it exchanges "recyclable" materials, and in this it has exploited a cynical shift presided over by US industry in the late 1980s: the extension of "recycling"—a legitimate practice when it comes to old steel or used paper, which had been getting traded around the world for decades—into a range of material sectors where it cannot be said to truly or safely work. "Manufacturers can claim anything is recyclable, even when, as a practical matter, there is no place to recycle it," Consumer Reports, the NGO dedicated to investigating American products, warned as early as 1991.[6]

The rampant invocation of recycling—pushed in marketing campaigns and school curricula and community boards across the 1990s— amounted to, yes, a deliberate deception foisted on residents of rich countries, who were led to believe that being a prodigious consumer was compatible with being a planetary steward. Larry Thomas, former president of the Society of the Plastics Industry, later characterized this corporate strategy: "If the public thinks the recycling is working, then they're not going to be as concerned about the environment."[7]

But the real victim of this campaign was to be found half a world away. Because, while today's waste flows through the same North–South networks carved out by the 1980s toxins trade, most of it isn't getting surreptitiously dumped in North African mineshafts or strewn along Caribbean beaches. You don't need to spend weeks driving around a colossal Guatemalan jungle attempting to find it. It travels thousands of miles to be "recycled" or "processed," terms that seem to imply efficient methods of resource conservation or renewal, but in the context of the global waste trade are better understood as the myriad, occasionally deadly ways that communities across developing nations attempt to source something — anything — of value from the trash rich countries send them. Plastic gets sorted and washed and shredded — or burnt. Broken TVs get shucked of their precious metals — then burnt. Automobile tires are almost invariably burnt.

Exporting waste to developing nations is often still defended as better for the planet than landfilling it within rich countries themselves. But is it? The result of sending waste to poorer countries is a malignant, monstrous, epically magnified version of what happened to İzzettin Akman's citrus trees in Turkey in late 2016. Today, trash travels thousands of miles and then — far from becoming unsightly though inert piles of refuse — gets broken down into its constitutive chemical elements, disintegrating, melting, leaking masses of toxins and contaminants that have already condemned great tracts of the Global South to barren futures of eternal despoliation. It's a world that Rachel Carson would have exposed if only she had known to do so: The toxic dangers latent within seemingly innocuous household objects such as Styrofoam packaging and plastic cutlery and electronics have only really become clear in the last fifteen years.

"Waste traders are Fred Flintstones who look at waste as solid, inanimate objects to move around with their dinosaur shovels," Jan Dell told me. Dell is the founder of an NGO called The Last Beach Cleanup who has worked more than thirty years in the chemical engineering industry. "They are living in the 1970s before dioxins and microplastics were measured and understood to be harmful to humans and ecosystems."[8]

Think about it: After a great international campaign stretching from the Bahamas to Benin was able to reignite the embers of the anticolonial movement, after a flurry of anti-waste-trafficking legislation was hammered out over years and years, and after the apparent successes of Greenpeace and other NGOs to finally draw public attention to the depravations of pollution export, more trash was getting moved across more borders than ever before.

The hazardous waste trade hadn't just survived. Cloaked in the rhetoric of saving the planet, it had metastasized. It had democratized. And, perhaps most importantly, it had become increasingly illogical and insane. Products designed to take seconds to consume became the objects of months-long journeys from one side of the world to the other. To attempt to cycle tiny pieces of carbon back into production streams, megatons of carbon would be unleashed shipping them from one continent to another. Communities scattered across the equator with little access to technology or economic mobility or clean water were enlisted to receive broken computers, retired cruise ships, discarded water bottles, for the simple reason that we made too much of it all — then balked at being held accountable for its fate.

The tragedy of the waste trade isn't that it started happening. It's that we knew it was wrong and failed to stop it.

PART TWO

E-Waste on the Odaw

11
STATE AND SLUM

> A great deal of the uranium for nuclear power, of copper for electronics, of titanium for supersonic projectiles, of iron and steel for heavy industries, of other minerals and raw materials for lighter industries — the basic economic might of the foreign powers — comes from our continent.
> — Kwame Nkrumah, president of Ghana, 1963[1]

FROM AN AIRPLANE window, the city of Accra resembles a great brown-gray mass tacked to the littoral of West Africa. Veiny streets of ochre dirt cut through disorderly chunks of mudbrick and corrugated iron. At the city's southern edge, where Accra brushes up against the Gulf of Guinea, are a handful of thick granite monuments. They were erected in the early 1960s to commemorate Ghana's independence, sub-Saharan Africa's first successful decolonization movement. Shortly after midnight on March 6, 1957, a young revolutionary named Kwame Nkrumah — the son of an Asona goldsmith — proclaimed the release of the former Gold Coast from the British Empire. "At long last, the battle has ended," Nkrumah told hundreds of Ghanaians who had massed in the center of the world's newest capital city. "We have awakened. We will not sleep anymore."[2]

Today, the monuments to Ghana's independence jut out from the

Atlantic coast and exude a vacant pageantry. There is Black Star Square, a tarmac parade ground that stretches out into sunstruck emptiness. There is Independence Arch, built in 1961 to greet the arrival of Queen Elizabeth, Ghana's erstwhile ruler. There is Accra Sports Stadium, a sepulchral coliseum that rises forth from the sizzling pavement like a stale wedding cake. It's unusual to find anyone else in this section of Accra, no matter the hour of the day, apart from a few bored guards armed with assault rifles and tasked with removing anyone caught dozing on benches. On either side of the monuments, along peninsulas stretching into the Gulf of Guinea, strongholds that once held enslaved people cast their shadows over the thrashing waves of the Atlantic.

Accra grows denser and denser as it pushes inland — so dense, in fact, that in the decades after Kwame Nkrumah oversaw his country's independence, a succession of "ring roads" were constructed in repeated attempts to encircle Ghana's capital and delineate its outer limits. One highway was paved in the mid-1970s, another in the late 1990s, both in vain: For half a century now, Accra has kept on pushing its way through the bush, undeterred. A city that sixty years ago boasted a population of some three hundred thousand is today a megalopolis of more than three million, a bead along the belt of West African urbanization stretching from Abidjan to Lagos that is poised to become, by 2100, the most populated coastline on the planet.

Tucked deep within the bowels of this unruly mass sits the market slum known as Agbogbloshie. It's a notorious place, a household name among security analysts, cybercrime experts, and environmental agencies the world over. For the US State Department, Agbogbloshie has proved a headache for security and intelligence safekeeping ever since it made itself known to Western authorities nearly two decades ago; in 2008, a team of researchers from the University of British Columbia traveled to Ghana's capital and chanced upon Northrop Grumman military contracts amounting to $22 million on hard drives rusting in an Agbogbloshie market stall that looked much like any of the thousands of others that crowd the slum. In the years since, dozens of other

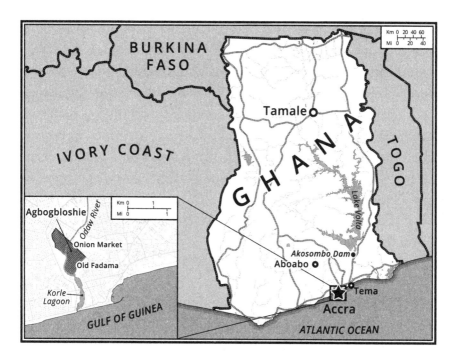

sensitive and precious documents — NASA mockups, Homeland Security memos, TSA spreadsheets, Defense Intelligence Agency files — have been recovered off secondhand computers discovered in the marketplace. It's probably a fraction of what's to be uncovered at Agbogbloshie. A recent Harvard study estimated that the value of the data that ends up in electronic waste streams could be as high as $13 billion a year.[3] No small chunk finds its way to Accra.

That's the valuable material Agbogbloshie receives — accidentally or otherwise. Then there's the valuable material it *takes*. For the FBI, the slum is a den of identity thieves and financial catfishers. Their culprits tend to be enterprising young men in Ghana who have spent their lives rummaging through the piles of keyboards, desktop monitors, and smartphones that waste brokers in rich countries have shipped to Agbogbloshie; they are seasoned at restoring these busted electronics back to life — and, on occasion, using them to conduct epic long-range fraud against residents of the countries that sent them.

In 2017, after hundreds of thousands of dollars failed to reach the accounts of a Memphis-based real estate agency, its realtors came to the belated realization that dozens of their email addresses had been spoofed and used to siphon off brokerage payments from recent home buyers across Tennessee. The culprit: a group of fraudsters from Ghana.[4] Other thefts are of a more intimate nature. Communities of Ghanaian teenagers, so-called browser boys, spend their nights luring Westerners into sending them gift cards, new phones, inheritances. In 2021, a seventy-seven-year-old man in Annandale, Virginia, matched on a dating website with a woman who claimed to be a widow in her thirties. A little more than one year and $500,000 in wired cash later, the man contacted local police with concerns that the widow might not exist. She didn't; instead, he had sent the money to a network of scammers in Ghana whose total earnings over the previous two years, concluded Virginia prosecutors, likely exceeded $40 million.[5] Indeed, even if you have never heard of Agbogbloshie, it's possible that someone in Agbogbloshie has heard of you. If an email has ever landed in your spam folder informing you that you've been designated as a handler of a West African prince's lost banking fortune — *I can't transfer this fund to my personal account rather I want you to assist me!* — there's a good chance it was dispatched from one of the several thousand corrugated-metal shacks that make up the residential sprawl of Agbogbloshie.

As for environmental agencies, Agbogbloshie is a byword for ecological ruin. The New York City–based Pure Earth is an NGO that tracks toxin levels across impoverished countries; it routinely lists Agbogbloshie among the most polluted places on the planet. Chemical analyses of Agbogbloshie's chicken eggs have determined they are probably the most poisonous on Earth.[6] According to the World Health Organization, a child who consumes a serving of poultry in the slum there will absorb 220 times the European Food Safety Authority's daily limit for intake of chlorinated dioxins, chemical compounds that can prove highly damaging even in minute quantities.[7] "One can assume they" — the residents of Agbogbloshie — "will be looking at a drastically reduced life expectancy," Matthias Buchert, a chemist at the Institute for Applied Ecology in Darmstadt, Germany, has claimed.[8] During afternoons spent in Agbogbloshie, it's not unusual to observe cows destined for the slaughterhouse chewing through last meals of yams mixed with plastic. According to researchers who have studied a similar phenomenon in India, the olfactory system of a cow — the vomeronasal organ — struggles to differentiate plastic coated in food from food itself.[9]

Yet the most confounding thing about Agbogbloshie may just be its location. One would expect a place of so many competing notorieties to be tucked far from the center of a bustling national capital, at sanitized remove from the UN agencies and World Bank offices that, on the face of it, make it their purpose to prevent such dire places from existing. And yet Agbogbloshie is *right there*. It sits at the beating heart of Accra, a mere mile from the daunting granite monuments to independence that seemed to celebrate the birth of a Ghana bent on never becoming the object of Western exploitation again. Indeed, driving toward Agbogbloshie is an almost hallucinatory experience. A string of highrise luxury hotels offers guests acres of manicured gardens and womb-temperature swimming pools. Gradually the road gets bouncier. The air gets smoggier. The buildings get shabbier.

A bridge is lined with men hawking secondhand clothes. Below them is the Korle Lagoon, a stagnant channel that may take one a moment to realize is actually a body of water. Across its surface stretches a thick canvas of garbage. It jiggles now and then as pristine white egrets touch down upon floating chunks of Styrofoam, sending ripples through the flotsam and a surge of black water splashing into a mushy bank that scampers with rats gnawing through shopping bags in search of scraps. The Korle Lagoon doubles as a latrine. From the Okai Street bridge, next to a vendor selling toothpaste and brushes from a red pushcart, men unbuckle their pants to piss while nearby a line of wobbly makeshift outhouses has been constructed out of wormy wooden doors.

A cramped tongue of land extends for more than two miles into the

Korle Lagoon. The residential section of Agbogbloshie, known as Old Fadama, is home to approximately sixty thousand Ghanaians who have largely reached Accra from their country's desert northern fringes. Shielded from the sun by thousands of overlapping corrugated roofs, hundreds of passageways weave through Old Fadama, across the peninsula, forming a beehive of humanity tucked out of sight from Accra itself. There are tattoo parlors and seamstresses and public latrines adorned with murals depicting women squatting on toilets. Small mosques bristle with minarets that pierce through the metal-roof expanse like arrows through armor. In secluded nooks, shirtless boys gather around televisions that play karate movies and soccer matches. A refuge from the sunbaked city beyond it, the warrens of Old Fadama can be wandered for chunks of the day without seeing the same shanty twice — or the sun at all. On three sides it is surrounded by the waters of the lagoon. Along the other, eastern edge, where Old Fadama runs back up against Okai Street, one finds the largest food market in all of Ghana, a great open-air bazaar known as the Onion Market. It sells not just onions but also yams, juices, coconuts, palm leaves, cookware.

Why have tens of thousands of Ghanaians relocated hundreds of miles from their ancestral villages in the north to live in such a place? The answer is to be found in the narrow strip of land that separates Old Fadama from the putrid waters of the Korle Lagoon.

Every morning in Agbogbloshie, with the exception of Sundays, thousands of male residents of Old Fadama trickle out from their shanties and make their way toward the Korle Lagoon's banks. There, they hammer and shuck their way through a perpetually replenishing reservoir of secondhand electronics that have reached Ghana from the richer countries of the world.

First there is the phone dismantling station, Agbogbloshie being the destination for huge quantities of foreign and domestic smartphones that, owing to exhausted batteries or cracked screens, or any number of other problems, no longer work. They may be "condemned," as broken electronics are known in Agbogbloshie, but they still possess worth in the form of precious metals and parts. Those valuable elements must now be extracted and separated. At the phone dismantling station, men are arranged in circles in groups of five or six. One strips printed circuit boards from old Android phones and tosses them onto a glittering pile of green silicon. He then passes what remains of the phone to the man to his right, who uses a pair of tweezers to unpick its camera, delicately dropping it into a plastic water bottle on the ground beside him, where, I notice, several other such bottles are arranged in a neat row, each filled to the brim with amputated smartphone cameras. I stay for an hour to watch the men process the phones round the circle. The devices lose more and more of their parts as they get farther and farther, until they arrive at the final man, whose role is to toss their plastic encasements atop a growing anthill of smartphone carcasses teetering along the edge of the lagoon. Then, just as the piles of phones requiring dismantling appear to be dwindling, that day's work nearly done, men aboard motorized wheelbarrows — the Nigerians who run much of the electronic waste market, I later learn — pull up to deliver dozens more rice sacks full of condemned devices. They toss them onto the ground as though they are bags of dirty laundry.

The dirt ribbon of track plods deeper into the peninsula, beyond the smartphone dismantling stations, toward the more complex and dangerous technological autopsies. Some fifteen feet wide, the path is an unrelenting vehicular torrent, juddering with motorized tricycles and pickup trucks and retrofitted tractors and choking with their comingling exhausts. Every morning, their drivers ferry hundreds of tons of mangled appliances (ceiling fans, washing machines, motorcycle engines, refrigerators) into Agbogbloshie; every afternoon they motor some of the world's most precious materials (cobalt, copper, gold, platinum) out — a ten-hour turnaround that extracts highly valuable material from Agbogbloshie, and eventually Ghana, and leaves little behind save a hazy mass of pollution and pittance wages. My feet crunch over shards of computer-screen glass and cracked iPad covers and stray Hewlett-Packard mice; a pink bra has been stamped into the mud by so many thousands of footsteps to the point of resembling a fossilized crustacean. Itinerant barbers bearing white plastic stools roam around doling out buzz cuts for ten cedis, or a buck. Women shimmering in tribal dress meander through the morass, selling juices out of plastic laundry bins. The constant clank of hammers laying into electronics pulses through the scene like a heartbeat. *Clank! Clank!*

About halfway down the peninsula, the appliances that have arrived on those motorized tricycles and pickup trucks are being bashed to pieces. Hundreds of young men — known in Agbogbloshie as "dismantlers" — sit in dozens of small circles straddling gutted microwaves and disemboweled computer monitors. Most are wearing knee-high colored dress socks and open-toed sandals, which poke out of great rats' nests of electrical wiring crowding the ground. The dismantlers have one job: For eight to nine hours a day they pound fat gavel hammers into the seams of old ceiling fans, motorcycle mufflers, speaker systems. The work has a factory-line monotony to it, only it is the exact opposite of assembling. It is a *de*-manufacturing line, reducing all the amenities of our modern world — the air conditioners that cool our offices, the refrigerators that preserve our food, the motors within the mowers that shear the grass of our lawns — back into their constituent elements. It's a juxtaposition that, even after weeks spent in Agbogbloshie, never ceases to be jarring: The work of the dismantlers may be pre-industrial and backbreaking, but what lies beneath the strokes of their hammers tends to be some of the world's most advanced technology. And while a streamlined process of automation might have manufactured most of these products, it is human labor — of an almost unimaginably archaic kind — that remains one of the few ways to get rid of them.

I watch as a Japanese refrigerator engine gets obliterated in the span of several seconds with rhythmic discipline, its dismantler capable of locating the seams in its sides without hesitation and, three or four hammer smacks later, cracking its torso in two. Gathered in the center of his circle are mounds of filthy gadgets. Smashed TV screens abut cordless electric teakettles. Webs of wiring seem to lasso the piles of junk together like spiderwebs. Along the edge of the circle, a young dismantler is halfway through the work of disassembling an industrial Epson office printer when one of its cartridges sprays his socks with firecracker bursts of pink and turquoise ink.

I leave the appliance dismantlers and keep walking. On the side of the path facing the Old Fadama slum, men are taking their breaks, lounging horizontally along banks of interconnected steel chairs that appear to have been lifted from an airport or bus terminal. A couple are putting back Club beers. As the path continues, after every thirty feet or so, I see one huge Alpha & Omega–brand scale after another, each of them capable of weighing up to 500 pounds of copper or aluminum. All around me, Frafra tribespeople — identifiable because they tend to be shorter than most other residents of Agbogbloshie — work fastidiously, hunched over at one-hundred-degree angles, plucking sparkling shards of emerald silicon circuit board off the ground and depositing them in nylon sacks.

Finally, after nearly two miles, the Korle Lagoon path rises to a five-story mountain of unsorted waste that lofts high above Agbogbloshie like an acropolis, a trashy summit befitting a society that owes its existence to the processing of refuse. Lording it over the slum, taller even than the minarets of its mosques, the garbage mountain bakes in the harsh beating glare of the sun. At its base, men doze beneath tarps held up by wooden stilts, battalions of flies buzzing around their eyelids. Around its peak wander the belated entries to the great scavenging hunt: gaunt cattle and bony dogs, rooting through pastures of plastic in search of scraps of yam.

12

TO THE QUAYS OF TEMA

> I put all my luggage in a taxi and told the driver to take me to Old Fadama. The driver turned to me and said, "Sir, the area is now called Sodom and Gomorrah."
> — Stephen Atta Owusu, Ghanaian foodstuffs importer based in Scandinavia, 2015 [1]

YOUR FIRST CELL phone, the VCR player you gave away after the advent of the DVD, the DVD player you donated to Goodwill after the arrival of Blu-ray, the Blu-ray player you never used, the college laptop you tossed away because it was ransacked by viruses — it all may very well have passed through Agbogbloshie, submitted to the stroke of a hammer and shucked of its valuables, the last chapter of a journey (What *did* happen to your childhood Game Boy?) you've probably never paused to contemplate in the first place.

Photos of Agbogbloshie are invariably enlisted to demonstrate Ghana's grim fate as one of the world's greatest recipients of Western electronic waste. But the reality is more complicated and, in certain respects, darker. For Ghana was never meant to turn out like this. It was never supposed to become a dumping ground for foreigners' unwanted electronics. And contrary to many descriptions of Agbogbloshie, not a single country or company on Earth ships, or has ever shipped, broken phones or busted televisions to the place as a matter of policy. No, none of this arrives in Ghana as waste *per se*.

What foreigners do send — and this is not merely legal but incentivized by global institutions such as the World Trade Organization and the International Monetary Fund — are old electronics they claim *do* work. When recycling firms or waste brokers in countries like Canada or Germany ship millions of broken cell phones or ceiling fans to Ghana, it's probable they may not think they are outsourcing pollution to West Africa. They may really believe they're bestowing the tools of enlightenment and progress upon a poor corner of the world's poorest continent.

How did any of this start happening? How did Ghana — a country that scarcely possessed a functioning computer a generation ago — emerge as the recipient of thousands of tons of busted electronics and appliances every year?

Early one February morning I took a taxi east of Accra to the faceless port city of Tema, whose construction along the prime meridian was overseen by President Kwame Nkrumah in the early 1960s in a bid to streamline the export of gold from Ghana. The highway ran parallel to the Atlantic, past noxious cement plants whose new Chinese owners had studded the grassy coastline with ranks of drab casinos; their blacked-out windows soaked in the rising sun. After half an hour we entered Tema, where an informal marketplace had latched onto the entrance of a gated cargo container terminal. Fanning out from the quays, hundreds of market stalls had been fashioned out of sawed-in-half steel cargo containers that still bore the names of shipping companies: COSCO, MSC, Maersk. Within each, masses of old electronics were for sale.

There were scattered arrivals. I could buy a secondhand ceiling fan from Italy. Refrigerators from South Korea had warning labels still visible in Hangul. A secondhand blender from the United Kingdom was available for 45 cedis, or about $3. But most of the products had been shipped in bulk and en masse. Stacks of Dell computer monitors bore yellow stickers identifying the elementary school in Hanover, Germany, they had once outfitted; a municipal post office in Norway had sent along its outdated landline phones, sequentially numbered 977, 978, 979.

None of this was getting "dumped" in Ghana. It had been shipped for the purpose of getting purchased by Ghanaians.

Dozens of containers packed with thousands of electronic devices of one sort or another reach Tema every day. Some are sent by waste brokers in Western countries who specialize in collecting secondhand electronics from recycling centers or dumps; others are donated by hospitals, universities, NGOs; others are sent by expatriate Ghanaians who, during the famine and tribal conflicts of the 1990s, relocated to the great metropolises of the north — London, New York, Toronto — and now wander their streets in search of old appliances piled on sidewalks that they can ship to relatives who work the street bazaars of Accra back home.

By the time they are packed up and shipped off to the port of Tema, one thing is true of all these secondhand blenders and ceiling fans and desktop computers, regardless of who sent them. Study after study has revealed how at least one-quarter of them don't actually work — and within three years of their arrival in West Africa, the majority of those that *did* work no longer do.[2] Someone in Canada or Slovakia was determined for good reason to dispense with that blender or fan or computer in the first place. They are, if not trash already, trash on the make, albeit trash packed with valuable materials. A crude calculus gets them from the quays of Tema to the market stalls of Accra. In West Africa's secondhand electronics business, one can purchase a ton of untested products for approximately six hundred dollars. But for *half* that amount one can purchase a ton of *untested* electronics, meaning cell phones and TVs that have been imported from Europe or the United States but are not necessarily guaranteed to work. A TV at twenty bucks instead of forty? A consignment of desktop computers at five hundred dollars instead of a thousand? Most Ghanaian vendors are willing to take the risk.

Several days later, just before dawn along Abeka Road in Accra, I watched as three men unloaded a pickup truck bed full of old flat-screen TVs that had reached Tema earlier that week. They placed them on a table. Over the course of the next hour, as the sun ticked up into the sky, they plugged the televisions in, one after the next. Those that turned on stayed atop the table. Those that didn't — approximately a third of them — were put on the sidewalk to be sold to scrap dealers, who would bring them south toward Agbogbloshie later that morning.

By the time an old television or laptop or cell phone reaches Agbogbloshie, it is deemed neither technology nor trash; it is a tangle of raw materials that happen to have been soldered and glued and drilled together, and that now must be separated.

Take the smartphone in your pocket, which contains more than seventy different metals and minerals, many of great value, comingled into tiny layers. Its speakers contain nickel. Its battery is composed of lithium and cobalt. There is silicon on the motherboard, scant but precious quantities of gold and silver, loops of copper wiring. The point of Agbogbloshie is not just to be a destination for "condemned" phones — not to be a "dump" in the conventional sense — but to separate and extract as much of these inner materials as possible, as quickly and cheaply as possible. It is difficult work. Beyond the long-term health consequences of operating a great scrapyard in the midst of sixty thousand people with negligible access to healthcare, there are reminders all over Agbogbloshie of the dangers of shucking and hammering broken electronics for ten hours a day. The slum is full of hands missing fingers, feet shorn of toes, limbs pocked with burns, and the occasional one-eyed dismantler.

13

TREASURE

> The ruler of Ghana is the wealthiest king on the face of the earth because of his treasures.
> — Ibn Hawqal, Arab chronicler, AD 988

MOHAMMED AWAL MOVED as he spoke: in jolts of gleeful, almost unmanageable excitement. We met just as he happened to be bringing a sledgehammer down on a filthy white block of metal. "Is that a coffin?" I asked, incredulous. "Refrigerator!" Awal shot back. He stopped sledgehammering and glanced up at me. A short but muscular man, he wore a couple of black Oxford dress shoes that were, seemingly always, frosted in red-brown grime. "Let's walk there like soldiers," he told me after agreeing to show me his home in Old Fadama, sweeping his arms up and down in imitation of a goose step.

Over the next three weeks Awal became my guide to Agbogbloshie. It took two of those three weeks for me to become convinced that he was really twenty-three. Awal's face was that of a tired, almost elderly man. Beneath his right eye, above a tribal mark carved into his cheek, a tattoo of a teardrop had been so poorly drawn that Awal had asked the artist to attempt another one beneath his left eye; a third tattoo — a pair of crossed knives mounted above the words DON'T TRY!, which Awal insisted was a popular taunt among American wrestlers — ran across his left biceps. Farther down, on his left wrist, a plastic watch no longer worked. "But what if you need to know the time?" I asked him. He pulled out his phone and pointed to a cracked screen.

At first Awal proved difficult to reach. Sure, he possessed a second-hand Huawei cell phone — three of them, actually. But their batteries were so depleted, none could last more than a couple of hours at a time — requiring Awal, over the course of any given day, in a slum literally piled with mobile phones, to transfer his SIM card from one to the next in order to be contacted by phone. After a few days struggling to pin him down, dialing a number that rarely picked up and often didn't ring, I settled on an easier way to locate Awal: showing up at his front door.

Awal's house, a single-room corrugated-iron shack where he lived with his wife, Mala, and one of their two sons, sat along Agbogbloshie's scrap thoroughfare, one of hundreds of shanties that extend for miles along the bank of the Korle Lagoon and look out on its garbagey waters. The shanty's front yard, if you could call it that, featured a huge mound of rusted scrap metal covered by a green canvas tarp; laundry hung on a rope slung above it. Behind the house sat one of Agbogbloshie's communal latrines, the runoff from which sluiced into a small channel parallel to the shanty itself; by nightfall, after hours absorbing sunlight, it had the smell and humidity of a bog, and seemed to spawn its own undeterrable population of mosquitoes. Tethered to a pole along the sewage channel, a white goat belonging to Awal's neighbor was being fattened on old yams.

Awal nevertheless required a working phone to earn his living in Agbogbloshie. For his job was not to extract circuit boards from VCR players. It was not to pluck the cameras off "condemned" smartphones. It wasn't even to chisel chunks of iron off old Japanese motorcycles. Mohammed Awal's job was to gather the worthless remains of all these things in a pile and set them on fire.

A complex tribal hierarchy dictates who does what work at Agbogbloshie — and who doesn't have to work at all. Over his eight years in Accra, Awal—partly through personal charisma, partly out of desperation—had come to occupy a unique position in this system, functioning as a one-man intermediary between two groups at opposite ends of the electronics dismantling process. At one end are the Konkomba, a powerful tribe who specialize in ferrying electronic scrap into Agbogbloshie from across Accra and beyond. At the other end are the Frafra, a weaker tribe who, to make their money, burn the PVC plastic off the cords of old electronics to extract their internal coils of copper wiring.

Awal is neither a Konkomba nor a Frafra. He is something else, an ethnic Dagomba. And as such—a third party, so to speak—he is in an advantageous position in Agbogbloshie to negotiate the terms of the relationship between the Konkomba and the Frafra.

It falls to Awal to deliver flammable material to the Frafra on behalf of the Konkomba. Over the previous four years he had emerged as an informal leader of what are known as the "burner boys." There are not many of them in Agbogbloshie, perhaps two hundred, most of them teenagers, almost all of them ethnic Frafra, and the majority operate in groups numbering ten to fifteen. But their existence is indispensable to how the slum—and the global e-waste trade at large—operates. The burner boys are the engine that allows the carousel of foreign electronics entering Ghana to keep twirling. They make all of what can't be repaired or dismantled disappear, clearing room for more cartloads of busted electronics to arrive. The plastic that encases television and computer screens, the Styrofoam that lines refrigerators, old tires—in a slum as crowded as Agbogbloshie, these can take up a lot of space. Every afternoon, the burner boys heap it all into piles, mounds of flammable trash they call *bola,* then set those piles alight, clouding the sky above Old Fadama with heavy billows of smog the color of tar.

But that's not all. The *bola* fires have another purpose. Trapped within virtually every gadget or appliance that reaches Agbogbloshie are spools of copper wiring. Be it a forty-year-old air-conditioning unit from Singapore or an iPhone discarded last year in New York, copper is the conduit that transmits electricity. In Agbogbloshie it's known as *capasta,* "treasure," and it is not just everywhere — after plastic, copper is probably the most abundant material found in the slum — but one of the few things that can be directly exchanged at scale, on-site, for cash.

Now here's the reality of the situation. There does exist a "proper" way to extract copper from electronics: You feed cords into an industrial cable-stripping machine known as a granulator, which shears the polyvinyl-chloride encasement from the metal, rendering more or less pure quantities of copper. Several of these granulators can be found in Agbogbloshie, donated to Ghana in 2016 by the New York–based NGO Pure Earth. BURNING IS BAD, read placards mounted across the machines. PROTECT YOUR HEALTH.

The other way to extract copper is to bypass the granulators and incinerate the polyvinyl chloride away by simply torching the cables, using the great abundance of *bola* — plastic trash — as kindling. This has never stopped being the preferred method of copper extraction in Agbogbloshie, for it has two advantages over the granulator system. First, it helps get rid of the mountains of plastic crowding the slum. Second, and more importantly, it is quicker.

Enter Mohammed Awal. It is Awal who brings the *bola* to the Frafra to burn; and it is he who brings the *capasta,* the copper treasure that it renders, back to the Konkomba, who then sell it to networks of Nigerian scrap dealers, who in turn hustle it out of Agbogbloshie to the port of Tema, where it is next sold off to Chinese businesspeople, eventually making its way to Northern Europe or East Asia.

Six days a week, Awal starts work around six in the morning. His first few hours are spent roaming Agbogbloshie's scrap thoroughfares on the hunt for knots of cords and wires the size of soccer balls. Here's where a working cell phone — or three phones that sometimes work — comes in handy. Calls come in from Konkomba scrap dealers. A few kilograms of wiring needs to be picked up from a smartphone dismantling stall; a shopping cart full of HDMI cables needs to be collected from another. Awal directs the Frafra boys working under him to fan out across the slum and fetch it all. Over the course of any given morning, cohorts of them can be seen approaching Awal, on bicycles, clutching rice sacks full of wires.

After a few hours, it is time to render the *capasta*. Awal typically waits until the afternoon to do this, the hottest part of the day, when many of the residents of Old Fadama are inside their homes resting, and when the smoke that rips across the slum will theoretically prove least bothersome.

A little after noon the day I accompanied him, Awal led a band of burner boys out to a stretch of wasteland on the far bank of the Korle Lagoon, where they dumped out their rice sacks, piled the contents into small heaps, then set the heaps alight with matches. Instantaneously, the electronics — laptop chargers, USB cords, car alternators, VCR cables, iPod headphones — were enveloped by fire. To increase the intensity of the heat, a great mountain of gutted TVs and Styrofoam refrigerator insulators had already been gathered along the bank of the Korle Lagoon; this is the *bola,* the trash fuel. The burner boys applied it generously to the flames.

Within moments, dozens of fires were sizzling along the lagoon. Each Frafra boy tended to three or four of them. They had begun by tossing wigs and dried-out coconut shells into multiple conflagrations at once, which crackled and danced and turned emerald green as thick clouds of leaden smoke spewed toward the Onion Market. When the fires grew larger, the burner boys switched to heavier kindling, feeding disemboweled televisions and refrigerator insulation foam into the flames. Every few minutes they used iron rods to lance out chunks of copper, gradually shorn of their PVC plastic encasements, dexterously lassoing the metal around their spear prongs and examining its state. All the while, the flames continued to get hotter and hotter. At one point an unearthly, gunshot-like noise sent a white blob careening into the distance: A gust of hot air from one of the bonfires had popped an automobile airbag resting atop a distant mound of trash. None of the burner boys took any notice; next to me, one discharged a snot-rocket onto the ground.

TREASURE

It took about ten minutes for the plastic to be segregated from the "condemned" cell phone chargers and computer cords, which had puddled into a congealed blob of frozen black goo. Their copper innards, the *capasta*, the treasure, were gathered by hand after the dozens of fires had been extinguished with small plastic sacks of water the burner boys had ripped open with their teeth. As we left the scene, I turned back to the dying flames. They still simmered. And in the wider landscape, I counted innumerable fumes rising out of the folds of garbage mountains. It was a scene all the more terrifying to witness at night, which I later did. For only in the darkness do you realize that the landscape around the Korle Lagoon is not just smoldering but perpetually *on fire:* The leaching of chemical solvents into the soil from old electronics, combined with the accumulated heat of each new day, results in the spontaneous self-combustion of patches of earth all around Agbogbloshie, all the time. The smattering of conflagrations stretched into the distance like so many tent fires of an encamped army. Indeed, parts of the Korle Lagoon landscape have been burning longer than many of Agbogbloshie's residents have been alive.

We headed back across the lagoon toward Old Fadama to take stock of the spoils. A huge Alpha & Omega scale registered three pounds of copper, earning Awal eight cedis, or about fifty cents. He and the burner boys repeated the process about half a dozen times that afternoon and evening. The final tally — some three dollars — is roughly what Awal has been making each day in the seven years he has lived in Accra. It is less than what the ethnic Konkomba make buying and selling scrap. It is more than the ethnic Frafra make collecting silicon and burning *bola* to acquire copper. It is enough to get by, never enough to prosper, enough to keep living in Agbogbloshie, never enough to leave it. And in this the burning turns out to be fitting symbolism for the slum writ large. In so many ways it is not really an electronic waste dump at all — not a place where phones and TVs simply get shoveled into a landfill — but a locus of exploitation, an economic purgatory whereby residents of rich countries, determined to recycle their unwanted electronics, have been led to believe that doing so is helping a place like Ghana, when in fact this is forcing the country into a kind of dependency stagnation, reliant on an import that may provide a number of middle-class Ghanaians with some working computers but does so at the cost of inflicting upon *far* more poor Ghanaians huge quantities of toxic material that never gets totally incinerated away. After a year or two, even the best of those electronics cease working; the toxins left in the wake of their dismantling are nevertheless bound to saturate the soil, contaminate the water, and poison the air for the rest of time.

"Burning *bola* is good. It makes an area clean so that we can start new work the next day," Awal explained when I once asked him if he had ever considered the prospect of not burning mountains of garbage to render copper.

But what about the pollution that burning *bola* leaves behind?

"It doesn't concern me. I pray to God every day to stop the burning. But for now I need it — and I need *bola* to do it. It doesn't concern me when I leave here whether there are problems or not."

This is a common sentiment in Agbogbloshie: Because it is a place most of its residents are not actually from, because it is essentially a foreign land, perhaps even because it is located within the capital of a state that has proven remarkably unwilling to help formalize their work or recognize its value, there is little reason to care one way or another about the long-term health of Old Fadama or its environs.

Awal had spoken to foreign journalists before about his life in Agbogbloshie. European filmmakers had come to interview him about the years he'd spent burning their electronics; they'd won awards at prominent festivals for documentaries that put his misery up for Western inspection. Little had changed for Awal. His life in Agbogbloshie required a deeper intervention. By the time I met him, Awal was resigned to the fact that burning *bola* constituted his existence; and until he died (average life expectancy in Agbogbloshie is less than that of Ghana itself), this seemed to be what it would remain.

Eight years earlier, he'd come to Accra from a village on the outskirts of Tamale, a city four hundred miles to the north. He was fifteen. His father was dead. His mother sold bags of charcoal to support the family. To help bring in more money, in late 2014 Awal headed south to Accra on a sixteen-hour bus ride with nothing more than a school satchel packed with a change of clothes. The bus dropped him off a stone's throw from Agbogbloshie, which in his eight years in Accra he has never left. The great clamor of the capital now around him, he entered the Onion Market that first morning and called a childhood friend, a fellow Dagomba named Raulf, who found him a home with four others; Awal lived there for three years, until he found his wife, Mala, and the two of them moved into their shanty along the edge of the Korle Lagoon.

Awal liked to tell me that he wanted only two things in the world. The first was to make enough money to construct a second story atop his one-room house so that Mala, a seamstress, might have a space of her own to sew dresses for her clients around Old Fadama. "She has too much talent," Awal told me, closing his eyes and shaking his head. "Too much talent." Awal's other ambition was to make enough money to leave Accra altogether and return to Ghana's north, where he and Mala might purchase some farmland and harvest yams. The absurdity of this second ambition — that in order to save up enough money to purchase a plot of land and do the same thing his ancestors had probably done for millennia, in the exact same place they had done it, Awal first needed to spend a decade or longer hundreds of miles from his birthplace, burning electronics that arrived in Ghana from across the Atlantic — was not lost on him. "What would you do if you were me?" he once asked me.

I didn't have an answer for him. In any case, Awal's biding concerns tended to be more immediate than the loftier questions of why he was in Agbogbloshie or how he might leave it. Eight years of burning electronics had pocked his hands and feet and neck with scars, the result of shards of scalding iron or plastic leaping forth from bonfires and singeing his skin. And like many of the residents of Old Fadama,

Awal could tell — with uncharacteristic, almost eerie dispassion — stories about what the *bola* fires seemed to be doing to his health over the longer term. Days passed, Awal claimed, when he lost almost all movement in his arms or legs; they were stuck in place by a burning sensation. And there were nights when Awal had woken up coughing blood.

For Awal and countless others in Agbogbloshie, one recourse was proffered to address such problems. In 2016, the German state of North Rhine-Westphalia donated €5 million for a field clinic to be constructed at the edge of Agbogbloshie's scrapyard.[1] A brick building encircled by a chain-link fence, outfitted with a large green placard crediting the generosity of its German funders, it was constructed to alleviate the respiratory problems caused by the burning of electronic waste — several tons of which, on any given day, are no doubt German in origin, perhaps even from North Rhine-Westphalia itself. A nurse at the clinic told me that an average of six boys enter the clinic every day to complain of chest problems. The majority shun it, and for understandable reasons: Any moment not spent collecting *bola*, or burning it to capture *capasta*, is generally considered lost time in Agbogbloshie. Awal told me the nurses gave him cold water before sending him back home after a night of bloody coughing. "They said there was nothing else they could do," he explained.

14

LOGGING ON

Now the most important thing is education. Education, schooling, the acquiring of knowledge. We are so backward, so backward! I think that the whole world will come to our aid. We must be the equals of the developed countries. Not only free — but also equal. But for now, we are breathing freedom. And this is paradise. This is wonderful!
— young Ghanaian accountant to Polish journalist Ryszard Kapuściński, 1958 [1]

AGAIN, IT ALL provokes the question: Why Ghana?

Not so long ago, Agbogbloshie was nothing like what it is today. There was no slum, no fuming scrapyard piled with mangled electronics, no mountains of trash crisscrossed with emaciated cattle, no burner boys reliant on stockpiles of TVs to burn rice sacks full of cell phone chargers, no landscape sizzling with chronic fires. There was just swamp.

You can still meet Ghanaians old enough to recall when the largest tributary of the Korle Lagoon, the Odaw River, emptied out into an estuary presided over by ethnic Ga fishermen. In local Ga mythology, the Korle was not just sacred but a deity; for centuries the Ga attributed the honor and protection of their tribal lands to the goddess Korle. In the 1660s, in response to the Portuguese — who arrived from the west in galleons and were mesmerized by the sight of gold gleaming along Ghana's coasts — it was Korle who enticed the Akwamu tribespeople to send a message to the foreigners and burn their encampment at Accra to the ground. In the 1820s, when the Ashanti Kingdom invaded the Ga lands from the Guinea interior, it was Korle herself who rose forth from the lagoon's waters to send the would-be conquerors scurrying back to the hinterland.

It was the British who, wresting control of the Gold Coast from the Dutch in 1872 and designating Accra its capital five years later, first struck upon the idea of using the sacrosanct Korle marshlands as a dumpsite for their refuse. They constructed an incinerator that by the interwar period was burning tens of thousands of tons of feces and rubbish a year.[2] Colonial edict, meanwhile, required European settlers to segregate themselves "one mosquito flight away" from the lagoon's waters to stave off the threat of malaria. A shrine to the goddess Korle still adorned its banks as late as 1927, the year Governor Ransford Slater connected the marshlands to the Atlantic with a canal that today sluices thousands of pounds of plastic and electronics out into the Gulf of Guinea every April with the first of the spring rains.

During the Second World War, American and British soldiers stationed in Accra were determined to put an end to the city's recurring malaria epidemics. As historian Jonathan Roberts has observed, in the Gold Coast, where no Germans were to be found, mosquitoes became the enemy.[3] In 1942, a team known as the Inter-Allied Malaria Control Group conducted a series of experiments intended to eliminate larvae from the Korle Lagoon. To test the effectiveness of various chemical combinations, the Allies first constructed eighteen shacks along the lagoon outfitted with special netting that allowed mosquitoes to enter the structures but prevented them from leaving. And as bait for the mosquitoes, the Allies imported tribespeople from the Northern Territories of the Gold Coast colony — the same lands from which most of Old Fadama's current residents, including Mohammed Awal, now hail — whom they proceeded to spray with different combinations of toxic agents. Long before Agbogbloshie was poisoned by dioxins and bromides from cell phones and televisions discarded by Western consumers, it was despoiled by antimalarial agents at the hands of Western soldiers.

The heavy granite monuments that rise forth from the coast of Accra today may commemorate the first successful independence movement in sub-Saharan Africa. But in many ways the history of Ghana after 1957 was to prove a story of foreign domination by other means. In 1961, in a bid to counter the Soviet Union's creeping influence throughout West Africa, the Kennedy administration—along with the United Kingdom and the World Bank—funded the construction of Akosombo Dam across a stretch of the Volta River sixty miles north of Accra. The largest investment project in West Africa's history, converting 4 percent of its landmass into an enormous reservoir, Lake Volta, visible today from space, the dam's purpose was to provide cheap electricity for Ghana. There was only one caveat. The dam would also be required to provide energy to a huge, US-owned smelting plant that would help convert the former Gold Coast, already a massive exporter of gold in the 1960s, into a titan of the emerging global bauxite and aluminum industries.

It never happened. As with much of the developing world, Ghana's future prospects came crashing down in 1973. That year's spike in oil prices drove the costs of harvesting cocoa ever higher just as recessions in developed countries clipped demand. Over the course of the decade, Ghana's export revenues collapsed in half.[4] Public-sector wages fell by 74 percent, GDP by a third. Debt skyrocketed. Over a million Ghanaians left their country. By 1980, Ghana was poorer than it had been in 1960.[5] As for the mighty Akosombo Dam, touted as Ghana's future, it had devolved into a national disaster. Accra had been promised a glut of discounted electricity, but 70 percent of the dam's electricity ultimately went to powering the operations of Kaiser, the American mining company that had all but stopped excavating bauxite in Ghana when it discovered greater troves closer to home in Jamaica; all the while, Ghana, heavily indebted for the cost of the dam's construction, was forced to import electricity at crippling rates from neighboring Ivory Coast.[6] By April 1983, two years into one of the worst droughts in a century, Ghana was on the brink of bankruptcy.

That summer, Ghana — the first nation in sub-Saharan Africa to liberate itself from colonial rule — became the first to undergo "structural adjustment" at the hands of the International Monetary Fund. Ghana is still regarded as one of the more successful economic interventions of the 1980s — the "star pupil," in the eyes of the IMF; the "success story" of West Africa, according to the World Bank and the US State Department — for its willingness to follow the blueprint of liberalization, privatization, and deregulation without question or hesitation.[7] And yet it's an intervention that in almost every way set the pieces in motion for the desperate existence of someone like Mohammed Awal decades later — a man who has spent the last eight years in Accra amid piles of busted electronics packed with invaluable metals and minerals, some of which were certainly extracted out of the subsoil of his native land. The irony is grotesque: In a place so blessed with mineral wealth it was christened the "Gold Coast," tens of thousands of Ghanaians are now forced to scrounge together a living by "mining" electronic waste hustled in from countries thousands of miles away with little to no mineral wealth to speak of.

To help restore Ghana to solvency, its creditors focused on boosting exports to the Global North. Capital went into propping up Ghana's natural resources sector to make them more attractive to privatization, even as that investment did little to alleviate poverty in those parts of the country and among those populations where it had become most dire. Extraordinary steps were taken to ramp up cocoa production at — by the IMF's own admission — the expense of agricultural sectors such as maize, millet, and yam, which feed Ghanaians themselves. Nowhere was this clearer than in Ghana's savannah north, the region of Awal's birth, once an exporter of rice, now reliant on rice imports from countries like Vietnam and India. By the late 1980s, overall poverty in Ghana may have declined, but inequality yawned: In the cocoa industry alone, 32 percent of farmers emerged from the years of structural adjustment bearing 94 percent of the profits.[8]

Still, the greatest intervention was directed toward Ghana's legendary gold-mining sector. The IMF oversaw the sale of 80 percent of the industry—Africa's second largest exporter of gold throughout the 1970s, when more than half of it was state owned—to firms headquartered throughout the West. The majority were Australian or Canadian, lured by IMF-brokered promises to eliminate foreign-exchange taxes, quash export levies, cancel import duties on equipment, and slash corporate tax rates.[9]

Following the privatization of the mines, no measure was spared to force great numbers of Ghanaians off their ancestral lands—thirty thousand ethnic Akan evicted from the district of Tarkwa alone—often without compensation, occasionally through threat of violence.[10] The effect was to turn independent Ghana into one great conveyor belt of mineral wealth from the ground beneath the country to multinationals located outside it. Indeed, by the time two-thirds of the country had been handed over to mining concessions, resulting in the turnover of huge tracts of maize and yam farms to exploratory extraction, even a prominent mining journal had to lament the sorry state of affairs whereby "everywhere you go, you see huge cavities in the ground, discarded pits where thriving villages once stood and where nothing now grows."[11] And while the volume of gold excavated in Ghana may have increased fivefold after 1983, less and less of the proceeds went to Ghanaians themselves. Today, according to the Bank of Ghana, less than 2 percent of the returns on its gold exports cycle back to the land of their extraction, and more than 70 percent of earnings are whisked away to offshore accounts.[12]

It was in the wake of such developments, in January 1994, that two farmers—an ethnic Konkomba and an ethnic Nanumba—entered a street market in the bazaar town of Nakpayili, near Ghana's border with Togo, approximately one hundred miles from Awal's place of birth. In one of the market stalls, the men spotted a black pheasant-like bird known as a guinea fowl. They began bidding against each other to purchase it. The bidding turned heated; the two men eventually came to blows; within weeks, the dispute over the bird had spread beyond the marketplace, beyond the town of Nakpayili, out across the entirety of northern Ghana, erupting into a vast intertribal conflict over ancestral rights to swathes of increasingly unproductive farmland and grazing territories.

Virtually everyone you meet in Agbogbloshie today has, like Mohammed Awal, reached Accra from northern Ghana. The older among them are refugees of what became known as the Guinea Fowl War, a six-month conflict that pitted dozens of tribes against one another and resulted in the razing of some five hundred villages, only to be quelled when busloads of soldiers dispatched from Accra resorted to pacifying the belligerents by dousing them with hoses spraying scalding water. Over the next few years, the Konkomba and Nanumba—followed by the Dagomba and the Frafra and more than thirty other tribes—began streaming south into a boggy section of central Accra, crowding the banks of the Korle Lagoon. Together, the onetime enemies cut down trees and filled in the marshlands with the resulting sawdust, which still pads the alleyways of the great labyrinth of Old Fadama today. The northern tribes began living side by side in the south of Ghana in a state of shaky codependency, relocated from the epic desert expanses of the Sahel to a tiny spit of swamp soon cramped with the Onion Market, the slum of Old Fadama, mosques, herds of cattle, a bus station. The Konkomba began selling yams. The Dagomba began collecting pieces of scrap metal. The Frafra began scavenging for bits of silicon and copper. One of the striking things about Agbogbloshie is that for all its outward misery, the fact that it exists in the first place is little short of a miracle: For a generation now, with negligible support from their state, forty tribes with generational histories of violence have lived virtually on top of one another in near-frictionless interdependence.

Other countries across West Africa sought out foreign waste flows in the 1980s to raise cash that could help reduce their debt or build infrastructure. Ghana was different. At a time when the Republic of Benin was negotiating to take thousands of tons of radioactive waste from France and acres of cattle pastureland in Nigeria were getting despoiled with PCBs from Italy, at a time when hundreds of drums of hazardous material from Ohio were getting flown to Zimbabwe and mercury from New Jersey was coursing through South Africa's Mngcweni River, Ghana was putting itself forward as the African exception. The waste trade, according to Jerry Rawlings, Ghana's president during the late 1980s, was an "evil and degrading practice."[13] And yet within just a few years of Rawlings's claim — by which time African governments had successfully and improbably united against the import of Western toxic waste — what was entering Ghana would prove nearly as devastating and poisonous as any of the waste flows that had ever entered Africa. Ghana may have rejected toxic waste imports in the 1980s, but by the late 1990s, it had allowed itself to become a destination for trash of a different, though often no less dangerous, kind: electronic waste.

How?

Ghana achieved independence earlier than many other states across Africa, but the problems layered into its independence — a crushing debt, a stuttering agricultural sector, an exploding population requiring housing and education and healthcare — were as formidable as in any other place on the continent. And on top of all those well-documented problems, a new conundrum presented itself as the millennium drew to an end.

In October 1999, the UN Economic Commission for Africa gathered together the continent's leaders at the African Development Forum in Addis Ababa, Ethiopia. "The world is perhaps moving to a division, not between the 'North' and the 'South,' but between the 'fast' and the 'slow,'" the commission explained to some 600 representatives hailing from all 53 nations of the continent.[14] The Africa that had once produced raw materials for European industry was losing ground to new export markets emerging out of Eastern Europe and Southeast Asia. Even that Africa, the continent divvied up by European states and laced with railroads to shuttle resources out to rich countries, was a colonial legacy that needed to be relinquished, argued the commission. It had produced disarticulated zones of resource specialization — why should Ghana export cocoa and neighboring Togo export phosphates and nearby Benin export palm oil? — and impeded much-needed economic cooperation and advancement.

No, continued the commission. In the years to come, the peoples and economies of Africa needed to be fused together. And there just so happened to exist a remarkable new tool to help them do it.

The internet held the promise of an "information revolution" that could solder the continent together after the depravations of colonialism. Workers from Nigeria could find fellow English-speaking workers from Kenya. They could integrate into global economies not by digging out or chopping down raw materials for rich countries but by creating complementary economies that contributed — and assimilated — knowledge and engaged in mutually beneficial commerce. "Africa has witnessed its already tenuous position in the global economy deteriorate even further," concluded the commission. "Globalization is a reality.... Countries that do not facilitate this information revolution will likely fall further behind."[15]

The power of the internet was also such that one could theoretically bypass earlier stages of economic development. With the internet, one need not enter the industrialized world, as states like Ghana had so desperately attempted in the 1970s. One could "leapfrog" straight into the digital one, skipping over the nineteenth and twentieth centuries and landing headlong in the emerging cyber economy of the twenty-first century, achieving something akin to what Kwame Nkrumah, Ghana's first president and himself an early apostle of Pan-African unity, had meant when he said that a national investment in technology might allow his country to accomplish "in a decade what it took others a century."[16] In the 1990s, after a 1980s widely regarded as Africa's "lost decade," at a time when average wages had in fact *decreased* since the late 1970s, the arrival of the internet seemed to finally offer an opportunity to push Africa into a long-awaited modernization orbit.

Just consider what the internet could do for Ghana specifically, a country in which fewer than one in five citizens had any formal employment or received any kind of regular salary. The internet could provide jobs that would formalize the economy, broaden the tax base, connect an overwhelmingly English-speaking population to the rest of the world. And it could lure Western capital to a country deemed disproportionally safe for investment in West Africa — Ghana had few coups and relatively few tribal conflicts — which, unlike investments in mining or cocoa, would not go toward privatizing Ghana's wealth and siphoning it out of the country but toward training its population to excel at using the tools of the future. Ghana's population was young, 60 percent of it under the age of twenty-five, and growing, predicted to double in size by the year 2020. "If we Ghanaians fail to take advantage of information technology, we will be further marginalized in the world," warned Clement Dzidonu, chairman of Ghana's Committee on National Information and Communications Technology Policy and Plan Development.[17]

And so it was Ghana, more than any other country in Africa, that took up the idea of getting ahead by getting online. "You must remember, this was the period of emergence for Africa," Nii Narku Quaynor, the Ghanaian computer scientist widely credited with getting his country on the internet in the 1990s, told me. "Mandela had just come out of prison. The last hold on freedom was removed. There was a certain excitement around the continent about our ability to level final barriers to our development. And many of us did see the arrival of the internet through a development angle. It wasn't only about building a cyber economy. There was now a chance that if you're going to build a hospital, it will be a better one, a safer one. If you're going to build a school, it's now going to be more advanced than a traditional school. We could rid ourselves of the baggage of a legacy system."[18]

Under Quaynor's guidance, Ghana began liberalizing its communications sector, leading to the World Bank noting with approval in 1997 that it had become "the first developing country to introduce privatization and competition in all areas of services, in all parts of the country."[19] At a time when the entirety of Africa still possessed less bandwidth than the city of São Paulo in Brazil, Ghana became the second country in sub-Saharan Africa to reach full internet connectivity.[20] It made good on its pledge to install a public telephone in every town of five hundred people or more.[21] It expanded internet access — Ghana may have had a single internet café in 2000, but two years later it could boast more than six hundred — and it began sending students, the vanguard of what it called its "generation of technology entrepreneurs," for computer training to India, which in turn helped fund the construction of Accra's Kofi Annan Centre of Excellence in ICT, soon home to West Africa's first supercomputer.[22] By the mid-2000s, US companies such as Aetna and Liberty Mutual were relocating their call and tech centers from South Asia to the Ghanaian capital.

The only problem with Ghana's grand future as the cyber-savvy economy of Africa? Bringing the internet to Ghana was one thing. Bringing millions of ordinary Ghanaians to the internet was another. Huge numbers of computers, phones, keyboards, routers, mouses, and chargers were required. And in a country in which a majority of the population lived on a dollar a day or less, this in turn required a peculiar policy of national self-advancement: incentivizing the importation of cheap old electronics no one else in the world wanted anymore.

Literary theorist Sarah Lincoln has called Africa a "remnant of globalization" owing to its persistent status as a "waste product, trash heap, disposable raw material, and degraded offcut of the processes that have so greatly enriched, dignified, and beautified their beneficiaries."[23] The continent isn't just a destination of choice for Western trash; it's also the repository of what Lincoln likes to call "excess capital." Used clothes, the undesirable sections of chickens, expired batteries, secondhand buses and cars and motorcycles, even tranches of developmental aid — each is an example of a supposed magnanimity that has managed to turn an entire continent into the recipient of that which richer countries overproduce — or no longer want — under the guise of helping Africa chart a course out of chronic underdevelopment. The secondhand electronics trade is in many ways the cruelest iteration of this phenomenon. For, at the turn of the millennium, the promises of technology were widely perceived as guarantors of Africa never needing to become an object of such "charity" again. "We paid the price of not taking part in the Industrial Revolution because we did not have the opportunity to see what was taking place in Europe," Francis Allotey, the first Ghanaian to receive a PhD in mathematical sciences, explained to the Ministry of Communications in Accra in 2000. "This time we should not miss out on this technological revolution."[24]

In the early 2000s, to avoid missing out on the technological revolution, Ghana's governments felt compelled to incentivize the import of old electronics into the port of Tema every which way they could. A clause tucked within the Basel Convention, which theoretically should have prevented many of these old electronics from crossing into the developing world, facilitated the exchange: If an object is being shipped for the purposes of "direct reuse," it does not constitute a form of waste. It's an ambiguity that Western exporters promptly exploited, often insisting they were generously donating their old electronics to Ghana.

"There's just a lot more junk going to Africa now," Robin Ingenthron, onetime president of the World Reuse, Repair and Recycling Association, observed in 2006. "In Asia, the buyers tend to know more about the material than the sellers. But in Africa, it's the other way around."[25]

Other countries quickly recognized the pitfalls of such policy. "We either safeguard our environment or pay an immense price in the future," Uganda's finance and economic planning minister claimed shortly after banning secondhand electronics imports.[26] But Ghana kept importing, becoming the only country in Africa to lift all duties on old computers, in addition to levying no customs duties on monitors or printers.[27] And, for their part, Western multinationals — firms from Dell to Cisco — had every incentive of their own to direct their old products to Ghana. For a Western company to hand its electronics over to a waste broker, and for that waste broker to then ship it to the port of Tema, not only circumvented costs of disposal; there also was money to be made. To recycle a *single* computer monitor in the United States in the mid-2000s cost $15; to send to Africa a 40-foot container packed with 20 tons of electronics, the equivalent of some 2,000 desktop computers, cost $5,000, fees typically covered by a Ghanaian importer who was *also paying for the electronics inside.*[28]

By 2004, old electronics had begun arriving in Ghana from more than 145 countries. Per each of Ghana's 3 million citizens, nearly 20 pounds of electronics were getting unloaded onto the quays of Tema every year; certain Western universities, such as William & Mary, even began "donating" their "decommissioned laptops" to Ghana as a matter of policy.[29] Some of the arriving devices worked. Many of them didn't. Eventually none would. One scholar has estimated that by 2009, of the 215,000 metric tons of electronics that entered Ghana that year, up to 70 percent were in fact waste.[30]

And herein is one of the major problems with most understandings of Agbogbloshie. It is not some outlaw dump, an ungovernable slum that insists on operating beyond the bounds of legality. Agbogbloshie came to exist because of an entirely lawful and incentivized process whereby a poor corner of a poor continent, in its desperation to catch up with the rest of the world, believed importing thousands of cargo containers of old electronics — bound to become mountains of toxic waste though they were — to be worth the elusive promise of accelerated development.

"You must remember how difficult it was to refuse this stuff," Nii Narku Quaynor told me. "You have kids who have never seen a computer before — and then you get someone out of nowhere who says, 'I can send you one.'"

15

TECHNOLOGICAL TINKERING

> There are more Apple devices being used than ever before, and it's a testament to the ongoing loyalty, satisfaction and engagement of our customers.
> — Tim Cook, letter to Apple investors, 2019 [1]

WHY WOULD "SOMEONE out of nowhere" send you a computer? Ghana's whatever-it-requires scramble to become a cyber economy is only half the story of how it emerged as one of the world's great repositories of unwanted electronics.

The other half involves a Western tech industry that by the late 1990s was making ever-expanding chunks of its profits off a policy of exorbitant overproduction. Vance Packard, the mid-century American journalist who pilloried the rise of an economy built on torrents of new products phasing out perfectly good old objects, never lived to see the full flowering of the electronics industry. But one suspects he would have registered little surprise. For the cell phone and television and computer industries picked up right where the automobile industry had begun in the 1920s and the plastics industry had elevated to absurdist new heights in the 1950s: imposing a throwaway culture on consumers through a strategy of *deliberately* rendering their own products obsolete or useless.

When it came to technology, planned obsolescence proved unusually profitable — and all too easy. A product like an automobile, one might recall, required what Packard called "artificial stimulation" to convince a customer to prematurely purchase a new one. A new model — of marginal or even zero improvement — needed to be paraded on a showroom floor, its advancements emblazoned in advertisements, its virtues convincingly preached to consumers. Tech companies, however, could engage in a more cunning degree of corporate deviousness to boost their sales: They could make old products difficult to repair, or simply hardwire perfectly good products to stop working.

As it happens, the earliest known instance of "planned obsolescence," a term not coined until 1932, is found in the tech sector. In December 1924, in a concerted bid to "emerge from a period of long life lamps," executives from the major light-bulb companies in the United States and Europe — General Electric, Philips, OSRAM, among others — convened in Geneva, Switzerland, and colluded to artificially shorten the lifespan of their products and overcome what one manager called the "mire" in sales resulting from their own highly effective goods.[2] Across the world, almost overnight, bulbs that had once burned for two thousand hours started working for just half that time. Sales soon skyrocketed.

Now fast-forward to the smartphone and consider all the ways its manufacturers and sellers tempt, goad, and press you into purchasing a new one earlier than you had anticipated. There are security systems that fail to download on "older" devices. There are batteries that artificially decline in longevity owing to redundant software updates. There are specialized screwdrivers required to access those dying batteries. There are chargers that routinely change in design. There are replacement screens that purposefully reduce a phone's brightness settings. There are mobile phone contracts that, before you have even finished buying one phone, schedule the purchase of your next one within three years or less.

The more you think about it, the more absurd the premise of the smartphone is: a thousand-dollar-plus computer that fits in the palm of your hand and that you've become accustomed to replacing should its screen crack or its battery life start to wane. And the price of that new phone is not merely the dollar amount you pay for it. There's also the environmental toll. For the electricity that powers a phone accounts for only 14 percent of its total carbon footprint over the course of its lifetime. More than 80 percent of its carbon output is emitted during its production, meaning that every time you are prodded into purchasing a new phone, there's a vast additional price to be paid, one you never really see but that is devastating at the planetary level. For to produce a new smartphone requires hundreds of pounds of fossil fuels, dozens of pounds of chemicals, and at least one ton of water. Weighing the equivalent of two apples, a smartphone needs approximately a car's worth of materials in weight to manufacture.

And that's just the physical phone. There's another, deeper corporate marionetting that occurs, too, a machination entirely invisible to the eye. A generation ago, a crucial shift occurred within the US manufacturing sector. Historically, physical products were governed by patents, which protected, say, the design of a particular car from creative infringement by rival automobile companies. In the 1980s, physical products increasingly became reliant on software — functions managed by microchips and processors — which are governed by copyright. Copyright is different from patent. It protects "intellectual property" or "authorship." Attempting to fix your smartphone or refrigerator yourself amounted to, in legalistic terms, a kind of plagiarism. You were violating the law. If you wanted to fix your phone, your only legal recourse was to seek out expensive repair options often offered exclusively by the manufacturer itself — that or buy a new phone.

In the 1980s and into the 1990s, as more and more of our world came to be operated by software-reliant devices, our relationship with our purchases shifted in a fundamental, though often overlooked, way. Suddenly the phone you bought was not *quite* yours — or, at the very least, all that was inside that phone and allowed it to function was beyond your writ to touch or tamper with. You were not allowed — on any legal or practical level — to fix your device yourself; you could merely *use* it for periods of time that were themselves subject to corporate whim.

It all added up to a peculiar reality. Sure, you may talk about "getting a new computer" or "buying a new printer," but one might say that for the last thirty years you have not been purchasing these electronics so much as you have been renting them. And even as you have tended to pay more and more for them, those "rental" periods have been provably decreasing in length. The average lifespan of a computer manufactured in 1997 was six years; by 2005, the year that more than three hundred million were discarded in the United States alone, it was fewer than two.[3] The average lifespan of a desktop printer is now just five hours and four minutes of printing time.[4] The average lifespan of a smartphone is now less than three years.[5]

"There was once a time when you brought extra Nokia phone batteries with you in case your first battery died. Now people have no idea how to even open the back of their phone," Nathan Proctor told me. Proctor is senior director of the Campaign for the Right to Repair, a part of the Public Interest Network, which works to pass legislation in the United States preventing companies from blocking consumers' ability to fix their own devices.[6] "Any time a new technology arrives, the consumer is required to reset expectations of tinkering and repair."

This is all by design. It boosts tech profits. And while the drift over the last generation from patent to copyright might sound like a semantic shift, its material consequences have become colossal. As the idea of reusing and repairing technology became unfeasible, and as you were increasingly forced to "buy a new one" rather than fix your old one, the waste started to mount. By 2018, at least 140 million cell phones were being discarded each year in the United States alone, the equivalent of 416,000 phones tossed per *day*. At a global level, the picture was to prove more striking. In 2021, tech companies sold 1.5 *billion* new phones worldwide, or one per every five humans — more than fifteen times the number of automobiles sold that year and nearly fifty times the number of Bibles.[7] In three or four years, hundreds of millions of those phones will be faltering or outright unusable, by which time several billion *new* ones will have been sold. It's an endless enfilade of discarding and purchasing, producing and selling.

For the last thirty years now there has been a discussion in the United States and Europe about how its consumers have become the victims of the tech industry's policy of planned obsolescence. And yet the millions of Westerners who have been pressed into purchasing a new smartphone or television earlier than they had anticipated are only one group of victims of a rigged system of corporate enrichment. And they are, on the whole, a privileged set of targets at that. The other victims are those on the receiving end of many of those electronics once they have been rendered prematurely useless or unworkable. Phones blocked by activation locks, outdated VCR players, laptops with depleted batteries, printers with cartridges disabled by microchips — *these* are the electronics that go to Africa, typically under the guise of "charity" or "direct reuse" or "bridging the digital divide." And while some of them *may* continue working in Ghana, the odds of any of them working very long are slim. In the grand scheme of planned obsolescence, Ghana is the last and final stop, the country that reaps the minimal benefits of technology and gets stuck with the disproportionate — and aggregate — material burdens of a devious profit model. And in this, in their willingness to *pay* for cargo containers full of such faltering gadgets from the Global North, importers in a country like Ghana perform an invaluable service for the world's tech sector, offering up an outlet for electronics that would otherwise be messily accumulating in the countries where they were originally purchased — and leading even more consumers to ask why it really must all work this way.

Now, to be clear, only a fraction of this technological superabundance goes to a place like Ghana. But even the tens of thousands of old computers that do arrive every year have had, at least in Agbogbloshie, a monumental impact, reshaping the lives of hundreds of thousands of Ghanaians for what has now been a generation.

Not long after the refugees of the Guinea Fowl War began arriving in Accra in the early 1990s, motorized wheelbarrows started pulling up to the banks of the Korle Lagoon piled with Western electronics that had ceased working not long after reaching West Africa—or hadn't worked at all. The problem in Ghana was to prove an exaggerated version of the dilemma that would play out elsewhere across the Global South. Though the World Bank and International Monetary Fund would incentivize the breakneck urbanization of developing countries, slashing much-needed agricultural subsidies that eventually drained countrysides of farmers, they failed to provide the resulting waves of migrants very many opportunities in those cities. In Agbogbloshie, hungry for work, thousands of displaced northern Ghanaian yam and rice farmers began living on the margins of society in the beating heart of their capital, collecting scrap, dismantling it, extracting its value, and, sure enough, torching what remained.

16

THE FLEXIBLE MINE

> These mines will differ from any now to be found because they will become richer the more and the longer they are exploited.... [T]he same materials will be retrieved over and over again.
> — Jane Jacobs, *The Economy of Cities*, 1970

WE LIKE TO think that the growing preponderance of office work in Western countries toward the end of the twentieth century ushered in a cleaner world. Ranks of laborers in grimy overalls filing into belching factories gave way to analysts and coders entering sleek, antiseptic office parks. But the tucked-from-sight fate of electronic waste — armies of Konkomba whamming your discarded Wi-Fi router with three-pound gavel hammers — offers one corrective to that view: There *is* a huge environment toll from the tech revolution, only one that someone else tends to pay.

Electronic waste is currently the world's fastest-growing type of garbage. According to the *Global E-Waste Monitor*, an annual United Nations report on the state of electronics disposal, nearly 50 million metric tons of old cell phones and computers and televisions enter disposal streams every year, the equivalent of 125,000 jumbo jets of materials that require — though scarcely ever receive — careful handling. That amount is expected to double by 2050, by which time the production of electronics will be responsible for nearly 15 percent of *all* global carbon emissions.[1]

Consider it: Something that hardly existed in 1980 now constitutes a huge chunk of our waste outputs, the material corollary of the internet's unrelenting takeover of global economic activity itself.[2] And currently less than 20 percent of any of it gets "recycled" in any proper or systematic way. Twenty years ago it was believed that approximately one-quarter of the electronics discarded in the United States got shipped to impoverished countries to be dismantled by hand. More recently US environmental groups have been inserting tracking devices in electronics designated for "recycling" and discovering something worse: Roughly one-third now get exported overseas, the majority to developing countries.[3] Equivalent quantities are sent from the European Union, with at least a thousand used or broken TVs being shipped to Ghana from Europe every day.[4]

At the same time there's an important twist here, one that makes exporting electronic waste fundamentally different than, say, dispatching drums of toxic chemical ooze to developing nations. And that twist is climate change — and how to slow it.

Twenty years after an internet revolution that resulted in the production of so many billions of smartphones and laptops in the first place, a new global revolution beckons. This is the green energy transition. It is a shift that will require the historic predominance of one kind of material extracted from beneath Earth's crust (hydrocarbons such as oil and natural gas) to give way to the extraction of another (minerals such as terbium and metals such as lithium).

To achieve this transition, however, will first require a seismic, almost epochal restructuring of our current economies of mining. Take the Paris Agreement, for instance, signed by 196 parties in 2015 in a bid to reach zero net global carbon emissions by 2100. To even *approach* meeting those goals, the world will require, over the next twenty years alone, in comparison to current rates of retrieval, 40 percent more copper and rare earth elements, 70 percent more cobalt and nickel, and almost 90 percent more lithium — the extraction, in other words, of more metals and minerals in the space of two decades than humanity has excavated in nearly all previous history.[5] It will necessitate unfathomable quantities of water and will unleash — initially, at least — an extraordinary amount of carbon into the atmosphere. And it is a future in which the considerable geopolitical and financial muscle wielded by oil-extracting petrostates over the last seventy years — Saudi Arabia, Russia, Iran — will give way to a new topography of metal-extracting "electric states," most sprinkled along the equator, many currently the recipients of huge quantities of Western waste. Indonesia and the Philippines boast great quantities of nickel, necessary for the production of solar panels and wind turbines. Chile sits atop epic reserves of lithium, a crucial component of electric vehicle batteries. The Democratic Republic of Congo possesses more than half the world's known stocks of cobalt, vital for storing renewable energy. And then there's China, home to many of the world's more obscure rare earth minerals, ranging from germanium to magnesium.[6]

Agbogbloshie may be a place of unrelenting destitution. It may be a slum with little running water, dreadful sewage facilities, flickering lights, and pitiful — or even nonexistent — access to healthcare. It may be a place where Mohammed Awal makes three dollars a day inhaling *bola* fumes by afternoon and coughing up blood by night, and a place he dreams of leaving to go harvest yams back in his ancestral land in Ghana's north.

But it is also something else.

Agbogbloshie is *packed* with many of those mouthwatering resources that have the potential to stave off planetary devastation. All those phones and ceiling fans and refrigerators and air conditioners that get motored into Old Fadama every morning are infused with metals and minerals whose excavation is largely confined at present to the subsoils of those few countries I just mentioned — Indonesia, the Philippines, Chile, the Democratic Republic of Congo, China. Any future in which we continue to connect to the internet, accumulate and exchange information, work remotely or otherwise, power our cars, harness the energy of the sun — and continue our lives as we currently know them — cannot afford to overlook a place like Agbogbloshie.

And that's not all. Agbogbloshie is also the *most sustainable* way to get those materials.

Industrialized mining is a deeply toxic, carbon-intensive process — and a crudely inefficient one too. Once development of a mine begins, it can take fifteen to twenty years for it to start rendering ore. The material it does start removing from Earth's subsurface contains, on average, only 5 percent mineral content and requires tedious, energy-demanding processes of floating and refining and smelting to be rendered usable. The rest of what conventional mining extracts — the vast majority of what gets lifted out of the ground — is waste rock that requires careful, expensive disposal.

Consider, by contrast, the material that gets sent to Ghana. The gold, platinum, germanium, silver, palladium, and copper within the electronics that get hustled into Agbogbloshie every morning are, in two fundamental ways, different from those same metals when they get extracted from the ground: Pound for pound, they are much richer and they are much greener. The average ton of e-waste contains approximately fifteen times more metals and minerals than a ton of excavated ore.[7] More crucially still, those metals and minerals slotted throughout e-waste have *already* been excavated: The carbon-intensive heavy lifting — the floating and smelting and refining — has already been done. All that's left to do is remove and sort the metals and then send them to an industrial processor.

In other words, Agbogbloshie's residents receive a few dollars a day to handle materials that are not merely essential and precious inputs for clean technologies — a recent estimate puts the *annual* worth of discarded electronics at $57 billion, greater than the value of American Airlines, Burger King, and Adidas put together — but also the most sustainably attainable versions of those materials.[8]

So, what is the problem here? If they are so valuable, how have electronics become the fastest-growing waste stream on Earth? And why are the world's richest countries—countries that are currently courting poorer nations for the metals and minerals required to accelerate the green transition—sending material of so much apparent value so far away?

Geographer Julie Michelle Klinger has argued that it is a myth that rare earth minerals are rare. Yes, *certain* rare earth metals can only be found in the subsoil of a select few countries. But the majority exist in abundance. The problems owe instead to the difficult, often dangerous ways they must be dug up and then turned into workable material. It is not just energy intensive. It is often highly toxic. Most rare minerals—in addition to many vital metals—are thinly spread across reams of Earth's crust. Whenever sizable concentrations of such materials are found, that crust must be leached away, a process requiring dozens of chemical procedures, ranging from acid baths to controlled heating, and leaving behind huge quantities of hazardous material, some of which can be radioactive.[9]

The resulting toxicity is why the history of such activity, as Klinger notes, has tended to be one of "frontierization," of pushing the process of not just mining but also metal and mineral processing as far as possible from populated areas and meddling regulations. Indeed, despite the fact that many rare earth minerals and metals actually exist across much of the world, "the geography of their production" has historically been confined to some of the remotest places imaginable. The Soviet Union territorialized production to Kyrgyzstan, China to the Mongolian steppe, the United States to the deserts of the California-Nevada border, Brazil to the innermost depths of the Amazon. Such places—remote landscapes inhabited by sparse, marginal populations—have been turned into the sacrificial zones of modernization and technological progress.

This has long been the foremost problem of mining: not locating ore, no, but finding a way to excavate and process it at profit without inflicting irreversible ecological calamity on surrounding areas. Even recently, over the last twenty years, highly profitable mines have been shuttered from California to the Rhine Valley precisely because of the unacceptable environmental damage thrust on nearby communities. Indeed, it's no great wonder that the future the mining industry often feels compelled to propose for itself is one in which its messy toil has been marginalized to almost ludicrous extremes. When the tech barons look to the future of how their electronics will get mass-produced, it is rarely to Earth. Minerals and metals will be dredged out of the ocean floor, mined from the moon, or dislodged out of asteroids. Extraction will be handled by robots, all to the benefit of a planet that can finally be handed over to unencumbered human habitation. Amazon's founder, Jeff Bezos, has spoken of a future in which *all* industrial production will one day be outsourced beyond the stratosphere. "This sounds fantastical, what I'm about to tell you, but it will happen," he assured a journalist in 2021. "We can move all heavy industry and all polluting industry off of Earth and operate it in space."[10]

The mine that exists "in the most favorable location," where you need not worry about stringent environmental regulations or endangering local communities or despoiling entire landscapes, where you can extract what you need as quickly and as cheaply as possible — as the environmental scholar Freyja L. Knapp has argued, the so-called flexible mine, a trove of resources that you can position *anywhere you want*, then plunder for profit — has been the industry dream since the 1970s.[11] And here is where Agbogbloshie comes in.

Agbogbloshie is the dream come true. It is a mine situated in an impoverished slum in West Africa not because geographic necessity dictates that it must be there but because — owing to cheap labor and zero environmental regulation and a general Western resignation that Africa is the place where such calamities tragically, though inevitably, exist — it is profitable and convenient to Global North interests that it exists there.

Excavating and processing rare earth minerals leave behind huge quantities of toxic material. The work done in Agbogbloshie —

stripping the discarded remains of electronics — is every bit as dangerous. Rare earth minerals can corrode the skin and cause organ damage if ingested; five of them can lead to radioactive poisoning: promethium, gadolinium, terbium, thulium, and holmium — minerals that often go into lasers and batteries and magnets and that invariably find their way into accumulations of electronic waste.[12] Clouded with cancerous smoke, encircled by acres of poisonous dirt, Agbogbloshie is the frontierization of the de-manufacturing process, one that, were it to be conducted by hand in the Global North, would be prohibitively expensive and labor-intensive and unacceptably polluting. It is a sacrificial zone of *disposal* to which ever more abundant quantities of electronics can disappear, get shucked and sorted, their precious metals removed to the side, their plastic encasements torched away, and their toxic elements dissolved out, leaving behind the most poisoned eggs in the world and deadly air quality, all under the guise that it is helping West Africans.

While the working conditions in Agbogbloshie may be medieval, they are often defended as necessary for global salvation, sparing the rest of the planet the emissions that would come with retrieving new metals from the ground. But it is a situation that, examined from the perspective of the last seventy years, would be grimly ironic, were it not so downright tragic. The Global South's drive to *bridge* the digital divide has resulted in its *circling back* into the material binaries of colonialism. Countries like Ghana — denied any fair chance at becoming an industrialized nation in the 1970s, and determined to make up for it with a cyber future in the 2000s — are turned into stockpiles of valuable resources whose extraction still does little to benefit them.

17

START-UP CESSPOOLS

> In this building, there was no Basel Convention, no dioxins and furans, no acid fog, no water whose lead content exceeded the safe threshold by 2,400 times, no soil whose chromium concentration exceeded the EPA limit by 1,338 times, and of course nothing about the men and women who had to drink this water and sleep on this soil.
> — Chen Qiufan, *Waste Tide*, 2019

ONE DAY I asked Mohammed Awal what happened with all the copper he handled.

"They make jewelry out of it in Europe," he replied.

If only. Most of the copper extracted by men like Awal at Agbogbloshie is exported to either Scandinavia or China, where its fate is likely to be put *back* inside new electronics, phones, and TV remotes, which, one may rightly suspect, could end up *back* in Agbogbloshie in a few years' time.

If the story of colonization in Africa has historically been that of Western extraction, a new picture seems to be coming into view with the rapid rise of electronic waste: a picture of parallel assembly and disassembly lines running out of and into the continent, the one extracting the metals and minerals that go into electronic devices — cobalt and tantalum from the Democratic Republic of Congo, gold from Ghana, iridium and platinum and rhodium and ruthenium from South Africa — and whisking them away to Northern Europe or East Asia, the other ferrying faltering old electronics back in. In this scenario, it is Ghana, the former Gold Coast, that begins to look a lot like the knot at the center of a globalized figure eight. New metal gets dug out of its hills; old metal gets motored into its slums.

The situation in Nigeria is just as dire. Electronic waste deposits in Lagos are believed to "employ" as many as one hundred thousand people.[1] "We recently analyzed cow's milk at one of the big new electronic waste dumping sites in Abuja," chemist Oladele Osibanjo told me. Thirty years ago, Osibanjo was tasked by Nigeria's government with testing drums of Italian toxic waste that had been dumped in the coastal farming town of Koko. "And then we gathered the villagers around us and asked them who among them drank the milk. They all raised their hands. And I remember thinking, *God help them.*"[2] Meanwhile, on the other side of the continent, in Tanzania, makeshift dumpsites of solar panels, many the product of a rising renewable energy sector in neighboring Kenya, are despoiling the outskirts of the capital, Dar es Salaam.[3]

Nor is it just an African phenomenon. E-waste scrapyards are popping up with greater and greater frequency in Asia — or, better put, getting "discovered," in much the same way that Agbogbloshie "existed" years before it came to Western attention in 2008, when a group of Danish journalists, attempting to document the inroads of electronic waste elsewhere in West Africa, embarked upon an impromptu detour to Ghana.

"I was in the Philippines with Greenpeace in 1993 when we heard about a ship arriving from Australia. And the data said that it was full of hazardous waste," Annie Leonard, one of Greenpeace's field operatives in the 1990s, told me. "So we decided to block the ship as it arrives into Manila. It becomes a big to-do. We motor out and chain ourselves to the ship's anchor. It was an unbelievably dangerous thing to do. And eventually the captain agrees to let us on board. This ship was so big it had an elevator. It felt like being inside a shopping mall. And after making a big fuss, the captain agrees to show us the cargo in question. By this time some press had arrived and started filming the scene and I'm getting pretty nervous that we could have made a mistake. We get to the containers. And it wasn't hazardous waste inside at all. It was something even worse. It was electronic waste. Just electronic shit everywhere. It looked like fucking hell. No one had ever really looked into electronic waste before that point. And, if you can believe it, it was literally heading for a facility in Manila called CRAP: Computer Recycling Area Pacific."[4]

The world's most notorious e-waste entrepôt is probably the city of Guiyu in China, where two-thirds of its 150,000 residents are estimated to work in the electronics dismantling industry—in China cables are not burnt, as they are in West Africa, but dissolved in baths of acid. The processing of e-waste has ruined the surrounding landscape to such an extent that potable water now must be trucked in from more than twenty kilometers away.[5] Many of the electronics come from Japan; some enter China with the help of North Korean smugglers, who receive the e-waste from traders in Hong Kong, then traffic it into Northeast China via the Yalu River, via which it then travels south by truck to Guiyu.[6]

The much-heralded start-up culture of Tel Aviv—generator of some 6,000 newly patented technologies per year—has its corollary environmental calamity far closer than the slums of sub-Saharan Africa; Israel's "Global South" is the West Bank, where a bartering economy deals in the 130,000 tons of old computers and phones and televisions discarded by Israelis every year. To process a ton of electronic waste in Israel costs ten times what it takes to smuggle a van of busted laptops and TVs by night east of the Green Line, where a collection of Palestinian scrapyards purchases the material. That which is usable is resold; all else is stripped for valuables, then torched. For hundreds of Palestinians who lack formal work permits, this informal waste economy—the tossed-away scraps of those who are continually encroaching on their lands — has become a last lifeline.

"We have little land left for agriculture, and what land and water we have is contaminated," a waste picker named Abu Jheisha told the *Times of Israel* in 2022. "The scrap industry creates some prosperity and self-reliance." The dangers are nevertheless apparent; rates of cancer in Palestinian scrap villages in 2022 were four times those of other towns in the West Bank.[7]

18

A NEW AGBOGBLOSHIE?

> My hotel's dining room, when I went to breakfast, was full of more of those whites — discussing Africa's untapped wealth as though the African waiters had no ears.... [I]t had been her human wealth the last time, now [they] wanted Africa's mineral wealth.
>
> —Malcolm X, in Accra, 1962 [1]

IT IS AN oversimplification to say that those who live and work in places like Agbogbloshie are mere victims of the unequal global geographies of extraction and disposal. The problem runs deeper. For all its misery and depravations, dismantling electronics and setting them alight brings in *just enough* money that Ghanaians like Mohammed Awal are willing to relocate hundreds of miles to undertake it — and, even after years of unrelenting misery, remain generally unwilling to consider any other type of work. The average dismantler at Agbogbloshie makes three dollars a day: a pittance by Western standards, but an earning roughly equaling the 1,280-cedi monthly wage (about US$100) of the average public-sector worker in Ghana. Awal understands as much as anyone the trade-off between the dire conditions in which he lives and the relative benefits Agbogbloshie is able to afford him — or, if not him, his children. "I burn *bola*," he liked to tell me. "But my son will become a soldier or a football player."

One of the reasons why Agbogbloshie remains a slum, pitched up with hastily constructed shanties, is that most of its residents refuse to consider it their permanent home. It is a work camp, a place to earn money for a few years before returning north. And yet the work they deem temporary — extracting metals from Western electronics — may very well prove more reliable in the long-term than the subsistence farming to which they are saving their money to return. There will never be a shortage of Western gadgets to burn. Harvesting yams, meanwhile, is hostage to government fertilizer subsidies, global competition, and a warming climate.

And in this it is hard not to detect a sweeping shift in humanity's relation with the planet: In a trend I would witness from Turkey to Tanzania, one first has to generate enough money off other people's trash in order to go back and make a living off one's own land — if, indeed, such a living is still feasible. And it is often the case that those forced off their land at least in part by upheavals in the climate have little option but to work in an industry — sorting trash, collecting it, burning it — that is itself accelerating environmental breakdown.

Until that time comes to go back and farm their ancestral land, if it ever does, the concentration of squalor at Agbogbloshie and the immiseration of those who live there is providing an invaluable service for the global electronics industry — as well as the Ghanaian state. During the three weeks I spent with Mohammed Awal, three unremarkable weeks of his life in which I witnessed him set hundreds of piles of *bola* ablaze, I watched as he concluded each of his days the same way. Just a few feet from Awal's shanty was an umbrella with the words MTN MOBILE MONEY stamped in black letters along its weathered yellow canopy; beneath it, a few men in sunglasses sipped on Club beers and handled wire transfers for the residents of Agbogbloshie. Every afternoon, Awal approached them for the purpose of sending half that day's wages three hundred miles north to his family's village outside the savannah city of Tamale. His mother could not get by otherwise.

Men like Mohammed Awal are not the only ones financially reliant on work that is poisoning them. So are the families they left behind in the desert north of Ghana, extensive tribal networks amounting to tens of thousands of Konkomba and Dagomba and Frafra who couldn't get by without the cash flows that the dismantling of European smartphones and American DVD players hundreds of miles south in Agbogbloshie generates. The material redistribution that men like Awal facilitate in Ghana — one that is attempting to reverse decades of IMF-induced capital flight out of the country's north — is a lifeline for great swathes of the nation today.

And yet Ghana's government has rarely seen it this way. When Agbogbloshie is acknowledged to exist at all, it's largely brushed aside as a Western press fabrication. "There is no dumping of e-waste in Ghana," claimed a deputy minister in 2008. "There is no need for people to be concerned," an environmental officer claimed later that year. "The situation is not as scary as the media is making it look like."[2]

In 2021, after years of insisting that Agbogbloshie was nothing worth worrying about, Ghana's government sent in the bulldozers. Early on the morning of July 1, with zero forewarning, battalions of soldiers and police circled its scrapyard bearing machetes and machine guns. They began leveling the dismantling stations, hacking down neighborhoods, tearing markets and shops and mosques to the ground. Shanties were ripped up, their inhabitants still inside them. "Read my lips: No one is coming back," Henry Quartey, the Greater Accra regional minister, would later pronounce. "You'll see a new Agbogbloshie that we will all be proud of."[3]

As a piece of terrorism, it was a master class. Virtually overnight a bustling scrapyard — an ecosystem comprising some forty different tribes that had delicately pieced itself together over the course of a generation and come to stretch to the size of a hundred football fields — vanished. Who had ordered it? What would be destroyed next? None of those caught in its crosshairs had much idea. According to residents I met who witnessed the destruction, anyone caught filming the demolition was marched to a truck bound for a nearby police station and ordered to delete their videos on threat of a prison sentence. But it never occurred to most residents to film the destruction; they were too busy attempting to protect the shanties they had come to call their homes. "The first thing I did was run," Awal told me. "I had no time to think of anything else."

Before long a preposterous plan began surfacing in the Accra press. In August 2021, six weeks after it razed Agbogbloshie's scrap market to the ground without forewarning, Ghana's government announced its intention to erect a great hospital complex atop the bulldozed land. Mock-ups of the Ablekuma Central District Hospital featured a hygienic campus comprised of soulless off-white buildings interconnected by trim stone paths cutting through well-watered courtyards; to demonstrate his commitment to the vision, President Nana Akufo-Addo himself arrived by limousine to cut the first pallet of grass sod that would be layered into the future campus, which his government touted as the "biggest investment ever" in Ghana's healthcare industry. No mention was made of what would be done with the dirt beneath that grass and the future Ablekuma Central District Hospital, dirt deemed by some scientists to be as toxic as the soil around Chernobyl. And no mention was made of what role, if any, the sixty thousand residents of Old Fadama would play in such a promising future.

By the time I arrived in Agbogbloshie — a place whose existence no one in Ghana still has much interest in acknowledging — one-half of it lay in a theatrical state of destruction. You might be forgiven for thinking it a war zone. Bony cattle grazed the evacuated dismantling grounds, pawing through rusty wires and the trashed remnants of mosques in search of stray scraps of yams, a scene resembling a freakish rendition of one of those bucolic seventeenth-century Dutch paintings. Craters speckled the bulldozed terrain, the result of expelled residents having returned by night to dig up valuables they'd stashed beneath their dirt floors at some earlier date. As for the *bola*, the fodder of what prosperity could still be attained in Agbogbloshie, the only place to burn it was now along the cramped banks of the Korle Lagoon, a stone's throw from the Old Fadama slum; the government — in its attempt to impose on the community the biggest investment in Ghana's healthcare industry ever — had only managed to subject tens of thousands of its residents to some of the most carcinogenic fumes ever measured on Earth.

And even then, even after the razing of so many neighborhoods and the upending of so many lives, the state's campaign against Old Fadama's residents hadn't relented.

One late January afternoon I hopped out of my taxi on Okai Street to find several dozen soldiers in camouflage fatigues emerging from an olive-green bus parked along the shoulder of the Korle Lagoon. Small daggers were holstered in their belts; some carried rifles. An officer marshaled them into a column, their rifles now bayoneted, then ordered them to goose-step the length of the demolished scrapyard. They fanned out across the plain, marching their way toward the bonfires of electronics where, a kilometer or so in the distance, Awal and the burner boys were settling into their daily ritual of setting piles of *bola* alight. After a few moments, the burner boys realized what was happening; they extinguished their fires, collected their knots of wiring, then retreated into the distance, before eventually starting the work again, only now burning their *bola* in the folds of the stagnant garbage mountains at the farthermost edge of Agbogbloshie. By the inky plumes of green and black that tinged the gray smoke, one could witness even at great remove when the flames had chewed through the wires' plastic encasement and begun lapping their copper insides. From Okai Street, I watched as the soldiers eventually disbanded, too tired or bored to chase the burner boys any farther, instead retiring to the air-conditioned interior of their bus after an hour or so of hapless pursuit.

To destroy swathes of Agbogbloshie was one thing. To prevent its residents from continuing to make their living was another.

19

GOING FISHING

> Nkrumah liberated his people politically. I am going to liberate them economically.
> — John Ackah Blay-Miezah, world-famous Ghanaian con artist and scammer, 1966[1]

I MENTIONED EARLIER that Agbogbloshie is known for more than ecological ruin. There exists in the slum another group of boys who have devised their own ways of dealing with the streams of Western electronic waste piling up around their shanties. They aren't the burner boys. They are the browser boys. They live on the opposite side of Agbogbloshie from Mohammed Awal, approximately two kilometers south, along a separate stretch of the Korle Lagoon. Like Awal, they are also Muslim migrants from the Sahel, the sons of Dagomba yam farmers who had trickled into Accra over the previous decade and considered Ghana's capital little more than a temporary stop on the way to a life of prosperity back north.

One morning, Awal connected me with a man who went by "Pablo Escobar," an ethnic Dagomba in his early twenties who had a messy goatee and was almost only ever dressed in a fluorescent tropical shirt. We agreed to meet at one of Agbogbloshie's smartphone dismantling stands. Escobar introduced himself as a drug dealer who made a living selling hash to the browser boys. I told him I would give him two hundred cedi — approximately fifteen bucks — if he would take me to them.

"That's not enough," Escobar shot back with what seemed like fabricated disdain. "I sell drugs. I can make two hundred cedi anytime I want."

"Then why do you need my money?" I asked.

Escobar paused for a moment, then agreed to bring me to the scammers free of charge. He led me through the winding warrens of Old Fadama, down damp passageways that felt like the interiors of caves. We threaded our way past hundreds of minuscule one-room shacks that, like Mohammed Awal's, slept four or five on common floors of mud. Laundry hung from the rooftops; the clanking of cookery accompanied the clucking of chickens. At last, after twenty minutes, shafts of light poked through the corrugated roofs, and we emerged from the humid darkness onto the sunstruck southern expanse of the Korle Lagoon, which somehow was layered with even more trash than its northern stretch; its waters were entirely sealed beneath a three-foot-deep sheet of coagulate sludge comprised of takeaway containers and water bottles and plastic bags.

Almost every young male on this bank of the Korle Lagoon conducted internet scams for a living. Over the next week I met dozens of them. Here, the work done by Awal just a couple of kilometers away in the very same slum — the task of burning Western electronics for handfuls of copper — was held in disdain. The browser boys did not get dirty; their clothing was immaculate, often imitation luxury Western name brands. They did not huff trash fumes; they sipped Benylin cough syrup to get high and to allow their fingers to type out their solicitous messages faster and faster — so fast, one browser boy told me, they could anticipate "the replies of our clients before they even arrive." Browser boys do not work during the day; they rise from bed in the midafternoon, as time differences with the United States necessitate they extend their dealings deep into the Ghanaian night. Their leader was a burly man in his mid-forties, Osman, who went by "Boss of

Bosses," though mostly just "Bosses"; he wore a maroon tribal robe and tuxedo loafers and was blood brother to the neighborhood chieftain, a charismatic man in aviator sunglasses whom the boys called "Lieutenant Commander." Bosses was scarcely literate, but it was he who compiled some of the most successful "scripts" for the browser boys, which they had pasted into the Notes sections of their iPhones. "Browsing like fishing," Bosses once told me, imitating a reeling-in motion with his arms. "You be bringing in the big white fish slow-slow now."

"I'm Unemployed. I use to be an accountant and lost my job as a result. the manager of the company i work for wanted to rape me,i run away and never returned.i reported him to the police they pay less attention cause he has money and power so lucky i applied for a nursing school,am a final year student in the UNIVERSITY OF MARYLAND COLLEGE PARK here working on my RN degree to work as a nurse seen ..I have left with 1 month for me to start a final year exams to be come a RN.. "

Or so ran one of Bosses' scripts. The goal of such overtures is simple: to use the internet to make money in lieu of having to incinerate VCR cables like Mohammed Awal on the other side of Agbogbloshie. But browsing takes patience. First, you have to make Facebook profiles professing to be those of gorgeous American or European girls. The profiles are fake, of course, but the girls themselves really do exist. Their schools, ages, and professions are accurate. And the photos are all genuine. To anyone who has any doubts, trust me on this: Browser boys across West Africa—either through harvesting the hard drives dumped in their slums or through years spent soliciting foreigners—possess *vast* photo collections of American and European women, ranging from the perfectly innocuous to the provocatively clad to the outright nude.

After making the profile, browser boys fire off hundreds of friend requests to middle-aged men, mostly white, and, if accepted, attempt to strike up conversations: "I'm real dear sure. I'm name is Vivian." Weeks are spent exchanging saccharine messages ("am yours forever take care of me very well"), coy appeals ("hope you are not here for games or drama"), or desperate demands ("please Frank I beg"). Then, when enough trust appears to have been won from the client, there comes what Bosses liked to call the "reeling in." A unique form of cancer needs urgent care. The car needs to be filled with petrol. My iPhone broke; won't you please send a new one to... Ghana?

Browser boys in Agbogbloshie can be persistent. Many are willing to message a foreigner for weeks on end with no hope of response or reward. In this respect, the working rhythms of browsing are not all that different from burning: a tedious grind that rarely produces anything beyond five dollars after ten working hours. On rare occasions, though, the spoils of scamming can be life altering. A "big fish" can offer a ticket out of Agbogbloshie forever in a way that burning never can and never will. Browser boys you meet in Old Fadama like to talk about the mansions that exist somewhere in the outskirts of Accra and that they insist have been purchased with swindled Western fortunes. They like to discuss the divorced American men who, they claim, have penned browser boys into their wills. They swear there are European women who have sent thousands of dollars to Ghanaian boys in the belief that they are purchasing plane tickets for French Legionnaires caught in intractable conflicts in distant lands. Along with pictures of pretty Western girls, browser boys' phones are full of photos of their country's most successful scammers—legends like Joey Gucci and Pope Skinny, who pass for celebrities in certain quarters of Ghana, and who can be seen on Sunday afternoons partying at Labadi Beach, east of Accra, where they sit at tables full of gorgeous girls and sip energy drinks. ("We are Muslim," Bosses once told me. "No alcohol permitted.")

A far cry from the "internet-based economy" that importing secondhand electronics into Ghana was supposed to generate. Or is it? For of course browsing has proved in many respects the predictable progression of a secondhand electronics trade that for more than two decades now has inundated West Africa with our faltering mobile phones and laptops: A scam generosity has given way to an economy of scammers.

British historian Stephen Ellis has argued that graft and theft in nearby Nigeria — where a scamming culture has flourished longer than in Ghana, originating as far back as the 1920s — emerged as responses to the machinery of colonial rule. The nature of British control in Nigeria, notes Ellis, was indirect, mediated through localized tribal systems for whom "Nigeria" was an unknown construct to be distrusted and, when possible, undermined. When independence came, rural chieftains balked at answering to any centralized authority. Instead, they built patronage networks and slotted trusted kin throughout the new state. For Ellis, the crucial shift came in the late 1960s, when these traditional forms of corruption — hardly unique to Nigeria — became financialized. On the eve of the rest of the world experiencing huge hikes in oil prices, hydrocarbon discoveries off the coast of Delta State would prove a gigantic windfall for Nigeria. State revenues quadrupled over the next decade, generating a great pot of money at the epicenter of a weak bureaucracy no one had trusted to begin with. Instead of just infiltrating the newly constructed state, now Nigerian elites also stole money from it, a practice that spawned a rowdy range of scams and deceptions. So endemic was fraud to the history of late-twentieth-century Nigeria, Ellis contends, corruption might even be construed as an alternative form of patriotism, or even "Nigerianism."

In Ghana the situation was different. The closest thing it had to the liquid gold of oil was actual gold, much of which was leaving Ghana's borders and enriching outsiders. Scamming in Ghana didn't begin with the arrival of electronic waste, no. But no one can seriously doubt that the inroads of old phones and laptops into downtown Accra didn't have a cataclysmic effect on earlier scamming habits, accelerating them and supplying Ghanaians with a limitless deluge of just-working electronics — and plenty of fodder for resentment against the Westerners sending them.

The results? Check your spam folder.

If what was coming to Ghana was often not workable, it nevertheless succeeded in achieving one thing: Many Ghanaians got a firsthand look at the munificence that they appeared to be getting denied. In Agbogbloshie they were literally choking on it. Mohammed Awal told me that he would have loved to become a scammer, to wake up midday with clean clothes, to chat on Facebook for a living, to provide for his family from the safe and anonymous remove of the internet. The only problem? Mohammed Awal was *completely* illiterate. He couldn't scam if he tried.

On a sweltering Friday night in January, I sat at a picnic table with Bosses and a group of browser boys. Nearby, a bar called the Wembley Pub was playing reggae music and flashing strobe lights. Across the Korle Lagoon, that day's accumulation of *bola* was just getting torched. There were no wires or electronics within it, no *capasta* to be rendered; it was simply getting burned in lieu of any other way of making it go away. A fireball of flames raged along the bank, sending ripples of thick black smoke warbling into the humid night. It was approaching ten o'clock; five hours and five thousand miles away, on the East Coast of the United States, lonely American men were just leaving their offices for the weekend.

Boss of Bosses and the browser boys mixed bits of hashish into their cigarette tobacco and tapped their fingers across their phones. They passed around a bottle of cough syrup and took swigs. A little past ten, Bosses — operating on Facebook as one Constance Weill, a brunette from Lyon, France — befriended a seventy-nine-year-old American from Arkansas who called himself Popi. Within a few minutes Popi had accepted Bosses' request, then sent hand-wave and heart emojis.

"Hi," typed Bosses, followed by a hand-wave emoji of his own. "How are you doing"

The music emanating from the Wembley Pub began to thump louder. Boys in glittering shirts arrived in tricked-out sports cars, steering through clusters of exhausted goats dozing along the road, then parking with panache on the muddy bank of the Korle, opposite the trash bonfire crackling across the lagoon. These, I was told by Bosses, were the browser boys who had made it, the "real hustlers." Many Friday nights they returned to their old abode in Agbogbloshie to flaunt the riches that scamming had delivered them. They emerged from their cars with faux Louis Vuitton trucker caps and a girl or two on their arms, then strutted across the filthy road into the Wembley nightclub with heavy importance.

The conversation with Popi from Arkansas was progressing. "Baby! I am really looking for the right one! I do have a relationship developing now that is very promising! I am going to commit to that! I'm afraid I cannot be what you want now!" he wrote Bosses/Constance.

"Call me," Bosses typed back. And Popi called.

At any one time a browser boy in Agbogbloshie is operating dozens and dozens of bogus Facebook profiles, friending hundreds of men, and maintaining an array of conversations whose slow drip of personal details — a man's relationship status, his kinks, the name of his dog — require keeping careful tabs. Browsing is not always easy work. Of course, scammers must stay up until the early hours of the morning in Ghana, when it is evening in the United States. But they also must look up the weather in their alleged hometowns: A Ghanaian claiming to be Constance Weill from Lyon, France, must know that in January it is not hot in Lyon, France, and that sending bikini photos in January might excite a client but might also become cause for confusion. And browser boys must also have a range of tricks at their disposal to demonstrate the credibility of their existence to older men. Some browser boys like to Photoshop selfies of their clients into picture frames around their houses, photos of which they retrieve from Google Images: "inside beautiful american house." Boss of Bosses' preferred trick was the video chat, which he proceeded to demonstrate for me with some pride. First, he muted his video. Then he put his thumb over his phone's camera. Then he switched his Facebook video feed to a video of Constance Weill that Bosses happened to have in his possession: a video of Constance chatting into her phone's camera while doing yoga.

Popi proceeded to frantically point to his ear, making it clear that he couldn't hear Constance, while she intermittently changed yoga positions and continued to talk into the muted camera. After a few moments Popi, exasperated, hung up.

"I'm sorry we had a poor connection!" typed Popi. "You're absolutely beautiful! I want more pictures of you! Dozens more! You are so beautiful I can fall in love with you at first sight!"

Boss of Bosses and I headed into the Wembley, a dark, overcrowded room pulsing with reggae and cloudy with hashish smoke, then moved toward a rusty spiral staircase at its back. The second floor of the Wembley turned out to be the communal scammers' den of Agbogbloshie. "This is where we do business," Bosses told me as we surveyed benches lined with boys tapping into secondhand laptops. Facebook chats filled their screens, along with tabs of weather reports and Google Translate pages and lingerie websites; most of the boys looked dreary and bored. The columns upholding the room were lined with electrical wire sockets, each of them charging extension cords leading out to battalions of grungy smartphones. Cameras had been installed in the ceiling corners. They weren't for the police, who in Ghana are widely alleged to take a cut of the scamming profits in exchange for making zero attempts to stop it. They were to make sure the browser boys weren't getting too distracted. Distracted browser boys, Bosses liked to say, were not successful browser boys.

Popi, meanwhile, had sent another message:

"I want to tell you, I am not a jealous man! I cannot satisfy you sexually because I cannot have an erection! I would want you to enjoy your life with people your own age and I would have no problem if you took a lover. I would expect you to be discreet and that you both be respectful of me!"

And so it went. Two weeks later, Bosses called to tell me that Popi had sent him a twenty-dollar Victoria's Secret gift card, which Bosses was able to convert into Ghanaian cedis through a website called Vanilla. Not a big fish, no — but a catch all the same.

20

MAGICAL THINGS

> Traditional religion in Ghana is dying slowly. It started to die when the Europeans and Muslims came and saw us as pagans. Their superior technology killed us. We have witches who fly in the air. But when we saw aircraft we came to abhor what was our culture.
> — Ghanaian to author V. S. Naipaul, 2010

MY HOTEL IN Accra was a squat, graceless concrete building located along the edge of one of the loud ring roads that slice through the capital. It had one advantage for my purposes. It was a short walk away from the busiest intersection in all of Ghana, a hair ball of thoroughfares known as Kwame Nkrumah Circle, the market stalls of which were sprinkled with internet cafés where stray browser boys could be found working their clients into the early hours of the morning. Every afternoon after leaving Mohammed Awal, and around the time he was wiring half his daily wage to his family in the north, I would walk to Kwame Nkrumah Circle and tour its street markets. The stalls extended for many blocks out of every spoke of the Circle. At most, ranks of sparkling iPhones were on display in glass cases carpeted with sheets of artificial putting grass. The majority were brand-new, some the latest available models. And they weren't just cheaper than new iPhones one might purchase from an Apple Store in the United States or Europe. In many cases they were hundreds of dollars cheaper.

Here was an irony almost too surreal to be true. With the exception of a mansion in the outskirts of Accra, or a six-figure cash inheritance, the most valuable thing a browser boy in Ghana can acquire via internet fraud is...a new iPhone. Most of the iPhones on display at Kwame Nkrumah Circle have reached the country courtesy of Agbogbloshie's scammers, who sell the phones shipped to them by duped Western men to the marketplace's vendors in exchange for hard cash. In the haphazard world of browsing, it turns out, receiving a phone is the easiest way to get large amounts of actual money. Asking a client to send cash through a service like Western Union or MoneyGram is problematic: A browser boy is left to explain why the money needs to be sent to a man's name, and even if the money does get wired, most browser boys don't have government-issued identity cards allowing them to collect it upon arrival in Ghana. Gift cards, meanwhile, tend to be trifles: Rarely is a client willing to shell out more than twenty or thirty dollars for a Victoria's Secret or Amazon card.

But an iPhone. An iPhone! For such a small item, it is incredibly valuable. And that a girl needs an iPhone is perfectly believable. Beautiful girls break their phones all the time. And the client does not even need to send the device all the way to Ghana. He can be convinced to send it to the address of an expatriate Ghanaian in the United States or Europe, who in turn ships entire suitcases full of them to Accra every few weeks. The economy of broken iPhones has given rise to scamming, which has in turn given rise to a derivative economy centered around *new* iPhones.

You might be inclined to think of browsing as the ultimate twenty-first-century business: The all-powerful reach of the internet has allowed enterprising teenagers from Ghana to conduct serial identity theft and inflict long-range economic retaliation on the richer half of the world.

But in many respects the scamming culture in West Africa represents something much older: the activation of an ancient voodoo tradition distinct to the Gulf of Guinea. For the most important thing to understand about Boss of Bosses and the other browser boys is that they don't believe they are stealing anyone's identity. No. They claim they are inhabiting it.

One afternoon, Bosses and Escobar took me down to the coastline south of Agbogbloshie. Bobbing across the surface of the Atlantic, a panoply of wooden fishing boats hovered in the distance.

"Do you know the first thing white men brought to Africa?" Bosses asked as we sat down at a picnic table on the beach.

"Guns?"

"A mirror. This was the first time we be seeing ourselves. For the first time we understood what we looked like. And then they gone and took our children away from us."

This is how many Ghanaians justify the business of scamming: retaliation against the descendants of those who kidnapped their ancestors. And with the help of a local juju tradition, they enlist their ancestral spirits — the spirits of those who were deprived of their relatives so many hundreds of years earlier — to help them do it. The instruments of revenge are the electronics that show up in Agbogbloshie every morning. Photos are downloaded by the tens of thousands from Western devices and shared among the hundreds of browser boys. Broken laptops and phones are restored to a working state, if only for a few weeks or so. Cracked screens are replaced, new batteries inserted, all in the service of swindling the residents of the rich countries who sent it all.

One afternoon a browser boy called Ransford asked me if I wanted to accompany him to one of the juju shrines. Ransford — who often wore a white tank top and jean shorts, and had a USB cord curled around his waist in lieu of a belt — had been having little scamming luck of late. But in the past, success had invariably followed a visit to what was known as a fetish priest; over the course of a few months in 2017, after consulting a priest in the northern city of Tamale,

Ransford told me, he had made 15,000 cedis (about US$1,000), an almost miraculous sum, which he claimed to have invested in a plot of land in Kasoa, a city west of Accra. More recently Ransford's success had run dry. He had only a few hundred cedis left. It was worth giving most of it over to a fetish priest to help conjure up a new financial windfall.

In the taxi the next morning, Ransford detailed his struggles at greater length. For weeks now Ransford had been posing as Rebecca, an American brunette. Recently he had begun reeling in a man named Todd P., a fifty-four-year-old from West Reading, Pennsylvania, who, though married, had floated the idea of leaving his wife for Rebecca a few hours after accepting Ransford's friend request. A day earlier Todd had even sent Ransford — "Rebecca" — a twenty-five-dollar gift card. It seemed a promising start to what could become a much-needed new revenue stream for Ransford. He might even be able to purchase another plot of land outside Accra. And he had already struck up a plan, he told me. Rebecca had menstrual problems and needed money to purchase tampons. Ransford showed me photos of bloody female genitalia on his phone, which he told me he had sent to Todd to convince him of the urgency of the situation.

As he put his phone away, I noticed that Ransford's phone background was a photo of Todd P. getting a haircut, surrounded by his wife and children.

"Why did you make that your background?" I asked Ransford.

"I thought it was a nice photo," he said.

This fetish priest was to be found in a village called Aboabo, which sits in the bushland north of Accra. A taxi took us deep into the hills beyond the capital, leaving behind the parched plateaus of the coast and entering a cool belt of rainforest. The air turned misty, then unleashed a lukewarm downpour. Strips of cocoa plantation lined the roads; out one window of the car, the remains of the Gold Coast Government Railway, hammered across the hills by the British in 1923, had been reclaimed by the jungle and now lay ensnarled in vegetation, Ozymandian testament to a vanished empire. Eventually, after about an hour, we entered Aboabo, a dozen or so thatched-roof homes arrayed in no discernible order between clusters of tall palm trees. We headed down a dirt road, passing a funeral service for a local woman who had died some days earlier. Outside a small church, a group of shirtless boys were chopping up a goat for the funeral feast. Its intestines lay strewn on a mat of palm leaves as a teenager split its skull in two with a machete.

British and American missionaries had done their utmost to spread Christianity to these parts of Ghana. But behind the facade of conversion, the ancient practices still carried on. Slightly beyond the church was the older place of worship, a mudbrick structure that was overhung by a corrugated-iron roof, which Ransford told me had been donated by a browser boy who'd gifted a cut of his earnings to the voodoo priest who'd helped him swindle it. Dwarf goats wandered around the enclosure. On the ground in one corner were several gin bottles emptied of their libations. Abutting the structure were two buildings. They had wooden doors; feathers and guts were smeared across each. One of the buildings could be entered by visitors. The other one could only be entered by the priest; there, I was told by Ransford, "the ancestors lived." Around this second building, several cats had been crucified on the trunks of palm trees; they hung upside down, small nails hammered through their paws into the bark.

The shrine itself was at the end of the enclosure, a bare table piled with the carcasses of birds, porcupines, and field mice, and the wings of bats. Red cotton handkerchiefs layered the floor underneath. Nearby, several dozen jawbones sat in a tidy pile.

After a few minutes the fetish priest arrived on a motorcycle, dressed in jeans and a green T-shirt. He entered a distant building without greeting us, then reemerged several minutes later wrapped in nothing but a white bedsheet. He was neither skinny nor fat, had a shiny round head and a goatee with hints of red hair, and he proceeded to walk around the enclosure barefoot. He ordered Ransford and me to remove our shoes before entering the sacred area. We did.

Ransford explained the predicament with Todd P., briefly flashing Todd's Facebook page for the priest, who nodded in a way that conveyed understanding and concern. Money had been coming, Ransford said. But only sporadically, and not enough of it. What could be done?

Without replying, the priest opened a black plastic bag he had brought with him. He pulled out a piece of dried fruit known as an alligator pepper, cracked it in half with his fingers, and picked out small seeds. He handed them to us. Just keep them in your mouth, he said. We did; they had a bitter taste.

Next he ordered us to squat. He proceeded to pour gin along the edge of the shrine, then took up an iron instrument — it resembled a fire poker with three small bells on each end of a handle — and shook it as he called on the ancestors in the local dialect. The aim, Ransford later told me, was to lure the spirit of the *obruni*, the "white man," into the gin bottle, then trap him, where his will would prove easier to subdue.

Next the priest rolled out a green-and-white matted carpet. He sat on it with folded legs, then wordlessly took out a white sack full of shells, seeds, bottle caps, animal teeth, dried fruits, and rusty coins.

The priest proceeded to chant, pouring gin over certain stones around the shrine that had been smeared with blood at some earlier point. "Every stone here has a meaning," Ransford told me as we watched him. The priest ordered Ransford to produce a one-hundred-cedi bill, which he did; the priest rubbed it through Ransford's hair before spitting on it and adding it to his magical pile of shells and coins.

He kissed a hog tusk, then rubbed it in the sand.

The smell of gin permeated the air. The priest whispered to himself, appearing lost at times, illuminated at others.

He raised his eyes.

"Have you ever killed a man?"

Ranford told him that he hadn't.

"And how many have you killed?" the priest asked, turning toward me.

"None," I answered.

The priest cast out a chain of shells.

He put a stick between the toes of his left foot.

He cast out the chain of shells again.

He paused.

A white candle was to be burned that evening at exactly midnight, the priest explained, a doctor prescribing a cure. The candle represented Todd P. of West Reading, Pennsylvania. A white piece of lace was to be tied around the candle. It would lock Todd's mind. Todd was to wake up in the near future and, instead of deciding to give Ransford/Rebecca $25, it would strike him to give Ransford/Rebecca as much as $2,500. The idea might even come to Todd while he's in the shower, the priest said.

That next morning, Ransford was to go to the ocean near Agbogbloshie and whisper Todd's name three times, his date of birth three times, and explain what he wanted from Todd P.

Then Ransford was to find a small white marsh bird — the type that liked to land on the chunks of Styrofoam that drifted throughout the Korle Lagoon — and release it at the shore of the ocean, so that it flew out in the direction of the setting sun. The bird would eventually reach the other side of the Atlantic and convey Ransford's wishes to Pennsylvania.

Finally, Ransford was to sacrifice nine roosters — those, as well as the animal that a lion cannot kill. The priest was quiet for a moment before the creature's name came to him. "A porcupine." And then:

"Do this. And then finally you will ask Todd for anything and he will do it."

The remains of the nine sacrificed roosters and single porcupine were to be returned to the shrine sometime thereafter, broken up, then scattered in the surrounding bushland, the priest continued, along the banks of nearby sacred streams, where they would be eaten by wolves and birds of prey and other creatures in which the spirits of Ransford's ancestors had reincarnated themselves. And in so doing, in helping to nourish the older generations, Ransford would be nourished by the ancestors in turn and treated to great riches.

Ransford agreed to burn the candle, utter Todd's name, release the white marsh bird, sacrifice the roosters and the porcupine, which one could buy at most village markets outside Accra, then strew their remains across the forest floor to feed his ancestors. And then he would start messaging Todd P. once more with renewed commitment and the promise of richer returns.

Late that afternoon we headed back to Agbogbloshie. Our taxi took us south through the bush, back toward the edge of the Gold Coast, flushed with the pumpkin-orange light of the sinking sun. Before long the overgrown gray-brown expanse of Accra burst into view on the horizon, the monuments to liberation — monuments to a new country that would never be the object of Western exploitation again — just visible at the shoreline. Beyond them, the rippled sheet of the Atlantic unfurled into the distance, cluttered with cargo ships, their navigation lights winking like stars in the darkening ocean dusk, some of them no doubt bearing containers of old Western electronics — future miseries, future fortunes — somewhere in the innards of their hulls.

PART THREE

Aegean Abomination

21

GLOBAL JUNK HEAP

> If you don't have steel, you don't have a country.
> — President Donald Trump, 2018[1]

AT THE HEIGHT of the 2009–2017 financial crisis in Greece, it was possible to witness a peculiar sight on the streets of Athens, the country's capital: respectable-looking, middle-class Greeks pushing shopping carts piled to the brim with scrap metal of all imaginable varieties. There were automobile mufflers, copper wiring stripped from foreclosed apartments, rusty old clothes irons, boxes of cutlery, doorknobs, even the occasional sewage cap crowbarred out of the street. From morning to late afternoon these upstart junk collectors — onetime civil servants and bank tellers and waiters — could be seen working the sidewalks, frequenting dumpsters, abandoned buildings, shuttered factories, stacking their carts higher and higher. By night, they sold their pickings to a network of Roma dealers outside of Athens, who in turn arranged for the scrap to be shipped beyond Greece to İzmir and İskenderun and any number of the other great ports of the Eastern Mediterranean. During a crisis in which a quarter of Greece's GDP effectively disintegrated, and in which one in two youths had become jobless, the city of Athens was being cannibalized — ripped up and carted away and sold abroad — so that members of a capsized middle class of a bankrupted European Union state could continue to put food on their tables.

"People laugh but you can make a decent day's work from trash," a laid-off construction worker and father of two named Dimitris claimed at the time.[2]

Not that such scavenging isn't common elsewhere in the Balkans. Consider for a moment the country of Kosovo, Europe's newest state and a place that not so long ago was hailed as the emerging promise of the continent. Kosovo appears to have all the ingredients to become a successful country: one of Europe's youngest populations, tremendous mineral wealth, defense courtesy of the biggest NATO base in Southeastern Europe. But more than a decade after the recognition of its statehood, following repeated attempts to reboot its Cold War–era mining industry, scrap metal isn't just part of Kosovo's economy. It *is* its economy, or much of it at least, amounting to the country's biggest export. Europe's newest state is reliant on getting rid of all that is old; every year approximately forty million dollars' worth of metal junk — the bulk of it clandestinely stripped from old Yugoslav-era factories by Roma, Kosovo's most destitute population — gets shipped abroad, much of it ending up in Germany. The poorest people of one of Europe's poorest states play a small but crucial role in propping up the manufacturing sector of Europe's richest and most powerful country.

Geographer Mazen Labban has described the phenomenon of the "planetary mine": The fact that over the last two centuries the most astonishing transformation in humanity's relationship to its material surroundings hasn't been the invention of the automobile or the moon landing or the rise of the internet. No: It has been the seismic diversion of our planet's natural wealth — oil, ore, minerals — from under Earth's surface to the top of it. Bridges, skyscrapers, highways, bicycles, spoons — these are not just signposts of civilizational development and urbanization and modernization. In the rawest sense, in the most strictly material sense, they represent redistributions of resources, minute chunks of Earth's substrata — atomic slivers of iron and deposits of carbon and reams of cobalt—that have been dredged up from beneath the planet's crust and placed atop it. Cities, in Labban's formulation, aren't just loci of human habitation. They are epic accumulations of reallocated raw materials.

And this mine—the planetary one, the current and future detritus of human civilization—isn't like other mines. It's not stationary. Its material reserves don't deplete; they grow. The planetary mine is constantly being redistributed, disintegrating, breaking apart, getting rezoned or rebuilt or bulldozed or stripped apart by upstart scrap collectors wheeling shopping carts around a crisis-wrecked capital city. The average lifespan of a concrete building may be a hundred years, a car possibly fifteen, an outdoor cooking grill perhaps ten. But the lifespan of the steel that goes into that building and car and grill is infinite; it can be melted down and used over and over and over again.

Alan Greenspan, former chairman of the US Federal Reserve, once said that the most accurate barometer of global economic health isn't the stock market. It isn't the value of one currency or another. It isn't employment rates. It's the going rate of old steel. The scrap economy can nevertheless be a baffling, bottom-feeder world, one full of unexpected material consequences rippling throughout disparate places.

Across Afghanistan in the 1990s, construction became reliant on melting down old Soviet tanks and using the resulting rebar to rebuild infrastructure devastated by decades of conflict.[3] Later, in Iraq, the rapid disintegration of Saddam Hussein's army triggered a scattershot of scrap metal—helmets, troop carriers, the joints of tanks—which was loaded into pickup trucks and smuggled out to black-market junkyards across the Middle East, leading to a long, regional plummet in metal prices.[4] In June 2019, Russians in the remote Arctic region of Murmansk woke up to find that a twenty-three-meter steel bridge spanning the Umba River near the abandoned settlement of Oktyabrskaya had disappeared; it had been dismantled in the night by a band of metalworkers, sliced up and bartered off, perhaps the work of the same group that had taken down a two-hundred-ton bridge under cover of darkness in the city of Khabarovsk some years earlier.[5] In India today it is said that one can find a college, a car repair lot, and a textile showroom that have been constructed with some sixty thousand tons of steel that once formed New York's Twin Towers.[6] The city of Dubai is a veritable scrap heap refashioned into a sparkling metropolis; the world's tallest building, the Burj Khalifa, is constructed largely out of reforged old steel, some of which is said to have once graced the Palast der Republik in Berlin, onetime seat of the East German parliament.[7] In the United States, secondhand metal economies have even emerged as an index of urban degeneration more broadly, with scrap trades statistically predominating in cities like Portland and Seattle at times of rising drug use and homelessness.

A ragged underbelly economy, the scrap industry is nevertheless indispensable to the destruction of the old — and the construction of all that's to come. As the dictates of a warming planet demand that we evolve beyond our current extractivist economy, one in which new materials are pulled out of the ground through highly energy-intensive, carbon-emitting processes, the planetary mine will not just continue to exist as an alternative to most conventional mines; it will increasingly become the all-important mine, bearing out a prediction made by the American Canadian urban theorist Jane Jacobs half a century ago that "the richest, the most easily worked, and the most inexhaustible mines" of the future would be the city skylines rising before our very eyes.[8] And it turns out that incentivizing an efficient recycling economy for steel is not difficult. We already have one. Steel is not just the most recyclable material in the world. It's the most recycled one too. Less than 10 percent of the world's plastic gets cycled back into production streams, compared to nearly 90 percent of the world's steel.[9] In 1980, one out of every ten tons of fresh US steel was manufactured from recycled steel. Twenty-five years later, two-thirds of all new US steel was derived from old steel.[10] The percentage continues to climb. Globally, a full third of our global steel output now derives from metal that has first been thrown away.[11]

The planetary benefits are clear. Producing new steel from old steel requires a seventh of the energy and releases a tenth of the carbon dioxide as producing it from newly excavated ore.[12] To its promoters, the scrap metal economy is an environmental salvation that holds the key to a future in which humankind will continue to urbanize faster than ever before, under greater ecological strains than any in human history. How we got to a world of rampant plastic exportation — of the type that is decimating citrus groves in southeastern Turkey — in many ways owes to the fact that the recycling of materials like steel makes good sense.

It should come as little surprise, then, that in recent years, as the price of virgin steel and calls to curb global carbon dioxide emissions have simultaneously risen, the scrap metal trade has taken on increasingly geopolitical and protectionist dimensions. Here, alas, is something that one might think little of tossing away — a beleaguered bicycle, a busted toaster, a broken mini-fridge — but is nevertheless worth hoarding at scale at the level of nation-states themselves.

There's a long history of geopolitical contention over scrap. In the 1960s, the American Iron and Steel Institute lobbied Congress for a ban on junk metal exports to Japan, which at the time took in more scrap from the United States than any other place on Earth, only to in turn act as the biggest exporter of *finished* steel back to the US, a turnaround that badly undercut domestic US steel manufacturing.[13] But in the last five years the contention over scrap has become more deeply entrenched than ever. More than sixty nations have banned — or are in the process of banning — the export of secondhand steel.[14] The European Union, currently the world's largest exporter, is preparing to outlaw its sale to developing countries and turn itself into a future net *importer* of old metal in a bid to accelerate continental decarbonization. "The importance of using indigenous secondary raw materials such as scrap metal and other residues cannot be emphasized enough," the European Commission has advised member states.[15]

And scrap hoarding is happening in more unexpected places too.

Two weeks before I arrived in Kenya, in January 2022, President Uhuru Kenyatta announced a ban on secondhand metal exports after sections of steel guardrail and several hundred signposts were discovered to have gone missing from the Mombasa and Thika highways; stolen by bands of junk collectors, the scrap was believed to be making its way to the Chinese city of Guangzhou through local Chinese traders who had set up scrap export yards in African states along the Indian Ocean.[16] At just the time when Beijing had decided that thirty years of plastic waste importation had done unforgivable harm to the Chinese environment, it was doubling down on the import of secondhand metal, some of which was being sourced from construction sites across East Africa run by Chinese state companies themselves. In the months that followed Kenya's ban, Tanzania, Uganda, South Africa, and the United Arab Emirates would announce similar moratoria on the export of old street signs, electricity transmitters, copper wiring, and other such objects.

How? I remember thinking from Kenya, where police had begun threatening street vendors with three-year jail sentences if they were caught carting around accumulations of rusty junk. How exactly does this all work? How does the planetary mine operate? How do barter economies of waste scrap transform into national economies of steel, manufacturing, development? And what is the future of an industry that proffers itself as the solution-in-plain-sight to the problem of guaranteeing economic and urban growth while staving off planetary destruction?

To attempt to understand, I traveled to Turkey, for years now the world's greatest importer of old metal, to witness the most insane and improbable scrap trade of them all: the maniacal process whereby hundreds of steel ships of all sizes — oil tankers, fishing trawlers, naval frigates, passenger ferries, luxury cruise liners — are dispatched every year to a small handful of coasts across Asia to be beached, dried, ripped up, chopped to pieces, and made to disappear by huge armies of workers.

22
SHIPPING OUT

> If I kill a beast in the Argentine and sell the product of that beast in Spain, this country can get no tax on that business. You may do what you like, but you cannot have it.
> — William Vestey, English shipping magnate to a British royal commission, 1920[1]

ALL SORTS OF unpleasantries are built into the global waste trade. But let me present you with one of the more obscene ones.

What if I were to tell you that there exists an industry that claims it is alleviating the climate crisis by sending poison-laden cruise ships to poor countries in order that despondent communities of migrants can dismantle them by hand and — almost invariably — get killed in the process, either instantaneously, struck by a collapsing mass of steel, or slowly, through lung cancer or any number of illnesses contracted during the work? What if I were to tell you that some of the richest companies in the world insist that this hazardous waste trade — and, rest assured, it *is* a hazardous waste trade — is a salutary and necessary thing? It would be enough to make even the most sinister of the toxic waste exchanges of the 1980s — the act of dumping your toxic chemicals into, say, miles of tarmac in East Africa — seem almost harmless by comparison.

But it exists. It's called shipbreaking. It happens every day. And it rarely gets discussed because you aren't really supposed to know that it occurs in the first place.

You've likely never considered it before: the life and death of some of the world's most complicated structures, vessels heavier than apartment complexes and longer than New York City blocks that — in seeming defiance of physics — *float*. But how do they disappear — and where exactly do they go? How does the Brazilian Navy get an outdated aircraft carrier off its hands? What happens to a thirty-thousand-ton oil rig when Chevron determines that it's no longer seaworthy? A rusting ferry from the Adriatic? A decrepit tugboat out of the Red Sea? The shipbreaking process exists to make them all vanish. The industry that launches a thousand vessels a year to their death, allowing a seaborne hunk of metal under the legal writ of Liberia or Panama to become a shopping mall in Pakistan or a skyscraper in Doha, its material haul is vast, transferring the equivalent of 420 Empire State Buildings' worth of steel every year from sea to land.[2]

If the electronic waste trade is the escape valve of environmental responsibility that helps our tech industry expand and profit, if the plastic waste trade helps us lead lives of astounding convenience without needing to confront many of the material consequences, then shipbreaking may be the altogether more indispensable phenomenon. Here is the secret disposal outsourcing upon which virtually *everything* is reliant.

Before there was the internet, linking enterprising browser boys from Agbogbloshie to lonely American men from Pennsylvania, there was the world that modern shipping created, one that effectively eliminated geography as a determining factor in manufacturing — and disposal — processes. Gargantuanized in the 1960s, containerized in the 1970s, shipping, by the 1980s, had truly shrunk the world.[3] Today, it allows denim produced in India and a plastic button manufactured in China and a zipper made in Japan to be sent to Thailand and converted into a pair of blue jeans — then sold in the United States for as little as fifty bucks. Shipping makes it more affordable to send, say, bales of garbage from Los Angeles to the Marshall Islands than to truck that same waste a few dozen miles away to the interior of California. Over the last fifty years, the quantity of goods it has moved between continents has quadrupled. At least 120,000 cargo carriers or tankers or bulkers are currently traversing the oceans at any one time, upholding your living standards without you probably even realizing it.

Yet what shipbreaking is and how it works remains all but unknown to those whose lifestyles so heavily depend on these vessels.

For the moment, let's start here: When a ship is designated for "recycling," all the pageantry and pomp of its birth — the bottle of champagne cracked against the hull at a dockside naming dedication — melts away. The fleets that keep our world running die unceremonious, piecemeal deaths, at the hands of cash deals that dispatch them to noxious beaches, far from the eyes of any adoring public, to be gutted and cut up like slaughterhouse cattle.

Yes, shipbreaking offers up a reservoir of millions of tons of steel that can be melted down and put into new infrastructure in developing countries. But it's also something more than that. Shipbreaking is the process that allows the insanely lucrative shipping industry to expand its reach under the guise of "sustainability," acting as something akin to a built-in renewal mechanism in which aging and environmentally detrimental vessels — such as the thousands of single-hulled oil tankers constructed in the 1980s and revealed by accidents like that of the *Exxon Valdez* in Alaska in 1989 to be susceptible to splintering apart — are removed from the oceans at a profit. As those ships are recycled into buildings and bridges across Asia, new types of vessels, such as double-hulled tankers, are constructed in the shipyards of South Korea or China, requiring tremendous quantities of carbon to produce but purportedly conforming to stricter rules of environmental regulation and "greenifying" a shipping industry currently responsible for 3 percent of global carbon emissions — a figure expected to rise to 15 percent by 2050 if left unregulated — in addition to causing as many as 400,000 premature deaths worldwide every year owing to its use of dirty fuels.[4]

Shipbreaking manages to add yet one more layer of complexity to the geopolitics of trash and the question of who must take your detritus and why. We've seen already how legislating the waste trade has been a decades-long endeavor with less than stellar results to show for itself. And yet attempting to govern where old ships go to get demolished has proven exponentially trickier. For here is a significant sector of the global economy that has, since before it even began internationalizing our world, expertly unmoored itself from most standards of legal, financial, and environmental regulation.

Shipping is a slippery, secretive industry, run predominantly by families who do their utmost to make its inner workings obscure to the public and its obligations difficult to pin to any single location. Since the rise of the nation-state, it has been notoriously difficult for governments to effectively legislate or tax the fleets that have upheld commerce and been lifelines of global energy production. For ships are, by their very nature, not bound by conventional borders; they travel. Their owners were some of the earliest pioneers of tax avoidance anywhere, evading regulatory oversight by swapping out one nation's flag for another and stashing their earnings on the island havens their vessels frequented.

Over the last two hundred years, little has changed. Today, the shipping industry is regulated by something known as the International Maritime Organization, a United Nations–affiliated agency based in London that ostensibly exists to push the shipping industry into abiding by a host of UN legislations to ensure "safe, secure and efficient shipping on clean oceans" but which often acts to shield the industry from those very responsibilities. The most important thing to know about the IMO is that its policies are disproportionately dictated by "flag states": five scattered tax havens — Liberia, Malta, Panama, the Marshall Islands, the Bahamas — which, for legal and financial reasons, currently have their flags flying atop more than half the world's ships. Despite being the headquarters of almost no shipping companies themselves, and despite being places where few crews are recruited and no ships are built or, indeed, destroyed, these five states have emerged as the world's foremost arbiters of the rules governing how international shipping works, whom it profits, and who has to pay the — life-and-death — price for the expansion and renewal of its fleets. It is a duty flag states make negligible attempts to perform. It's not unusual, for example, for "Liberia" or "Panama" to send employees of private shipping firms to the IMO as their national representatives in the work of regulating... those same shipping firms.

It is a corrupt system that purports to be monitoring against corruption. And it's a major reason why, as late as 2009, shipbreaking, which is an ecological nightmare, was subject to almost no environmental oversight. To this day, the legal basis for its existence remains thin. The landmark legislation of the international waste trade, the Basel Convention, signed in 1989 by more than a hundred nations and parties, was crafted precisely to stop the movement of toxic waste from rich countries to poor. Its writ should have included ships, which are literal tubs brimming with hazardous materials, ranging from asbestos to polychlorinated biphenyls to polycyclic aromatic hydrocarbons to heavy metals to paints infused with arsenic. And yet it never happened.

"In the 1990s we pushed to legislate that ships destined for destruction constituted a form of toxic waste," Jim Puckett told me. Puckett heads the Basel Action Network, an NGO that aims to further the work of the Basel Convention in banning the international movement of hazardous waste. "And as soon as we even raised the idea, the shipping industry went crazy. They invaded every convention, every conference, every meeting. 'A ship is not waste! A ship is not waste!' And though the International Maritime Organization is a United Nations body, you would meet its representatives and they would present you with two business cards. One was from the IMO. The other was from their own shipping firm. These people were working in the business and claimed to be regulating it. As for 'human rights,' we weren't allowed to mention the term. I kept pinching myself. *This is the United Nations?"* [5]

Under mounting pressure, owing in no small part to journalistic exposure of the day-to-day realities of shipbreaking in the subcontinent, the shipping industry finally agreed to submit its disposal processes to international regulation. Only it didn't choose to answer to the Basel Convention. It crafted its own set of rules instead. "When the discussion on all this started, people in the shipping industry were confused. Many still are," Nikos Mikelis, who served as the IMO's head of ship recycling from 2006 to 2012, told me. "The pressure started against us. They were hell-bent on us joining the Basel Convention. They complained bitterly. They had demonstrations. 'You are killers!' But [the environmentalists] also implicitly recognized that Basel had too many problems and that to really apply it to shipping would be impractical. A ship is not a country. It moves." [6]

In 2009, Mikelis and the IMO unveiled something that's now known as the Hong Kong Convention. The aim was to demonstrate to the public that a degree of environmental and safety oversight was being brought to bear on the lethal shipbreaking process. Yet its mandates were minimal — not requiring shipbreaking yards to be responsible for the fate of hazardous waste once it leaves their environs, for instance — and nonbinding. Today, the Hong Kong Convention's writ extends to the less than one-quarter of the world's nations that have signed it. No independent body ensures that it is followed. Its greatest advocates are the dismantling yard owners it is supposed to be regulating. The landmark piece of legislation governing how ships are recycled, it is little more than a "legal shipwreck," according to the Brussels-based NGO Shipbreaking Platform, which tracks the fate of the world's dismantled vessels. It is violated virtually every hour of every day.

Hazardous waste movement is exactly what shipbreaking has always been and will continue to be: the profit-driven transfer of toxic vessels from the ports of wealthier countries to what sociologist R. Scott Frey has termed the "waste frontiers" of the global system writ large. The most vulnerable in this system are not necessarily the world's most destitute nations but the "semi-peripheral" ones that possess the structural means of receiving huge volumes of scrap — harbors, highways, electric arc furnaces that specialize in melting old steel — along with, most crucially, pools of labor desperate enough to relocate hundreds of miles to undertake the dangerous task of wrenching those ships to pieces.[7]

Accessing these "waste frontiers" is difficult. Permissions are typically denied, journalists sometimes detained, cameras occasionally confiscated. Until 2018, when China deemed shipbreaking too unacceptable an environmental calamity and shuttered its facilities, there were five principal ones. Now there are four. Farthest to the east is Chittagong Ship Breaking Yard in Bangladesh, the world's largest yard if measured by the shipping tonnage it "processes"; Chittagong provides feedstock for ten million tons of steel a year, the equivalent of roughly 250 of Dubai's Burj Khalifa towers, or 70 percent of Bangladesh's total production.[8] Moving west, to the opposite curve of the subcontinent, there is Alang Ship Breaking Yard in India, the world's largest yard in terms of the number of dismantlers it employs, some 60,000 migrants hailing mostly from the inland state of Uttar Pradesh — to say nothing of the regiments of truck drivers, security personnel, vendors, and scrap collectors Alang indirectly employs, a workforce said to exceed 200,000 in total. Three hundred miles farther along the Arabian Sea coast is Gadani Ship Breaking Yard in Pakistan, the world's third largest yard. And finally, three thousand miles beyond that, along Asia's westernmost edge, looking out on the European continent itself, there is Aliağa Ship Breaking Yard in Turkey — one of the world's fastest-growing yards, and the one that had brought me to the strange, gloomy city of Sivas.

23

INTO THE HEART OF ANATOLIA

The Commissioner's eyes filled with tears. Ah, the humble people of Anatolia, he thought, hospitable, noble, long-suffering.

— Yaşar Kemal, *Anatolian Tales*, 1969 [1]

SIVAS IS NOWHERE near any sea — or, for that matter, any ship. The town sits square in the dusty innards of Turkey, two hundred miles south of the Black Sea, three hundred north of the Mediterranean, nestled in an almost lunar landscape that I reached by plane, from Istanbul, one frigid April morning. The road from the airport into town took me past a billowing carpet of yellow-gray farmland powdered with snow and sprinkled with busted bullock carts; groups of Kangals — the massive blond shepherding dogs native to eastern Turkey — snarled and lunged at my taxi as we passed. Scattered across the highlands, small turquoise lakes had just shed their ice, creasing the hills below with rivulets of muddy runoff. Perhaps because the scenery bore such a resemblance to the Central Asian steppes from which they had poured out a century earlier, the Seljuk Turks paused their nomadic existence in 1174 to proclaim Sivas their capital, then used it to spring new waves of invasions across the rest of Anatolia. Long after the Seljuks have

gone, their mosques and madrasas still crowd Sivas's ancient downtown, stony relics of a stuttered transition to a settled life.

An hour east of Sivas sits the still-more-forlorn town of Zara. It feels forgotten, a once-formidable agricultural community whose population has been draining away for forty years now. Drab blocks of concrete apartment buildings rise forth from undulating red hills. Closed storefronts line vacated, tumbleweed-blowing-in-the-breeze streets. A statue of a bee at Zara's main intersection is a reminder that the place was, not so long ago, one of the foremost producers of honey in the Mediterranean. But like so much else in Zara, the honey industry has been on its way out of late, as the wildfires that have become annual fixtures of Turkish summers have begun to kill off its legendary bee populations.

I had traveled to Istanbul, then from Istanbul to Sivas, then from Sivas to Zara, in the hopes of meeting the Taşkın family. I eventually found them at Zara's outskirts, along the bank of the Kızılırmak River, in a small but tidy apartment that sat on the sixth floor of a crumbling concrete building with no elevator. A young bearded man named Enes answered the door. He led me into a living room layered with floral carpets and smelling of herbs that sprouted forth from old plastic yogurt containers lining the windowsills. After a few moments, Enes's mother arrived from the kitchen bearing a tray full of teacups and candies and hot scented towels, and over the next half hour, I explained to them why I had come all the way to Zara. They soon called up the other siblings of the Taşkın family, three brothers who arrived and seated themselves around the living room, each dressed in a black leather jacket and bearing the same narrow Anatolian face of the young man I had seen plastered across Turkish news websites.

Months earlier I had read about the unbelievable death of their brother. In July 2021, Oğuz Taşkın was scarcely thirty years old, a recent hire at the Aliağa Ship Breaking Yard hundreds of miles away on the opposite side of Turkey, when he descended into the hull of a ship in a halfway state of demolition and was met by an explosion that effectively roasted him alive. Rushed to Bozyaka Training and Research Hospital in nearby İzmir, Oğuz clung to life — beyond all expectation — for more than two days. According to doctors, 99 percent of his skin and insides had been scorched by the flames; a radio he'd been carrying at the time of the explosion was found melted into what remained of the bone structure of his right hand.

It wasn't just any ship that had killed Oğuz Taşkın. No, his was a tragedy deepened by the fact that it had occurred within the bowels of a vessel of almost cartoonish awfulness: a fourteen-story wedding-cake structure that had sailed the seas under the name *Carnival Inspiration*.

For a decade and a half the *Carnival Inspiration* had drifted around the Caribbean like a floating shrine to the stupefying self-indulgence of America's cruise ship industry. Pride of the Carnival Cruise fleet, it weighed the equivalent of seven Eiffel Towers. Its bow ran the length of three football fields. It emitted more carbon than a US suburb. Fourteen glass elevators disgorged two thousand passengers into ten stacked passenger floors, shuttling past a five-story bronze sculpture known as the *Birds of Paradise,* which twinkled beneath a glass atrium that emerged onto a sports deck—one of thirteen across the *Carnival Inspiration*—threaded with a three-hundred-foot-long corkscrew water slide. Guests—honeymooners, retirees, everyone in between—were tended to by a thousand crew members hailing from more than twenty countries. Their ship boasted twelve bars, not including the Serenity, an adults-only retreat featuring a couple of whirlpools tucked into the aft of the promenade deck. Golfer? The *Carnival Inspiration* had a driving range. Gambler? There was a casino. Theatergoer? There were two.

"*Carnival Inspiration*'s creative spirit sparkles throughout the ship," marveled its operators, based out of Tampa.[2]

Such was the creative spirit whose engine room would end up cooking Oğuz Taşkın alive a little more than a year after the ship's arrival in Turkey. In July 2020, after just fifteen years spent sailing the coasts of California and Florida and following "an intensive review of sustainable ship recycling facilities," the flagship *Carnival Inspiration* was dispatched by its owners to the Mediterranean for recycling. Its destination was the Ege Çelik shipbreaking yard at Aliağa, a facility whose credentials amount to one thing: It is a place where gruesome accidents of the sort that ended up killing Taşkın are vouched not to happen—by European Union inspectors, no less, who tour Ege Çelik's premises once a year to ensure that its workers wear proper uniforms, that the air they breathe isn't toxic, and that all hazardous materials are properly disposed of.

"Our highest responsibility and top priorities are compliance, environmental protection, and the health, safety and well-being of our guests, the communities we visit and our crew," Bill Burke, Carnival Corporation's chief maritime officer, claimed just weeks after Carnival Cruise Line dispatched the *Carnival Inspiration* to Turkey. "In addition to limiting our vessels' impact on the environment throughout their service time in our fleet, recycling our retired ships following the [EU] Ship Recycling Regulation ensures we are applying the highest global standards and contributing to a sustainable cruise industry."[3]

Inside the family's apartment, Oğuz's mother and brothers were somber, soft-spoken, and embarrassingly attentive. Did I want more tea? Did I like the candies? They prodded the floor with their eyes as I asked about Oğuz, almost as though they were imploring their surroundings to offer up some greater sense of what had happened to their son and brother and why. But over the course of a few hours, the story of Oğuz's death came out in pieces, and went as follows:

Oğuz Taşkın always had a knack for fixing things. He was that neighborhood boy who could get old radios running or restore TV reception by tilting the rooftop satellite dish back into working position. And, like many residents of Zara, he aspired to get out of it. For years he worked at a café in the city's downtown. But by 2020, he had a wife and a child to support. He needed money.

So, in early 2021, he left Zara, busing out west to the Aegean and the shipyards of Aliağa, where several of his cousins already worked and helped arrange his employment at its shipbreaking facilities. According to his brother, Oğuz Taşkın — who had a light frame, with an angular face covered by a trimmed beard — thought the work of shipbreaking might just suit him. Was this not just a more complicated version of restoring old radios?

Of course, what Oğuz was doing at Aliağa was of a ludicrously different order from anything he had ever attempted in his life. His position at the shipyards may technically have been that of "engineer," but there was nothing being designed or constructed; shipbreaking is the absolute opposite of engineering, an act of chaotic demolition, a construction job conducted "rapidly in reverse," as one industry executive told me, and a job often subject to frantic improvisation at that. And as his family now sees it, the very point of Oğuz's "training" was to make the dangers of what he was doing as obscure as possible. Within two hours of arriving at work on his first day in January 2021, they claim, he was directed by his bosses into the hull of the *Carnival Inspiration,* having first signed a paper that, so he later learned, certified his capabilities as a qualified asbestos removal worker.

At first it went well. The money was good: 5,000 liras a month (about $250), approximately half of which Oğuz sent back home to his parents and brothers, along with his wife and young child. Then, six months into the job, on a scalding morning in early July, Oğuz and a fifty-five-year-old Turk named Yılmaz Demir were tasked with climbing down into the engine room of the *Carnival Inspiration* to operate the evacuation pumps capable of discharging the water that had flooded the ship's stern during its beaching process. Oğuz entered the engine room carrying a blowtorch; an instantaneous reaction occurred between the torch's flame and the gas invisibly gushing out of the evacuation pump, igniting a catastrophic explosion that left Demir incinerated to death on the spot.

Not Oğuz. He managed to make his way back up five flights of stairs within the *Carnival Inspiration*, even as his body had turned into what was in effect a human fireball.

"Yılmaz is still downstairs!" Oğuz cried out to the other workers of the shipyard as he burst out onto the *Carnival Inspiration*'s taffrail. Then he collapsed. "Get the sun off me!" he was said to cry as he fell. So disoriented, Oğuz apparently believed he had been struck by the sun itself.

Reading articles about Oğuz Taşkın's death, and then hearing about his fate firsthand from his family, it was hard not to be struck by the sordid tale of globalization it told. A US-owned cruise ship, constructed in Finland for the purposes of sailing the Caribbean, registered under a Panamanian flag, had claimed the life of a Turk who hailed from the landlocked plains of Central Anatolia and had never glimpsed the sea before in his life. But the more I considered Taşkın's death, the more I wondered if it wasn't actually a story about the price enacted by systems of globalized recycling even when they *did* work. An industry premised on the ability to *salvage* old material was, in this case, reliant on tremendous populations of poor Asian migrants to do so — populations that evidently constituted expendable collateral in the service of famous shipping firms being able to publicly broadcast their environmental bona fides. Nowhere does this injustice reach more preposterous or schizophrenic heights than with the cruise ship disposal industry: Vessels that bring vacationers from one half of the world to luxuriant paradises bring misery and contamination — and, on occasion, death — to workers hailing from the other half of the world.

Consider for a moment the death toll of Aliağa in the three years leading *up* to Oğuz's death. Again: Aliağa's credentials amount to it being a safe shipbreaking destination. There was Ercan Yıldırım in March 2016, crushed to death by a collapsing piece of steel, then Cemal Doymaz three months later, choked by toxic fumes, then Ahmet Güleç shortly thereafter, who fell off a ship and drowned. Or consider just the three months leading up to Oğuz's death, when Musa Gezer suffered a heart attack while stripping the wood from a ship's deck, when Can Sünmez fell nearly twelve stories and drowned after an errant piece of balustrade struck his head, when Mustafa Coşan had his legs shorn off by a cargo truck, and when the life was knocked out of Turan Arslan by a twirling block of scrap steel.[4]

Or, crazier still, consider what happened *after* Oğuz was killed inside the *Carnival Inspiration* — the Turkish workers who went on to meet their ends at the hands of that very same luxury liner. Consider the fact that Oğuz had hardly been put into the ground back in his hometown of Zara when, in September 2021, a pulley system lifting Ilyas Bidid and Veli Bal up the starboard side of the *Carnival Inspiration* suddenly snapped, dropping both men to their deaths.[5]

Someone, in other words, has to pay a price for this exalted recycling economy committed to sustainability — to pay a price for the executives of a company like Carnival Cruise Line being able to proclaim that their ships are, contrary to public opinion, or common sense, net boons for the environment because they conform to the highest standards of responsible disposal, that all their glass and wood and electronics are repurposed for new ships, that all their hazardous material is meticulously handled, that "all health, safety and environmental measures are followed" until "the last piece of steel is brought to the smelter to produce new products."[6]

Which is to say that if you only read Carnival Cruise's telling of the story, you might be forgiven for thinking that the *Carnival Inspiration* would arrive in Turkey, then vanish, leaving behind nothing but truckloads of precious, reusable materials available for local consumption. Oğuz Taşkın offered up a one-man corrective to this argument, his gruesome death being a metaphor not just for the dangers of shipbreaking but for the countless human perils layered within the business of globalized waste diversion more broadly.

The longer I sat with the Taşkın family, in their herb-scented living room layered with floral carpets, the clearer it became to me that they didn't just have many questions still to be answered about what had happened to Oğuz. Almost none of their concerns had ever been addressed. In July 2021, shortly after the incident had occurred, Oğuz's brother and father traveled to Aliağa, only to find that the company that had hired Oğuz balked at the idea of taking them to the spot where he had been killed. It was still too dangerous, the shipyard insisted. Worse still, the company refused to acknowledge any liability in the matter. "They were trying to tell us that this was all Oğuz's fault," Oğuz's brother told me.

Soon after they returned from Aliağa, the Taşkıns were offered a onetime payoff of one million Turkish liras — roughly $50,000 — from Oğuz's former employer. They had no illusions about what the money represented. They were being paid to stop asking questions. But they were also in no position to refuse it. Oğuz had gone to Aliağa to be their financial lifeline. And now he was gone.

"We were told that it was a miracle if he would even live forty-eight hours after the accident," Enes, Oğuz's brother, told me as I prepared to leave the apartment. "But he did. My brother was a tough man. The next morning, we got a phone call from the hospital. He had died early that day." Enes took a thin gold wristwatch out of the pocket of his jeans. Its band was streaked with black marks, its glass face a disfigured oval of melted goo. Oğuz had been wearing it when he died. "It's still ticking," Enes told me, running his finger across the band and putting it to his ear. "It makes me think he's still with us."

24

DEADLY BUSINESS

> And I, who longed to find my grave
> in some sea of the Indies, far away,
> will have a sad and common death,
> a funeral like those of other men.
> — Nikos Kavadias, "Mal du départ," 1933 [1]

IN MANY RESPECTS the world of shipbreaking as we know it today attempts to trace its origins to a fairy-tale-like incident said to have occurred along a Bangladeshi beach in 1969, the year a vicious cyclone whipped across the Indian Ocean and flung a Greek-owned ship called the *MD Alpine* up onto a flat stretch of shore somewhere south of the great city of Chittagong. Numerous attempts to refloat the *MD Alpine* failed. And so it sat — for five years, or so goes the story, until a local Bangladeshi steel company bought it and, over the course of several months, directed a ragtag band of locals to chop it to pieces, then drag the resulting chunks of rusty metal inland. After a few months the *MD Alpine* had vanished from sight. It was as if the vessel had never existed in the first place.

A mysterious foreign ship that offers itself up to a desperate community along a distant shore of Southeast Asia — the tale of the *MD Alpine* is the foundation myth the shipbreaking industry likes to tell about itself. And it's completely false. Ships have been getting "recycled" for centuries — Tudor England reused the timbers of decommissioned galleons to construct roofs in homes and pubs — but modern shipbreaking dates to the early 1950s. And before it was offloaded onto poorer countries, it was conducted in the same countries, and often in the same ports, where ships were actually built: As late as the 1960s, there were huge ship dismantling yards across the United States, the United Kingdom, Japan, Taiwan, and South Korea. In the 1970s, as those countries got wealthier, as the drive to protect their environments and the safety of their workers became objects of more and more oversight and legislation, an irony presented itself: *Other countries' environments and workers appeared increasingly attractive and expendable.* Ships ceased being the responsibility of those nations that disproportionately reaped the value of the commerce they facilitated.

Instead, their owners began turning to sandy stretches of tidal beach along the Indian Ocean to discard vessels that, to disassemble in the United States or the United Kingdom or Taiwan itself, would incur extraordinary financial and environmental costs. The useful life of the ship was deemed finished; the burden of disposal and recycling was pushed downstream. As shipbreaking scholar Tony George Puthucherril has explained, the dirty work of dismantling was contradictorily presented as an opportunity to clean up the shipping industry's dismal environmental footprint. Only by the early 1980s, old ships weren't deemed "ships" at all, nor would shipping companies, determined though they were to dispose of them, countenance calling them "waste." They were hunks of steel, measured and sold by the ton and repackaged as "raw material" for the subcontinent.

Disposal, once expensive for the shipping industry, now offered one additional windfall of profit. By the 1990s, hundreds of the world's ships were directed every year toward South Asia for dismemberment. By 2007, at which time four out of every five ships undergoing destruction anywhere in the world were being wrenched to pieces along the shorelines of the subcontinent, the environmental toll had turned cataclysmic and irreversible.[2] Just take India. Today, the waters surrounding the shipbreaking facilities at Alang are inhabited almost exclusively by species of fish "tolerant to petroleum hydrocarbons," or so local marine biologists have determined.[3] Moving inland, hundreds of local wells have been abandoned owing to the poisoning of the regional water table.[4] "This is a place that has seceded from India at every environmental and legal level," Gopal Krishna, an environmentalist who has spent decades measuring toxicity levels at Alang, told me. "It has been colonized by the global shipping industry and their protectors in the United Nations. And it is run by a local mafia elite that are the de facto lawmakers and lawbreakers."[5]

In addition to the Hong Kong Convention, shipping companies have honed a range of tools for evading responsibility for what ultimately happens to their vessels. When a shipping company decides to sell its cargo container or tanker to the subcontinent, it often arranges that sale through a crop of middlemen — typically Dutch or British, sometimes based out of Singapore or Dubai — who broker deals in cash and proceed to deal with Bangladeshi and Indian and Pakistani conglomerates directly. For the shipping multinationals, the aim is to put as much distance as possible between themselves and the calamity of dismantling.

As the ships are passed off to these cash buyers, they are promptly "re-flagged," taking up the nationalities of even more recondite island states — so-called black flag nations like St. Kitts and Nevis, Comoros, and Palau — which offer brokers, in addition to the usual hodgepodge of legal benefits, zero pretense of environmental oversight. The Republic of Palau, an impoverished archipelago of 340 coral and volcanic islands in the middle of the Pacific Ocean that earns approximately $10,000 in exchange for handing its flag out to cash buyers, has little incentive and less ability — or, as the shipbreaking expert Nicola Mulinaris put it to me, possesses "lack of knowledge, lack of resources, lack of will" — to hold a shipping company licensed in Valletta, or a shipbreaking yard based in Bangladesh, responsible for any calamities inflicted on the endangered mangroves of Chittagong.[6]

Finally, vessels tend to be sold while anchored in international waters presided over by no single nation, making it more difficult for authorities to take account of, among other things, any hazardous materials on board.

Which is all to say that by the time a ship is getting hacked to pieces by thousands of South Asian migrants, it has an exorcised relationship with the Western shipping company that it just spent the last years, possibly even decades, enriching. What the shipping industry has accomplished amounts to a supercharged version of how Coca-Cola justifies its sales in the plastic-infested metropolises of Africa: assuring you that though they have profited off it, the devastation is not their problem.

"Globalization has helped fuel this rush to the bottom," the world's largest seafarers' union has claimed about the failures to hold the shipbreaking industry accountable for its environmental and labor-rights depravations.[7] And, true enough, to bring a semblance of responsibility to the shipbreaking industry would first require something exponentially more difficult: bringing a modicum of accountability to the international shipping industry itself.

At least one striking thing has resulted from the last forty years of offloading environmental hazard onto South Asia: The methods and means by which ships get dismantled have never really needed to improve. *Where* ships get demolished has changed since the 1970s. *How* they get demolished hasn't. It is still done by hand, with a blowtorch, through a process that first requires the beaching of the ship, a haphazard endeavor in which the lunar tide helps propel the vessel into subcontinental mudflats, where holes are then drilled into its sides, the effect of which is to wash out bilge and residual oils with the eventual onset of the high tide.

Next, the ship's most hazardous elements — asbestos, mercury, sulfuric acid, lead, polychlorinated biphenyls — are removed. In Bangladesh it's common practice to first drop a live chicken into potentially poisonous sections of a beached hull; if the creature manages to emerge from the hull alive, the material inside is determined safe enough to be handled by Bangladeshi laborers, shipbreaking's own twisted version of the canary in a coal mine.[8] And it is here that the bulk of the costs of dismantling are cut. In a country that has legally committed itself to environmental due diligence, disposing of a ship's toxic waste would incur approximately 80 percent of the costs of dismantling, requiring meticulous treatment and transport and specialized disposal. In India and Pakistan and Bangladesh and Turkey the process incurs remarkably little cost; asbestos and biphenyls and cable casings and glass fibers tend to be piled on the beachhead at dusk and torched.

Next, the innards of the ship — the wood decking, engines, refrigerators, radar systems, tables and chairs, dartboards, soft-serve ice cream machines, maps, nautical charts — are removed and sold off to local buyers. In India there exists a bazaar that stretches six miles long and is cluttered with the bric-a-brac of the eight thousand ships that have been dismantled at Alang since the mid-1980s.[9] In the case of Turkey, if you have ever vacationed anywhere along its coasts, in any of the gorgeous beachside resorts of Bodrum or Antalya, it is possible that everything from your mattress to your mini-fridge lived out a previous existence half a world away in the Caribbean Sea, adorning one of the "staterooms" of, say, the *Carnival Inspiration.*

Then, over the next months, sections of the ship's bridge, decks, and hull are sliced up by hand, with the use of plasma torches, and dropped — the so-called gravity method — or shifted by crane onto the tidal flats. Hundreds of gas canisters are transported *inside* beached ships to provide fuel for the plasma torches; outside the vessel, the hull's surfaces are prepared for cutting through the removal of external toxic substances — heavy metals and toxic paints — by scrubbing and brushing.[10]

Over time, the ship's sections are moved up the beachhead by contingents of bulldozers, which are lassoed to one another in a line, like carriages of a train, and proceed to collectively yank the steel inland; then the chunks are dismantled by torches into smaller slices of scrap roughly the size of automobiles, turning the coastline into what is in effect numerous deconstruction sites. When at last the steel has been sliced into pieces capable of being lifted by several men, it is stacked into the beds of trucks and sent to a steel plant to be melted down. Approximately two thousand experienced dismantlers can make a tanker the length of three football fields disappear in four months. The process is grinding. It is ad hoc. Photos of vessels being dismantled in Kaohsiung, Taiwan, in the early 1970s are interchangeable with photos of ships being dismantled in Gadani, Pakistan, today. Or at least almost interchangeable: Over the last fifty years, ships have grown exponentially larger, more complicated, and more dangerous to break.

One might briefly compare all this to the changes in the ship and the port itself, both of which have been ground zero for global advances in automation. Today, the largest ships in the world employ fewer workers than your average grocery store. Some two dozen crew members oversee tankers and bulkers — vessels of monumentally increasing size and complexity — that are becoming the domains of computerized navigation; not long ago in Norway, a cargo ship began undertaking voyages without a single human aboard.[11] More astonishing still is the transformation of the port. Since the 1990s, cargo hubs like Valencia and Hamburg have become increasingly devoid of humans. Once one of the great bastions of organized labor, a staging ground for strikes and even political revolutions, full of essential workers whose demands required careful consideration by bodies from the League of Nations to the World Bank, ports today are the front lines of robotic labor, a computerized choreography in which the human element has all but receded from view. On a recent trip to the Port of Los Angeles, I couldn't help but feel that I was visiting the vacated movie set of a harbor rather than North America's busiest container hub. Along quays stretching into the Pacific for miles, sparse clusters of workers could be seen, meandering between stacks of twenty-five-ton cargo containers that were getting snatched up and swung around like dominoes by cranes rising as high as Manhattan apartment buildings.

Shipbreaking yards, on the other hand, have never had very much incentive to become more efficient. This is partly because efficiency is financially undermining: Flooding regional markets with cheap steel from dismantled ships at greater rates would be counterintuitive, driving up supply and diminishing value. But something even more fundamental explains the near-technological stasis in the world of shipbreaking. The process of breaking apart a giant ship may take four months; it may take two years. The bigger the ship, the longer the effort of dismantling, and the more time the slowly retrieved steel becomes subject to price fluctuations in global commodity markets. The one thing that *can* be controlled—the lone mainstay in the volatile economics of shipbreaking—is labor and how much you pay for it. Bangladeshi and Indian and Pakistani dismantlers, all of whom are migrant populations within their own countries, are desperate enough to work for thirty cents an hour; Turkish workers make about ten dollars a day. "Poverty is the selling point," Syeda Rizwana Hasan told me. Hasan is the executive director of the Bangladesh Environmental Lawyers Association and a decades-long advocate for dismantlers' rights in Bangladesh. "Shipbreaking gets to move from place to place. The ground realities of the industry themselves never have to change."[12] The price of a "sustainable" business is, in this case, the sustaining of highly exploitative systems of underpaid labor.

And so shipbreaking remains a capricious, human endeavor—and an excruciatingly dangerous one at that. The fate of Oğuz Taşkın in Turkey is a tragedy but not an outlier: To pick up a blowtorch and enter a ship destined for destruction routinely ranks as one of the most dangerous ways of making a living in the world, statistically deadlier than mining.[13] The problems often begin before you can even clamber inside the vessel. Most shipbreaking yards in the world are reliant on the aforementioned "gravity method," shipbreaking speak for cutting up a vessel from the top down and allowing bergs of steel to drop as many as ten stories before crumbling into more "manageable" sizes upon striking the ground. In 2022, a Bangladeshi cutter named Ariful Islam Sujan was flattened to death by a falling iron girder as he was walking the beachhead, an accident his employer was accused of lying about with claims that the twenty-six-year-old had died of a heart attack.[14] Then there are the errant machines, the poisonous gases, the conflagrations. In Bangladesh, it is estimated that on average such "accidents" take the life of at least one worker every week.[15] In November 2016, an army of dismantlers in Pakistan was tearing down a Japanese tanker when an unexpected blast "incinerated" more than a hundred men "without a trace" and sent shrapnel flying as far as two kilometers away; it took three days to extinguish the flames.[16] Other dangers are of an even more unpredictable nature. In May 2020, a group of Indian dismantlers entered a tanker called the *King Kong 1*, only to encounter a wild leopard believed to have taken up refuge three weeks earlier in the ship's hull, where it sustained itself on the carcasses of stray dogs it captured along the beach by night and dragged back to the *Kong*'s bowels.[17] Finally, there is the longer-term damage wrought by demolition. It has been determined that when shipbreaking was conducted in Taiwan in the mid-1980s, approximately one in four dismantlers ended up dying of some form of cancer.[18]

A similar phenomenon can be expected to play out over the next decades in the shipbreaking states of the Indian Ocean semi-periphery. As it stands already, life expectancy for men in the shipbreaking industry in Bangladesh is twenty years lower than that of the general population.[19] In India, the lung damage accrued working at Alang, a doctor at its Red Cross hospital has estimated, equates to smoking "ten to fifteen packs of cigarettes" every day.[20] There, as elsewhere, workers who die prematurely of lung or tracheal cancer — diseases almost unquestionably resulting from years spent huffing diphenyl ether and other toxic fumes — are not deemed work-related fatalities. For good reason has *"Alang se palang"* become a common refrain among Indian shipbreakers: "Alang till death."

25

SCRAP SHEPHERDS

> The community would prosper, but their sores would never heal. The factory waste would alter the color of the earth, the howling wind would scatter, and murmurs would turn into screams.
>
> — Latife Tekin, *Berji Kristin: Tales from the Garbage Hills*, 1993

IN THE LATE 1980s, Turkey emerged as one of the world's great new destinations for ships bound for dismantling. Three hundred miles south of Istanbul, the Turkish government designated a stretch of state-owned coast in the city of Aliağa a "shipbreaking zone." Some two miles long, buttressing a promontory that juts out into a deep stretch of the Aegean Sea, the land was leased to a clique of men who mostly hailed from a distant inland city called Niğde. They weren't shipping magnates. They were scrap dealers; most had been shepherds until the 1970s, when they traded vast herds of sheep for scrap picked out of junkyards and construction sites in central Turkey. In the 1980s, they gravitated to the Aegean coast in search of ever-greater troves of old metal junk, often in the form of broken electricity poles and busted ovens imported from the nearby European continent.

Eventually they began buying ships. Four families emerged from the decade more powerful than the others; today, ownership of the majority of Aliağa's twenty-two shipbreaking yards is divided among them. Unlike the yards of the Indian Ocean, the Turks tend not to arrange the purchase of their ships through an intermediary class of cash buyers operating out of offices in Singapore or Dubai; they themselves scour the harbors of the world in search of vessels, not so much recipients of wasted ships as hunters of them.

Over time, the interests of the Turkish yard owners diversified. They might have started out as shepherds, then turned to scrap, but by the early 2000s they stood over a wide range of concerns. Some acquired the steel plants to which they sell sliced-up ship hulls, as well as the construction companies to which those steel plants in turn sell newly fashioned rebar. Leasing their land from the Turkish state, they reaped the benefits of discounted electricity and advantageous construction contracts. Needless to say, their ties to the inner circles of Turkish power — the ministers, local and national, who tender public building contracts — run deep, perhaps most grippingly revealed by an offhand anecdote delivered by Sedat Peker, an infamous Turkish drug lord currently on the lam in the United Arab Emirates.

In September 2021, Peker turned Aliağa into a household name in Turkey when he claimed on his wildly popular YouTube channel that its shipbreaking yards doubled as importers of South American cocaine, which arrived in Turkey welded into the hulls of ships from Ecuador or Venezuela designated for dismantling. The drug trade enriched no less than the highest echelons of the Turkish government, contended Peker, who unveiled photos showing the owner of one of Aliağa's shipbreaking yards and the son of Prime Minister Binali Yıldırım gambling side by side at a Singapore casino.

Shortly after the shepherds-turned-scrap-barons from the city of Niğde came the beekeepers and cattle herders from the towns of Zara and Sivas. Throughout the 1990s, tens of thousands of them bused west from the interior of Anatolia to the coast of the Aegean to undertake the dangerous work of dismantling. They typically relocated in networks. Brothers would bring cousins, who would then bring sons. Many had never seen the sea in their lives; they arrived to find themselves ordered inside vessels that had often not touched dry land in decades. So drastic was the migration that by the time I reached Zara, a full third of its male population was no longer there. It had relocated to the Aegean. Most, like Oğuz Taşkın, sent remittances back east. And most — either because the money in Aliağa proved too good or, in some cases, because they physically couldn't — would never return home.

As the historian Matthew Sohm has argued, Turkey has long played a crucial role in Europe's penchant for maximizing economic opportunity while minimizing environmental risk.[1] True enough, Turkey has always been a frontier, a crucible of conflict saddling the much-disputed cusp of Europe and Asia. But during the last decade of the Cold War, its chronic status as a geographic frontier began extending into corporate and legal realms as well, turning Turkey into a kind of regulatory borderland — and at times a vacuum of oversight — for enterprising Western multinationals. Turkey's advantage was that it existed next to Europe without *being* Europe. It managed to be endowed with certain benchmark standards of European prosperity, security, and respectability — a founding member of the Organisation for Economic Co-operation and Development, for instance, and a NATO stalwart — while being devoid of so many others, such as stringent environmental regulations and enforced labor rights.

In the late 1980s, at just the time when Western ships began drifting into Aliağa for dismemberment, a parallel exchange was attempting to shift large amounts of another kind of material from Europe to Turkey. Before there was Turkey's First Lady Emine Erdoğan, garlanded by the World Bank and the UN Capital Development Fund for her purported success in turning Turkey into a "zero waste" nation, there was Semra Özal, Turkey's First Lady from 1983 to 1989. Özal had her own waste campaign: She was bent on helping import great amounts of it into the plains of western Turkey from factories across West Germany. As Sohm has revealed, between 1987 and 1988 companies such as

Weber in Bavaria and NE-Metall in Saxony arranged deals to repackage their hazardous residues as "fuel," ship them east at a fraction of what it would have cost to dispose of them anywhere in Western Europe, then have them incinerated in, among other destinations across Anatolia, the steel and cement plants ringing the port town of Aliağa. West Germany sweetened the deal for Ankara by dangling an investment package amounting to "billions of dollars" that would include the construction of two large factories outside the southwestern Turkish city of Isparta and the creation of several thousand local jobs — a generous alternative, or so it appeared, to the more traditional forms of monetary lending that Turkey, indebted to Western institutions by the late 1980s, was in no easy position to refuse.

As Sohm notes, the *çöp tartışması*, or "garbage debate," of Turkey's 1980s was as much about the country scouring for ways to bring in hard currency and industrialize itself as it was about deeper historical questions of Turkey's dependency on Europe, an inequality with roots extending as far back as the trading capitulations the "great powers" had extracted from the Ottoman Empire in the seventeenth century. By the 1980s, the difference between "developed" and "developing" countries was not merely that West Germany possessed factories capable of producing millions of tons of hazardous waste and that Turkey was desperate to possess factories of any kind whatsoever; the difference was that a "developed" country like West Germany also possessed the power and capital to pick and choose where its hazardous waste ultimately ended up — and who had to take it.

How do the shipyards at Aliağa fit into this broader story of Turkey as Europe's environmental frontier? How did Western countries sending shiploads of hazardous waste to Turkey turn into them sending actual ships themselves?

Sending cruise ships from the United States to Turkey to be recycled is, yes, profitable: A company like Carnival Cruise Line can receive tens of millions of dollars for a ship its executives have decided to dispense with regardless. And sending a cruise ship to Turkey is, sure, more convenient than paying for it to get tugged all the way to the shipbreaking yards of the Indian Ocean: One need not pass through the Suez Canal. But neither cash nor convenience is why one sends a cruise ship from the United States to Turkey to be dismantled. In fact, there are plenty of reasons why only about one percent of all ships that have ever been dismantled have gone to the Aegean. The principal one is money. For while the shipyard owners at Aliağa do pay for ships, their purchasing power has never been very strong; Turks can pay approximately $200 per ton of vessel, or about half of what yards along the Indian Ocean can offer, facilities owned by even mightier conglomerates of state-connected companies with interests ranging from timber to diamond mines.

But consider for a moment what Aliağa *does* offer a publicly minded, social-media-savvy multinational like Carnival Corporation, the world's biggest cruise ship company, whose accolades — among the "100 Best Corporate Citizens," according to 3BL, a platform that tracks environmental issues, and one of "America's most responsible companies," according to *Newsweek* — were rolling in just as it decided to sell off eighteen of its ships in early 2020 owing to a coronavirus that had obliterated a third of the company's stock value in less than a month.[2] For Turkey is unlike India and Bangladesh and Pakistan in one major — and, one might say, all-important — respect. A handful of its yards have recently been "approved" by the European Union — meaning that after an extensive application process, which requires European inspectors to journey to Aliağa and observe operations up close, they have received official "validation" that they conform to European standards of labor rights and environmental due diligence.

In the 1980s, Turkey was attractive to European companies because it *didn't* conform to their own national standards of environmental and labor regulation. Today it is attractive because, at least according to Europe's own environmental inspectors, it *does* conform to them. Turkey's recent galloping success in the ship-scrapping market is its ability to put itself forward as the green, ethical alternative to the breaking yards of the subcontinent. "If you're a European shipping company and you really want to aim high, then Turkey is the destination for your ship," Konstantinos Galanis, chairman of the International Ship Recycling Association, told me.³ "You go to Turkey because it checks all the right boxes," Simon Ward, a ship broker, told me in his office overlooking Piraeus harbor in Greece.⁴ "More shipowners come to Aliağa every passing day," recently boasted Kamil Önal, president of the Ship Recycling Industrialists Association of Turkey.⁵

So explains the boom in activity—and, less publicized, the uptick in fatalities—that have overtaken Aliağa in recent years. At a time when green credentialism has become an important public relations exercise for companies with carefully manicured public profiles, here is a place that manages to do certain quarters of the global shipping industry the ultimate—and rare—service: making them look good.

There was, as mentioned, Carnival:

> In addition to limiting our vessels' impact on the environment throughout their service time in our fleet, recycling our retired ships following the European Ship Recycling Regulation ensures we are applying the highest global standards and contributing to a sustainable cruise industry.⁶

But also Mediterranean Shipping Company:

> We recognise our industry's end-of-life challenges and are committed to promoting the respect for labour and human rights and supporting the welfare of our supply chain partners, including those working at the yards where MSC ships are recycled.⁷

As well as the shipping giant Maersk:

SCRAP SHEPHERDS

We will ensure safe and responsible ship recycling globally to the benefit of workers, environment, responsible yards and shipowners.[8]

By 2021, all three companies were dispatching vessels to Aliağa, splashing their commitment to human welfare and environmental responsibility across their web pages and quarterly memos.

The Finnish American law professor Anu Bradford has coined the phrase "Brussels effect" to describe the way certain European standards of labor rights, data privacy, and environmental regulation can be extended beyond the continent with the help of market forces. European companies operating in the Middle East, for instance, can usher in admirable standards of gender equality. In Central Asia they might bring new protections on data privacy, or help curtail corruption in East Asia by introducing elevated standards of monetary transparency. And at first glance, Aliağa appears to be an example of such European values in export. It is a place that, after all, European investigators literally inspect firsthand, arriving once a year to examine uniforms and measure air quality and test for water pollution.

But there's another story here, another possible type of "Brussels effect" at work: the ways in which European standards of human rights can provide the imprimatur of credibility to certain industries that are just too inherently dangerous and polluting to be conducted at scale on the European continent itself yet are too indispensable for its companies to stop doing altogether. For the more you dig into the murky circumstances that resulted in the death of Oğuz Taşkın, the clearer something becomes: European inspectors may "approve" of some of the facilities at Aliağa, but they have little idea of how the work at Aliağa actually gets done.

One consequence of so many big new ships being sold to Aliağa in recent years is that there have been fewer and fewer places available to dismantle them. In early 2020, the *Carnival Inspiration* may have been sold to the yard called Ege Çelik, but upon reaching Turkey from the Caribbean it was transferred to a neighboring yard, a facility called Metas, which is owned by the same conglomerate that controls Ege Çelik. Between these two dismantling yards there was one major difference. Ege Çelik is "approved" by the European Union. Metas, where the *Carnival Inspiration* ended up, is not.

The tragedy surrounding Taşkın's death is not just that he was scandalously undertrained to be dismantling a seventy-thousand-ton monstrosity like the *Carnival Inspiration*. It is that he should have been nowhere near it in the first place: He was not working for an EU-approved yard. He had no EU-approved training. He had no EU-approved equipment. And it says something about the enforcement mechanisms of the EU approval process that a ship like the *Carnival Inspiration* — a ship five times as long as the Statue of Liberty is high and whose reason for sailing a quarter the circumference of the globe from the Caribbean Sea to the Aegean Sea was to be dismantled at a yard that had been subjected to a scrutinous process overseen by bureaucrats from Brussels — could be getting torn to pieces for an entire year at the wrong dismantling yard without anyone seeming to realize or care.

Perhaps the real point of Aliağa is not just to break ships within sight of Europe. It is not just to dismantle them cheaply. It is not just to boost the Turkish economy.

Perhaps the *real* point of Aliağa is to give European and US shipping companies a plausible degree of deniability: By sending their ships to Turkey, and not Pakistan or India or Bangladesh, they are at least attempting to live up to their own self-professed, highly touted commitments to environmental regulation and human rights. And alongside the familiar tool kit of financial and legal evasion, the shipping industry now has harnessed a new tactic: constructing bigger and bigger ships, reaping bigger and bigger profits, all through the justification that they are disposing of their old fleets by the most responsible means — and in the most responsible place — that they possibly can.

None of this is ending anytime soon. Stricter emissions regulations proposed by the European Commission earlier this decade dictate that by 2030, the world's oceans must theoretically be traversed by new fleets of more sustainable ships, necessitating the dismantling of huge numbers of current seagoing vessels. By 2028, global ship recycling volumes are expected to double their 2022 rates. By 2033, they will have quadrupled.[9] A massive glut of scrap steel is on the horizon — fifteen thousand ships full of it — and no place is more poised to cash in than Aliağa, the shipbreaking destination that has successfully positioned itself as the humane alternative to the tidal flats of the Indian Ocean.

As for Carnival Corporation: Does anyone seriously believe that its decision to scrap the *Carnival Inspiration* was undertaken to any net benefit of the environment?

Let me present to you the *Carnival Jubilee*, vaunted successor of the gutted *Inspiration*. For those unawed by the latter, here sails a vessel stretching one full city block longer and rising six additional stories into the sky. The *Jubilee*'s 6,600 guests are to be privy to something the executives at Carnival have termed the "Excel-class treatment": a roller coaster called the BOLT, six "zones of fun," an IMAX theater, a bingo lounge and karaoke auditorium, a casino, a golf course, an in-cruise televised game show called Deal or No Deal — all, rest-assured, emitting immensely more carbon than the lowly *Inspiration*.

One can already imagine the Turkish shipbreakers of the late 2030s, armed with the same blowtorches as so many decades of dismantlers before them, staring up at such a behemoth in disbelief and trepidation as they prepare to enter its hull.

26

SCRAP NATION

> Property developers and construction companies have been with him [Erdoğan] since day one.
> — Turkish financial analyst, 2018[1]

HUNDREDS OF MILES away from Zara and the depopulating towns of Central Anatolia, a new Turkey was rising. GDP growth ranked among the world's highest, surpassing almost all the countries of the European Union, which, by the mid-2000s, Turkey appeared to be on the brink of joining.

"The Turkish economy is on the right track," commended an IMF official in 2003.[2]

"The country's progress in recent years has been remarkable," lauded the fund's managing director two years later.[3]

"Opportunities for the Turkish economy are enormous," one of its annual reports concluded two years after that.[4]

By 2013, the year a massive corruption scandal convulsed Recep Tayyip Erdoğan's Justice and Development Party (AKP), the success story of Turkey was more astonishing still, with the country's GDP having fully tripled in size since the start of the millennium.[5]

The foremost drivers of Turkey's mesmerizing growth? Bricks, mortar — and steel. Under Prime Minister — and, after 2014, President — Erdoğan, construction had become the roaring motor behind Turkey's full-steam-ahead prosperity. Blessed with a range of perks, such as electricity subsidies, "state of emergency" dictates that allowed for the circumvention of local zoning laws, and routine "amnesties" that summarily legalized years of unauthorized construction, the building sector was helping Erdoğan to usher in not merely a new country but a new kind of Turk too. Over the course of the 2000s, one Istanbul slum after another was cleared away to make room for ultramodern glass skyscrapers, office parks, and ranks of concrete apartment buildings totaling more than 700,000 new homes for a rapidly expanding middle class; over the course of just two decades the city's population, already Europe's largest, would double.[6] Elsewhere, a land of bazaars was resurfaced into a nation of malls, with more than three hundred shopping centers — some built to resemble Ottoman army barracks — replacing the age-old street markets of Anatolia. What Erdoğan called the "New Turkey," a country he vowed would shortly "get into the top ten countries in the world" in democratic and economic indexes, was soon interconnected by more than nine thousand kilometers of new highways and three thousand kilometers of high-speed rail.[7]

And then there were the biggest construction jobs of them all, practically Pharaonic in scale. Since 2002, nearly a trillion dollars had been lavished on Erdoğan's so-called megaprojects, whose unveilings invariably coincided with the announcements of his reelection campaigns. More than two million trees were chopped down to lay the ground for a new airport outside Istanbul, the city's third and one of the world's largest.[8] Another forest was demolished to put up a glittering presidential palace thirty times the size of the White House. A new mosque capable of accommodating sixty-three thousand worshippers burst forth from the thicket skyline of Istanbul. A third bridge went up across the Bosporus; a new underwater tunnel conjoined Europe and Asia beneath it. The dredging of a new canal to connect the Black Sea and the Sea of Marmara was announced, allowing ships to bypass the Bosporus for the first time in history; next, proposed Erdoğan, a couple of new cities would be constructed along its banks. "When man dies, he leaves behind a monument," the president liked to tell his supporters at campaign rallies after promising to deliver them yet one more bridge or airport or mosque.[9]

But here's the strange thing. So much of what Erdoğan was portraying as the "New Turkey" was in fact reliant on what the rest of the world had deemed too old or rusty or worthless. Highway dividers from Florida, the scaffolding of Finnish coal plants, manhole covers from Athens, old Ferris wheels from the United Kingdom, shredded cars from the Balkans — Turkey had become a magnet for all of it. In addition, of course, to the million tons of steel scrap a year it was sourcing from ships dismantled at Aliağa by migrants like Oğuz Taşkın.[10]

For the last twenty years, Turkey has not just been the world's greatest importer of ferrous scrap metal. It has been the biggest importer *by far*, with twenty-five million tons of scrap imported in 2021 alone — four times that of the next highest importer, Bangladesh.[11] Turkey has been the greatest importer of secondhand metal from the European Union, itself the biggest exporter globally, as well as from the United States, taking in a quarter of all American scrap bound for export; in 2010, one in four pieces of scrap metal getting moved anywhere on Earth would find its way to Turkey.[12] Cargo containers bearing cracked hubcaps and dented refrigerator doors and limping shopping carts entered Turkish ports dotting the Mediterranean or Aegean or Black Sea, and were melted down to fashion new steel, then used to build apartment blocks, malls, hospitals — that construction mania otherwise hailed as the "New Turkey."

Why Turkey? In many ways it's a story that goes back to the foundation of its modern republic. Following the loss of its European territories with the dissolution of the Ottoman Empire, Turkey proceeded to engage in a massive concentration of investment in the — historically backwater — landmass of Anatolia. In 1926, in one of the early gambits of a modernization program that he vowed would turn Turkey into a "European nation," a secular state with a Western legal code and a Latin alphabet, Kemal Atatürk — father of the modern Turkish republic — brought in British experts to survey the country's hills in a bid to locate iron ore and develop a construction sector that could make Turkey look like the rich Western states it was so eager to emulate.

But Turkish steel production remained negligible over the next fifty years. Only three steel plants were constructed. Much of the country's steel had to be imported from Europe.

The problem? The British experts Atatürk had brought in to search for deposits of iron ore were never able to find very much of it.

In 1980, something changed. Following the military coup of that year, and guided by the vision of Prime Minister Turgut Özal, a former employee of the World Bank, Turkey underwent a stark pro-business pivot. During the years of "liberalization," in which the IMF helped oversee the streamlining of the country's banking sector and the opening salvos of its European Union accession courtship, Turkey started investing in a specific type of steel plant, something known as an electric arc furnace mill.

Requiring gargantuan amounts of electricity, the daily equivalent of that required by a US city with a population of some 150,000, the arc furnace is nevertheless considered a net boon for the environment.[13] For unlike a conventional blast furnace, which produces virgin steel from iron ore, an arc furnace gets fed old scrap steel, which it then heats up and melts down. What eventually results is fresh steel in the shape of rebar rods or flats or slabs or casts, which can be used directly in the construction of steel structures, such as bridges and skyscrapers, or as reinforcement in the foundations and walls of mostly concrete structures, such as high-rise apartments and malls.

Between 1980 and 1987, more than twenty electric arc furnaces were assembled across Turkey, the majority rising along its three coastlines. They acted simultaneously as producers of new steel and depositories of old steel from across Turkey and beyond. It proved an auspicious, almost incredible moment to enter the global scrap market. For the previous forty years, on the distant cusp of the Black Sea, the Eastern Bloc had been the world's largest producer of steel; by 1992, the Soviet Union had collapsed, not only resulting in the disintegration of the upstart Turkish steel economy's only real competitor in the region but also leading to a tremendous industrialized landscape itself devolving into scrap.

In the mid-1990s, ships chartered by Turkish steel companies and flying Turkish flags started ferrying off literal towns' worth of old steel and iron from the northern rim of the Black Sea, much of it stripped from factories of former Soviet republics such as Georgia, where, until recently, junk metal to Turkey was the country's biggest overall export, second only to used cars sent to Azerbaijan.[14]

"The transportation minister of Poland called me shortly after communism had collapsed," Nathan Fruchter told me. Fruchter led Eastern European scrap operations of the commodity house Marc Rich / Glencore in the 1990s. "They had a railway going from Warsaw to Gdańsk that they needed to get rid of. And a lot of old railway cars. I went to Gdańsk to see it all. And yes, we bought the railway. But that was just the beginning. You would go to ports in Eastern Europe in the 1990s and see cranes just sitting there. They weren't working. They had become scrap, waiting for someone to come and take them away. It was the same with their cars and their ships and their military equipment. All just waiting to be taken away. And I would go to Istanbul once a month to sell three or four cargo ships of it all to the Turks."

On the other side of Turkey, meanwhile, a separate opportunity presented itself: Stupendously rich markets emerged, to which all that old scrap, melted down and manufactured into new steel, could be sold and exported. In the 1990s, scrap from Israel and Yemen would be shipped to Turkey, melted down, refashioned into fresh steel, then sent back to those same states. Farther east, Turkey became the preeminent supplier of construction stock for places like Dubai and Doha just as they were gussying themselves up from backwater deserts into sleek international hubs of finance and commodities trading. And newly manufactured Turkish steel went farther afield too.

"We would load a ship in Philadelphia full of scrap, send it to İzmir, wait a few days, then load that same ship up [in İzmir] with newly minted steel and then send it on to South Korea," Fruchter told me. "In Turkey, the buyer of your scrap and the seller of the new steel were the same person. The country had become literally one giant steel factory for the rest of the world."[15]

It was an astonishing transformation: Old junk unwanted by the rest of the world had allowed Turkey, a country with almost no iron ore to speak of, to turn into one of the planet's foremost producers of steel. And at the same time, the electric arc furnaces played a crucial role in producing something rather unexpected: a country whose greatest economic activity was to ultimately prove its own pell-mell renewal. Today, the majority of Turkey's newly minted steel isn't going to Israel or Dubai or South Korea. It's not going anywhere at all. It's staying in Turkey, going into the construction of shopping malls and railway stations and, yes, mosques big enough to accommodate the populations of entire cities. Fueled by a sugar rush of cheap foreign credit, over the last twenty years construction in Turkey has increasingly become *the* economy, taking the place of manufacturing sectors that only a generation earlier had made Turkey a major supplier of, among much else, carpets and textiles and automobiles to Europe and the Middle East.

By 2010, Turkey's building sector had come to rival that of manufacturing as a share of GDP, doubling the sizzling growth of the Turkish economy more generally; by 2017, construction had overtaken manufacturing, with building and related activities accounting for, in the estimation of one analyst, nearly 30 percent of Turkish GDP.[16]

Scraps of metal tossed away in foreign lands had been recycled into steel rebar in Turkey, the buttressing of a new nation bristling with new thirty-five-story minarets and spanned by new mile-long bridges, but the metal had been turned into something else too. It had been converted into political power.

For it was not just the promise of new airports and palaces that kept Erdoğan in office one year after another. His ties to Turkish construction also operated at a grubbier, more subterranean level. Sure, Erdoğan — whom, owing to a series of telephone transcripts leaked to the Turkish press in 2013, the construction tycoons were known to refer to as the "big boss" — bestowed perks on the building sector. But there was also a clientelistic network of local construction bosses required to support the "big boss" in return, converting Turkey into a Matryoshka doll of construction-infused interests — contractors, local bosses, workers, scrap dealers, the millions of middle-class beneficiaries who had been lifted out of poverty by a building boom fed by foreign junk and cheap credit. This network vaulted Erdoğan into office again and again over the decades. By 2018, five of the world's ten biggest construction companies were Turkish firms, all enriched by huge contracts tendered by — and to — members of Erdoğan's inner circle.[17]

Aykan Erdemir, an anthropologist and former representative in the Turkish parliament with the Republican People's Party, told me, "Construction in Turkey is a mutual help association. It is one that takes skill, to accept bribes and pass on spoils and make sure that everyone is getting paid on time. Erdoğan didn't invent it. He perfected it. And under his rule, as the Turkish economy grew, so did the spoils."[18]

Your old car, your old clothes iron, your old cooking pans — it's not improbable that such items made their way to Turkey over the last twenty years, accumulating into so many indispensable props in Erdoğan's increasingly authoritarian clutch on a nation that he liked to claim with some, but not total, exaggeration he had built. Where the state ended, where the steel manufacturers and construction firms began — in Turkey it was impossible to say.

"My brothers," Erdoğan would ask his audiences after rattling off the successes of his building program, "could a corrupt government do this?"[19]

And yet where *exactly* the billions of dollars in construction contracts were all going — this, too, was difficult to say. But here's one thing that's clear about Erdoğan's Olympian-scale building spree: Behind the smoke screen of progress and modernization there festered corruption of equally gargantuan proportions. Under Erdoğan, construction has never just been about building a new Turkey and a powerful electorate. It has a seedy additional function, operating as a piñata of crooked deals and rigged tenders and kickbacks between factions of the AKP and their servants in the building trade.

"A zoning law says you can't build anything higher than twenty stories. So you build something that's forty stories. All those new rents fund the municipality, and the municipality in turn funds the political apparatus," Bülent Gültekin told me. Gültekin, the former governor of Turkey's Central Bank, presided over the country's housing development in the late 1980s. "Construction in the building sector tends to be corruption that everyone likes. It increases employment. It creates rents to be distributed. It raises land values. But construction under Erdoğan became a war chest with which to fund the political machine."[20]

What better way to swindle cash and amass personal power than through an awe-inspiring building program that, a hundred years after the dismantling of the Ottoman Empire, claimed to be restoring Turkey to imperial stature? In 2013, a corruption scandal momentarily cracked the patina of Turkey's dazzling renewal to reveal the underlying rot. Like a flashlight illuminating the interior of a dark cave, a series of wiretaps revealed, among other tawdry exchanges, how construction companies across Turkey were obliged to donate land to charitable entities with basement ties to Erdoğan's inner elite and members of his family.[21] The AKP wasn't just Turkey's all-powerful political party. It had turned itself into one of its biggest landowners — and landlords. Over the next years, the investigation into the pillaging of the public sector through artificially inflated construction contracts became a proxy conflict within the Turkish state, one that Erdoğan eventually won by tossing the prosecutors investigating his corruption into jail, in addition to hundreds of journalists, judges, and civil servants.[22]

So the building bonanza continued. And the biggest emerging

source of all its scrap steel? The lifeblood of construction, the generator of a new Turkey, the motor chugging in the AKP's roaring electoral machine, the lubricant of corruption?

Western cruise ships heading to Aliağa.

Even as the scandals heated up, even as Erdoğan's hold on power often appeared to be thinning to a breaking point, ships like the *Costa Victoria* and the MS *Astor* kept on chugging in from the Caribbean, hustling record quantities of steel into Turkey — and keeping Erdoğan's promises of a grand new Turkey afloat. For enormous cruise ships had an advantage over the battered bicycles and crumpled cars that were simultaneously getting imported into Turkey. Their retrieved metal often didn't need to be fully melted down. Like old organs transplanted into sprite bodies, the billboard-sized sheets of steel forming their hulls could be re-rolled and slotted directly into the edifices of new buildings rising across Istanbul and Ankara. And because they constituted a form of waste, and an extraordinarily toxic type at that, cruise ships possessed another advantage: Here were the biggest hunks of scrap steel on Earth, which, even in an age of metal embargoes and industrial re-territorialization, rich countries were hell-bent on getting as far away from their borders as possible.

Late one April afternoon I hopped into the passenger seat of a cargo truck in the ratty outskirts of İzmir. The truck's bed was piled with propellers the size of trampolines that had been sheared off ships newly arrived at Aliağa for dismantling; some of their blades were encrusted with barnacles from years, perhaps decades, spent spinning underwater. Over the next hour, the driver made his way from a coast carpeted with olive groves toward an ugly suburb east of İzmir called Bornova. It resembled a company town, only its business was funneling in the junk metal of the world. Trucks brimming with dilapidated objects of copper and iron rumbled over too-small streets lined with scrapyard upon scrapyard fuming with the stinging, intermingling scents of burning rubber and diesel exhaust. My own truck driver eventually pulled up to a yard called Temurtaş Metal Hurda, which, like most of the other yards in Bornova, featured the head of a hydraulic excavator poking above a perimeter fence of corrugated iron. Around the fence, lion-like sheepdogs native to Sivas had been brought in and chained up; they kept away the local Roma, who for centuries had been the biggest scrap collectors in Turkey and were now forced to make their living looting metal from the scrap barons who had migrated to İzmir from central Turkey over the last forty years.

Or so I was told by Mehmet Temurtaş, the twenty-eight-year-old owner of Temurtaş Metal Hurda, who had inherited the business from his father. Temurtaş was emblematic of the interlocking scrap interests that had emerged out of the interior of Anatolia and turned Turkey into the world's foremost mecca of junk metal. His grandfather, Hacı Mehmet, had arrived in İzmir from the interior city of Niğde in 1977. A shepherd, he'd sold his 400-odd sheep earlier in the decade to finance his pursuit of old streetlights, busted radiators, rusting sewage pipes: Hacı Mehmet bought it all, setting up a network of recycling yards across central Turkey before finally heading west to İzmir, where even greater hauls of steel beckoned from across the seas — and also from the seas themselves. Today, the entire Temurtaş clan works scrap, importing it or sorting it or melting it down, a single family with an iron in every fire of the great Turkish steel industry, including in the shipyards at Aliağa; Hacı Mehmet's brother founded the shipbreaking yard now known as Işıksan, an EU-approved facility where some months earlier a dismantler fell more than a hundred feet to his death after a twirling piece of steel struck his scalp.[23]

"If you go to Aliağa, you will see his yard," Temurtaş told me. "He brings in ships from all over the world. And then their propellers" — he pointed to the bed of the truck that had brought me to his scrapyard — "come to me."[24]

27

AT EUROPE'S EDGE

> You remember the harbor. There were plenty of nice things floating around in it. That was the only time in my life I got so I dreamed about things.
> — Ernest Hemingway, "On the Quai at Smyrna," 1922

I REACHED ALIAĞA in early May. The city sits five hundred miles west of Zara, along the Aegean Sea, in what can only be described as a separate civilizational zone from the interior of Turkey. The barren hills of Anatolia open out onto the sun-splattered coastlines of the Mediterranean. There are flabby tourists and posh hotels, yachts and beach clubs, palm trees and fish markets. The islands of Greece — of Europe — can be glimpsed across a thin blue strip of sea.

Aliağa rests along a fault line that in almost every respect globalization has struggled to bridge. The Eastern Aegean is a fraught place. Greece and Turkey, though both long-standing NATO members, lavish chunks of their GDPs on their militaries, largely against each other, in a shadow conflict that plays itself out every day in the waters directly off — and the air above — Aliağa. Fighter jets strafe its skies, unleashing sonic booms as they swoop from the airspace of Europe to Asia and vice versa. Warships, not just Greek and Turkish but American and French and Russian and British, ply a narrow maritime alley that has been crisscrossed by armadas from the Trojan War to the Battle of Gallipoli; Aliağa may be where modern ships go to die, but it was where the ancient ones went, too, with the Persian emperor Xerxes I having sent the remnants of his fleet to a bay adjacent to Aliağa after its shock annihilation across the Aegean at the Battle of Salamis. Before my visit, the coast had become host to arrivals of a newer conflict: Russian oligarchic yachts that had fled Monaco and Cannes for safe haven in the harbors of Bodrum and Marmaris, where European and US authorities, nominal allies of Turkey, had no authority to impound them.

More apparent still — more evidence of the curious globalization buffer zone that is the Greek-Turkish border — is the political crisis over refugees, which has turned the waters off Aliağa into its greatest, and maybe most lethal, flash point. Since the early 2010s, hundreds of Afghans, Syrians, and others have drowned while attempting to cross from the shores around Aliağa to the Greek island of Lesvos a few dozen miles to the west. As of March 2016, most refugees are required to stay in Turkey, the result of a financial deal arranged between Brussels and Ankara, and yet one more example of Europe externalizing a problem partly of its own making — toxic industrial revenues, plastic waste, migration — to its eastern neighbor. The crossings have nevertheless never stopped. Today, the Greek and Turkish borders — gorgeous coastlines disfigured into soiled landscapes of coast guard stations, strewn life vests, impounded smugglers' dinghies, makeshift graves — have come to resemble grim mirrors of each other.

The fact that among the few things able to cross unencumbered from one half of the Aegean to the other are ships fated for dismemberment at Aliağa is testament to the powerful breadth of interests that shipbreaking satisfies. Europe relies on Aliağa; Turkey needs the steel it attracts.

To get to the city, one takes a train that follows the coast up from the heavily industrialized port city of İzmir, winding like a snake through hills upon hills that have been doused in shoddy concrete apartment blocks. One passes a suburb called Gaziemir, sometimes dubbed the "Chernobyl of İzmir," where in 2011 more than 100,000 tons of nuclear waste was discovered beneath the site of a former lead factory; as Turkey possesses no nuclear power plants, the origin of the waste, and why it ended up in Turkey, remains unclear.[1] It's another blow in the environmental carnage that has turned the shoreline north of İzmir into a grimy clutter of cement factories, chemical-processing plants, scrapyards, plastic importers, and steel furnaces.

At first glance, the city of Aliağa appears to be a sanctuary from this hellscape. It comes into view perched along the edge of a bay so parabolic as to look artificial. Along the inner half of the bay's curve are hotels, combed beaches, grassy hills with playgrounds and potted palm trees, an asphalt path where Turkish couples like to go for evening strolls. Men fish along the shoreline. Old trawlers have been retrofitted into cafés. The ancient historian Herodotus claimed that the trees that grew around Aliağa, though they bore no fruit, emitted a pleasant smell. They still do, rising above water that is porcelain in hue, and sprinkled throughout with small toast-colored islands.

Aliağa might offer up an appealing holiday destination, were it not for what exists on the other side of its circular bay: a Gotham City of fuel tanks and oil terminals and petrochemical refineries. It reeks. Above the complex, a platoon of wind turbines twirls the noxious air south across the bay, toward the city of Aliağa itself, stinking up the *meltemi* wind that blows down from the Black Sea. The first thing that strikes you upon waking up in Aliağa every morning is not the salty scent of the Aegean, or the fragrant trees that enchanted Herodotus, but the sour scent of petroleum, which pervades the city like a fog, permeating beachfronts and hotel rooms and the interiors of automobiles. For good reason do measurements of Aliağa's air quality consistently rank among Turkey's worst.[2]

The Aliağa Ship Breaking Yard sits at the far edge of this petrochemicalized peninsula. From the city's center, your taxi circumscribes the bay, then takes you along a thin asphalt road that veers through the refineries, zigzagging up the mountain, toward a ridge where you first hear, then behold, a barreling procession of cargo trucks bouncing up from the sea. Some are carrying chunks of steel scrap, which jut out of the trucks like metallic ribs. Others are carrying mattresses. As your taxi moves across the summit, all around you the rocky green landscape begins piling up with the leftovers of the world's shipping industry. It is a nautical onslaught. There are wooden steering wheels and lanterns and propellers and foghorns and old batteries and buoys. Life vests are strewn by the hundreds, refulgent on a floor of black gravel quilted with stray mooring lines. An armada of orange lifeboats has been bleached a dull yellow. Chunks of mountain have been blasted away to make space for anchors, industrial freezers, dishwashers. Next to an ancient Roman sarcophagus sits a stray cruise ship whirlpool, replete with an instruction placard bearing stark all-caps: NO LIFEGUARD ON DUTY. HAVE FUN. BE SAFE. It is as though a once-in-a-millennium biblical tsunami has dismembered an epic fleet and upchucked its contents far ashore, more than half a mile inland and high into the Ionian hills, then left them there to dry out, splinter, collect rust, lose their glossy luster under the beam of the pitiless Mediterranean sun.

As I look down from the summit, toward the Aegean, my first glimpse of Aliağa's shipyards is that of an apocalyptic LEGO set sprung to life. Along two miles of shore, a phalanx of cruise ships and cargo vessels sits in a patchy state of demolition, their hulls sawed open, revealing a jagged cross section of boiler rooms and passenger decks and ballast boxes. Four-story cranes swivel to and fro, effortlessly plucking chunks of steel the size of mobile homes out of ships as though they are made of Bubble Wrap, swinging them inland and dropping them onto the beach and unleashing a *kaboom!* that echoes up from the sea and boomerangs between the mountainsides. Just visible from the ridge, hundreds of workers can be seen meandering around the beachhead, capped in white helmets: a termite army lining up to enter the ships and do the outlandish — and, I now know, unbelievably dangerous — work of cutting their hulls with handheld plasma sabers. Above the fray, a formation of filthy gray-brown smog coats the seascape, intermittently pierced with columns of black smoke that twirl forth from great bonfires where masses of steel are being lasered down into lesser blocks.

Descending the mountain ridge, I count more than twenty ships in competing states of dematerialization. They stick into the beachhead like nails hammered into a plank of wood. There is a Ukrainian oceanic research vessel and a Canadian cement carrier and a colossal natural gas drilling platform that resembles a space module. I see workers clustered around a dirty white ferry called the *Delphin*, which — or so I later read — began life in the Black Sea in the 1980s as the *Belarussiya*, shuttling passengers between various Soviet republics until the Soviet Union ceased to exist, at which point the *Belarussiya* was sold off to an Australian concern to run budget cruises in the South Pacific before sailing back to Europe and, rechristened the *Kazakhstan II* by new Italian owners, began moving passengers between Genoa and Algiers. Now the *Delphin* — a ship built by the Soviet Union and that outlived it — is undergoing its own dismemberment at Aliağa.

Next to the *Delphin* is the egg-yolk-yellow hull of the *St. Damian* — a Greek-owned passenger ferry that in April 2011, I later read, under the name *Ionian Spirit*, was responsible for evacuating nearly a thousand refugees from the besieged Libyan port city of Misrata in addition to the bodies of two photojournalists killed in that spring's clashes; nine years later, the *Ionian Spirit* was renamed the *St. Damian* and began a new life shifting hazardous waste from Southern Italy to Albania on behalf of a sanitation company called Delfini 1.[3] First passengers, then refugees, then toxic garbage — and now nothing. I observe for a moment as a bulldozer yanks the *St. Damian*'s enormous black anchor up the tarmac coastline, producing a scraping screech like that of nails on a chalkboard amplified to a hellish scale.

Not every ship has such a story to tell. But a surprising number do. Looking out over the line of vessel carcasses at Aliağa, one imagines the books that might be written on the renegade world of globalized shipping that can be exposed from a hillside trashed with its detritus. And, historically speaking, Aliağa has been the undertaker of some true outlaws of the high seas. There was the *Kuito,* a tanker from Angola discovered to have been packed to the brim with radioactive waste, and the *Otopan,* a tanker from the Netherlands that arrived in Turkey bulging with several hundred tons of asbestos. And does anyone remember the ABC television series *The Love Boat?* From 1977 to 1987, the MS *Pacific Princess* played fictitious host to passengers in pursuit of romance on the high seas, all of it acted out aboard a real 560-foot luxury liner, which in 2013 was sent to Aliağa to be dismantled; two Turks were killed — and six permanently maimed — when poisonous exhaust infiltrated the *Pacific Princess*'s engine room as it was being carved into pieces.[4]

Shipbreaking insists on presenting itself as something distinct from the waste trade, a greenifying industry that exists to recycle materials rather than merely re-territorialize and discard them on the cheap. But what's clear is that in many respects shipbreaking — more than either the electronic waste or the plastic trades — bears the most obvious parallels to the hazardous waste trade of the late 1980s: Tracts of the Global South have been handed over to chemical despoliation in exchange for the elusive promise of development.

For while there are ships such as the *Kuito* and the *Otopan* that have been caught blatantly shifting hazardous material on behalf of distant factories or firms, there's also the fact that no matter what ship it is, they *all* contain spectacular compilations of toxic material. Indeed, so many toxins cross international borders drilled and sprayed and infused into ships bound for breaking yards, one NGO has termed them "floating toxic waste."[5] There are oil sludges, toxic paints, polychlorinated biphenyls, heavy metals; one study calculates the amount of hazardous material that reaches Aliağa within "recycled ships" to be fifteen thousand tons a year — meaning that because it is a shipbreaking destination, and not even a particularly big one at that, a single city in Turkey is currently on the receiving end of as much hazardous waste as the United States shipped to the Global South in any given month of 1987, the disastrous and much-derided height of the international toxins trade.[6] And, again, Aliağa is *as good as it gets*. The figures in India, Bangladesh, and Pakistan are greater by multitudes.

My eye continues moving along the beachhead, past a mammoth ship called the *Megastar*, past a Balkan car ferry called the *Jadrolinija*. And then, jutting out of a mass of white fiberglass, there it is. I spot it. It's the *Carnival Inspiration*.

By the time I got to Aliağa, it no longer bore much resemblance to the sparkling spirit that had once twinkled around the Caribbean Sea. It had been decapitated. A breakfast buffet hall on its tenth story sat exposed to gusts of toxic air and drags of bonfire smoke. Nearby, the onetime RedFrog Rum Bar had been denuded of its red leather stools and bamboo ceiling fans. The engine room where Oğuz Taşkın had met his terrible end also appeared to be gone. Two more recent arrivals to Aliağa — the MS *Marella Dream* and the *Horizon*, both cruise ships — flanked the vessel on port and starboard, propping up the disemboweled pride of the Carnival fleet like a pair of pallbearers.

Aliağa is commonly described as "a shipbreaking yard." But it technically comprises twenty-two separate dismantling facilities. They line the same stretch of coast, extending the length of roughly forty football fields from north to south, and are divided from one another by chain-link fences. And the first thing that strikes you about so many yards operating side by side is the meaninglessness of the "EU approval" granted to the third of them that make Aliağa such an attractive option for shipbreaking — since they allegedly practice "safe and environmentally sound methods of dismantling ships" and ensure that "any toxic materials in the atmosphere are within permissible concentrations."[7] Here's the inconvenient reality: Air pollution doesn't abide by the see-through chain-link fences that separate an EU-approved facility from a "substandard" facility next to it. The smog that emanates from any one yard drifts into the next yard and vice versa. And such is the flow of the sea too. All the grime and pollutants disgorged at any one shipyard drift along the Aegean coast not just to neighboring yards but to the entirety of the coastline around Aliağa.

A shipbreaking destination can only be as safe and clean as its most dangerous and polluting facility. Indeed, one need only look at the exorbitant cancer rates of the city of Aliağa as a whole — higher than almost any other place in Turkey — to grasp that twenty-two shipyards aligned along a two-mile stretch of windy coast are a generalized emitter and spreader of toxins; singling out any particular yard as possessing standards "substantially equivalent to that prevailing in the EU" is like claiming that an electric vehicle in a traffic jam full of gas-guzzling eighteen-wheelers is absorbing no air pollution.

But acquiring EU approval nevertheless comes with many advantages. On a blazing May morning, I climbed a steel flight of stairs and entered the air-conditioned office of Orhan Demirgil, owner of Blade Maritime Recycling, one of Aliağa's twenty-two dismantling yards and one that is currently awaiting listing by the EU. The spotless office had been fashioned out of the former bridge of a container ship. Inside, Demirgil sat on a plush leather chair in a black polo shirt, craftily puffing away at a cigarette. A small globe stood on the desk; a Wi-Fi password had been printed out and taped to the wall behind him. "It's impossible to lose money in this business," Demirgil told me after shaking my hand.[8]

Demirgil is among the new generation of Turkish shipbreaking yard owners. He has never owned sheep. He instead moved into shipbreaking from the used car industry, which he insisted was not all that different from shipbreaking; in both industries you are in the business of buying steel on the cheap and passing it off to someone else at a profit. Where it gets complicated, however, is in the international dimensions of the shipping trade. He waved an application for EU approval in front of me.

"If you have the EU certificate, you can buy ships at better prices," Demirgil told me. Other advantages follow. Turkish banks offer lower interest rate financing. The purchasing process is streamlined. You can dismantle vessels bearing the flags of European Union states like Malta and Greece. You need not suffice on the lesser fauna of ferries and tugboats picked up out of African ports or Black Sea states. You can dismantle larger ships, like the MS *Megastar*—Demirgil pointed to the yard next to his, where a massive bulk carrier was being sliced up by workers ambling around a seafront curtained by leaden smog. And you can make a good name for yourself in European circles.

At the moment, Demirgil was dismantling six ships a year, which he measured in the volume of scrap steel they could feed into Turkish furnaces: 40,000 tons. He sold the scrap for approximately $400 a ton to the nearby steel plants. In all, it brought in about $15 million a year. "It's impossible to lose money in this business," Demirgil repeated to me, shaking his head like a man dumbfounded by his own strange luck.

Several days later, I spent an afternoon at a picnic table along Aliağa's circular bay, where several of the *Carnival Inspiration*'s orange lifeboats had been converted into fishing trawlers by locals. At the table sat four former employees of BMS Gemi Söküm, one of Aliağa's larger shipbreaking yards. Their fathers had all worked at Aliağa, having arrived from Central Anatolia with the first of the migrant laborers in the 1980s; three had died of lung cancer before the age of seventy. Their sons, seated before me at the table, had themselves put in more than four cumulative decades at Gemi Söküm — and emerged as leaders of a February 2022 wildcat strike in which they and fifteen hundred other workers from every yard of Aliağa walked off the job demanding better safety measures, higher wages, and industry acknowledgment that diseases like the lung cancer that had killed their fathers constituted workplace fatalities.

"Here we are talking about an annual business volume of $200 million. Shipyard bosses are breaking profit records," ran their statement. "We want a humane, livable wage."[9]

A day after the strike was announced, the four men were unceremoniously fired by BMS Gemi Söküm. Three months after that, by the time I met them at the picnic table along Aliağa's waterfront, they were still looking for employment — and agreed to explain to me how the work at Aliağa really got done.

There was no such thing as safety training, began Gökhan Çoban, a thirty-nine-year-old with cobalt-blue eyes and bruised fingernails and a bushy blond handlebar mustache who had led the failed strike. "Training" occurred at lunch breaks, when most workers were so exhausted they could hardly stay awake. If they objected to the dangerous nature of the work, they were confronted by their bosses with the unfortunate reality of their situation: An ever-replenishable supply of men desperate for work existed back in Zara and Sivas and would be glad to take their position. And because most men were forced to take out ten-year mortgages on the apartments they purchased upon migrating to Aliağa, shipbreaking amounted to a kind of indentured servitude, keeping workers trapped in a system that paid them just enough money to prevent them from returning to Zara, all while wrecking their prospects of a long life. Meanwhile, over the years, as the incoming ships got larger and more complicated, and more and more of them arrived in Aliağa, no attempt was made to recalibrate "training." Be it a 120-foot car ferry or a 900-foot cruise ship, they entered inside and sliced it up the exact same way.

For more than two decades Çoban watched as the older generation of workers passed away from brutal diseases, while the shipyards that employed them raked in the prestige of receiving "certification" from European Union inspectors. They got rewarded with bigger ships, bigger hauls of steel, international credibility — and more and more money. And while it was true inspectors did arrive from Brussels once a year to ensure that certified yards were meeting European "standards" of labor rights and environmental diligence, it was such days, claimed Çoban, when "the makeup gets put on." The dismantling yard owners were notified in advance of the inspectors' arrival. Workers were given proper uniforms and face masks. Ships were ordered to be cut slowly and methodically. No hazardous waste was torched on the beachhead at night. And though in recent years more workers were getting killed at EU-approved yards in Turkey than at non-EU-approved yards, this fact got conveniently forgotten as the observers from Brussels seemed to relish the ultimate answer to the messy problem of how to dispose of the fleets that enriched and supplied their continent: a place looking out on Europe but not within it, where pay was decent enough, and the fatality rate still less alarming than in South Asia.

"Europe exploits Turkey," Çoban told me. "And Turkey exploits us."[10]

28

GREEKS BEARING GIFTS

> Go forward, sons of Greece.
> — Aeschylus, *The Persians*, 472 BC [1]

TO GET SOME idea of the future of all of this, the future of a shipping industry powered by fuels that it must soon cease using, an industry that pulls its profits off transporting hydrocarbons and cheap plastic goods that must be phased out of existence sooner rather than later, one need not look very far.

One need only look out across the Aegean from the fuming shore of Aliağa, fifty miles or so, to the Greek island of Chios.

If Aliağa in Turkey is where an increasing number of Europe's ships go to die, just off its coastline is where some of the continent's richest and most fabled shipping dynasties first got their start. For centuries now, Chios has been a virtual incubator of shipping know-how and capital. It is a dry place, ringed by rocky shorelines and layered with poor soil, and the only way to make money there has traditionally been to board a ship and leave. It's been argued that Christopher Columbus had family roots on the island, a Genoese colony in the fifteenth century; crews of Chiot sailors were in high demand by the Ottomans who occupied the island after the Genoese and awarded it a special tax status in return for its cultivation of mastic resin, a delicacy worth its weight in gold under the sultans. Later, in the nineteenth and early twentieth centuries, it was the great maritime clans of Chios — legendary names like Andreadis, Lemos, Livanos — who launched fleets of merchant ships around the Levant and the Mediterranean, intermarried, merged fortunes, relocated to Piraeus and London and New York, and proved crucial in streamlining and expanding the petrochemical and bulk cargo trades in the postwar era. Today, Greeks control 20 percent of the global shipping fleet, and Chiots 40 percent of the Greek fleet — meaning that nearly one in ten ships in the world is currently owned by a family hailing from an Aegean island not much larger than Guam.

The Chiots of today bear another distinction. They — and Greece's shipping elite more broadly — rank among the world's most unethical ship dismantlers, a distinction that has earned them the accolade "world's worst dumpers" from NGO Shipbreaking Platform, the Brussels-based coalition that monitors the fate of the world's ships. "The Greeks have always been our biggest problem," Ingvild Jenssen, the NGO's founder, told me.[2]

In the bustling, boisterous port city of Piraeus, the bureaucracy of the storied Greek shipping industry can be found almost everywhere you look. The offices of shipping agents abut brokerages that abut legal firms specializing in international maritime law; the local headquarters of the China Ocean Shipping Company sits opposite the consulate of Panama, itself next to a crummy old shop selling erstwhile tugboat lanterns and cast-off Hellenic Coast Guard gear. On the fourth floor of an unprepossessing building overlooking the naval harbor of Pericles's Athenians, I met a fourth-generation Chiot shipowner whose family first made its fortune transporting timber around the Mediterranean. More than a hundred years later, out of a corner office decorated with glass-encased models of ancient triremes, he presides over seven bulk vessels, which mostly shift dry cargo — grain, sugar, scrap — back and forth across the Atlantic Ocean.

"Greeks mastered the shipping industry because they turned

conventional logic on its head," the shipowner, who has a clean-cut, patrician face and a tidy head of black hair, explained.

Shipping is a boggling business. From an economic perspective, the magnates of Piraeus do almost everything you probably shouldn't do. When profits are high, Greeks often sell off their ships, sometimes to the subcontinent for dismantling. When income threatens to run dry, they snap up new fleets in a flash. Greeks specialize in no single type of vessel. They master no particular form of trade. Their carriers and tankers and bulkers ply all oceans and can be found in every port of the world. This makes Greek shipping companies hard to list on publicly traded stock exchanges because, to a board of directors or a mass of shareholders, much of their strategy is logically indefensible; there is none.

Or so it seems. The Greek shipping industry has proved so resilient precisely because it eschews corporate rigidity. Greeks don't succeed because they have the most capital or the most ships or the most seats at the International Maritime Organization in London. They have a Slinky-like ability to expand and contract their fleets in narrow windows of time, a flexibility in which they are forever seesawing between possessing huge armadas and amassing huge piles of cash, and which affords them the ability to weather geopolitical crises and emerge ever more advantageously positioned on the other end.

The 1940s? Greeks lost many of their vessels to German torpedoes and confiscation, only to acquire new fleets on the cheap courtesy of the United States, which entrusted the emerging world of postwar commerce to the wise men of Piraeus. The 1950s? Greeks and their vessels became instrumental in allowing boatloads of US petroleum to overtake stocks of German coal as the dominant energy lifeline of Western Europe. The closure of the Suez Canal from 1967 to 1975? Greeks like Aristotle Onassis and Stavros Niarchos — who successively married the same heiress of the same Chiot shipping dynasty — endeavored to build new tankers, the biggest commercial ships ever constructed, chartered them out to oil companies, directed them around the Cape of

Good Hope, then used the revenue on those charters to fill out whole armadas. Cold War embargoes on states like Cuba or Lebanon or Rhodesia? The Greek financial collapse? Trade wars between the United States and China? The Greeks have treated all these events as opportunities with which to maximize their versatility and indispensability, expanding their fleets at moments of crisis and trimming them down at times of boon.

And climate change? I asked the Chiot. What about climate change?

He paused. It's another challenge, he said. Years ago he stopped buying tanker ships that transport hydrocarbons. Other Greek shipowners didn't.

"They must know something I don't."

The climate emergency, unlike geopolitical crises, is fundamentally trickier. A warming planet upends shipping lanes, threatens ports, breeds deadlier storms. But the magnate from Chios suspects that shipping will learn to maneuver its way around the challenges. Fresh maritime lanes facilitated by the melting of Arctic ice are, paradoxically, reducing travel distances and carbon outputs. New vessels are being engineered that can harness the power of the wind. And swelling freight rates can be passed off onto consumers. In three years, from 2020 to 2022, during a once-in-a-century global pandemic replete with clogged ports and free-falling oil prices, the shipping industry generated as much profit as it had in the previous six decades *combined*.[3] It's hard to imagine climate change — at least in the immediate term — posing much of a threat to shipping.

"I'm the fourth generation in my family to own ships," the man added as I prepared to leave his office, a note of concern registering in his voice for the first time during our conversation. "I want there to be a fifth."[4]

29

COMING HOME

Fun For All. All For Fun.
— Carnival Cruise Line company slogan

BACK IN ZARA, before saying goodbye to the Taşkıns, I asked if I might be able to visit Oğuz's grave. His brother Enes agreed to take me. We drove north along the gravelly bank of the Kızılırmak River, a short distance from the family apartment, toward a steep embankment of red cliffs. The cemetery itself sat a few minutes' walk from the road, in a small fenced-off enclosure that featured a dozen or so graves belonging to generations of Taşkıns stretching back more than a century. Plane trees rose out of a hillock; the babbling of the Kızılırmak rattled off their trunks. Each plot featured a small white marble gravestone, simple but dignified, though the grave of Oğuz bore no such slab. A wooden stick nestled into a patch of dirt marked the location of his body. Around it were slung his black prayer beads, which he had taken to Aliağa with him three years earlier, and which were said to have been in his pocket when he died. Atop the dirt, three lilies were just beginning to bloom.

PART FOUR

Pacific Plastic

30

A LONG JOURNEY

The Javanese is naturally a husbandman. The soil on which he is born, which promises much for little work, lures him to this, and, above all, he is devoted heart and soul to the cultivation of his rice fields.... But strangers came from the West, who made themselves lords of his land.

— Multatuli, *Max Havelaar*, 1860

TRAINS IN INDONESIA possess not just routes — Bandung to Yogyakarta, Palembang to Bandar Lampung — but names. The Sembrani, which runs from the capital of Jakarta in the west of the island of Java to the port city of Surabaya in the east, takes the better part of a day. You enter one of its nine rickety carriages and spend the next thirteen hours slowly rolling along a windy coastline that fumes with the occasional brush fire and bristles with the odd minaret. On one side of the Sembrani, the Java Sea dips in and out of view; on the other, a horizon of volcanoes bulks across the landscape, spiny scales rising out of a dark-green stegosaurus's back. Buzzing around the train, a torrent of Japanese motorcycles — steered by long-bearded farmers, Muslim girls in hijabs, teenage boys out for a ride — flows from one market town to the next, separated by sprawling rice fields presided over by women in conical lampshade hats sitting in wooden huts atop bamboo stilts.

When you start your journey, the sun rises in front of the train; by the time you end it, it's nearly midnight, and the sun — and the landscape it has flushed with its unsparing tropic light — has slunk from sight. All the while the Sembrani crawls unhurriedly across the northern rim of the world's most populous island. A city called Cirebon gives way to one called Tegal, which gives way to Batang, and later Semarang. The arrival of a noodle cart in your carriage every few hours is the little that exists to break up the monotony of the trip — a journey that, in my case, took a good deal longer than the plane ride that had brought me to Indonesia in the first place.

I had flown to Indonesia to witness the lunatic phenomenon of the "trash towns": villages nestled deep within the volcanic highlands of Java drowning in flows of European and American plastic waste. Photos on the internet had displayed malign hellscapes, confetti mounds composed of shredded dull-gray plastic sprouting incongruously out of fields of the most luxuriant green vegetation you've ever seen. Somehow, of all possible places on Earth, these towns — shrouded by thick jungles, miles from any sea — had become the recipients of millions of pounds of garbage from countries whose residents probably couldn't locate the island of Java on a map. Trash layered their streets in waves. It sat in piles that rose higher than houses. Children played in it.

How, I kept wondering from afar, could such places exist?

For about a generation now we have normalized ourselves to the our real logic of the cheap consumer goods economy. We have accustomed ourselves to the idea that a handheld pocket calculator can be manufactured in a factory in Southern China, shipped to the United States, placed on a shelf in a chain store in Nebraska, then sold for a dollar. A dollar! We have accustomed ourselves to the idea that there is a place called the dollar store where such objects apparently exist to be sold.

But *garbage*?

Garbage was not so self-evident. A country like the United States, possessing almost limitless expanses of land, had shipped the detritus of its citizens' consumption — cargo containers packed with yogurt lids and old Pepsi bottles and empty bags of Doritos — nine thousand miles away to a crowded archipelago of seventeen-thousand-odd islands whose rapidly expanding population, the fourth largest in the world, was producing unmanageable quantities of trash of its own. The narrow margins of profit had, one suspects, already been rinsed out of these single-use products. There was no longer any yogurt in the container or Pepsi in the bottle or Doritos inside the bag; they were just pieces of plastic that, had they not been deposited in a recycling bin, may very well have been tossed out of a car window.

Shipping plastic waste, I told myself as the Sembrani train lumbered eastward across Java, was not like shipping sewage sludge from a rich northern country to a poor southern one. Though useless, such toxic residue can nevertheless be greatly profitable to waste traders insofar as the cost of its disposal in one place versus that in another can present a considerable, even enormous, margin. Nor was shipping plastic very much like sending a broken DVD player to West Africa. Electronic waste contains material — rare earth metals and minerals — requiring significant outputs of carbon to extract in the first place, and boasting a lifespan not of minutes but of years, and is in turn reusable: Much of the material stripped out of old electronics can be reprocessed and slotted back into new electronics. And the plastic waste trade was nothing like the scrap metal trade. Junk steel has become valuable enough for countries to curtail its export. Stockpiles of highway fenders and crumpled cars are now the objects of national hoarding. In some countries, scrap steel traders are being jailed.

Shipping discarded plastic made less sense. The transportation costs, the canal and port fees, the import duties, the rent for warehouses, the fuel for trucks and the wages for crews and drivers and the tolls for highways and the innumerable other middlemen who exact their cut of profit at each step of that plastic's transit to make its journey worthwhile — presumably none of these proved great enough economic or logistical barriers to the idea of shipping cargo containers stuffed with waste across the Pacific Ocean. And, by all appearances, the garbage was just *sitting* there in Indonesia. The timescales alone were dizzying. A bag of Doritos that may have taken half an hour to consume had become the object of a months-long journey from one side of the world to the other, where it would then take tens of thousands of years to disintegrate alongside so many millions of other bags of Doritos or Fritos or Ruffles. Someone's afternoon snack in the US had become an almost geological event in Indonesia.

You might recall that in 1989 a group of businessmen from Washington State had hatched a strange scheme to send the West Coast's garbage across the Pacific Ocean to the Marshall Islands in exchange for tens of millions of dollars. Even then, at the height of the hazardous waste trade, the plan was deemed preposterous — and rejected by President Amata Kabua, desperate though he was for the money such an arrangement could bring in. And yet somehow, a generation later, after the passage of an international hazardous waste ban and decades of intense scrutiny of the waste trade and its perpetuators in the petrochemical industry, that nightmare — that dystopia of a developing archipelago in the Pacific Ocean piled with the detritus of a huge, rich, far-off land — hadn't just happened. Western countries and their exporters had somehow gotten importers in a country like Indonesia to *pay for* their trash.

Who had first ventured that shipping empty bags of potato chips could be a viable business? Who was profiting here? What were these "trash towns" that had emerged as receptacles for consumers who had little understanding of, and apparently even less concern for, where their garbage was going? And how could it be that to make so many minuscule pieces of carbon go away, tremendous quantities of additional carbon were being released into the atmosphere through the act of ferrying them across the planet?

I had flown to Indonesia — and boarded the excruciating thirteen-hour Sembrani train from the manic megalopolis of Jakarta in the west to the chaotic port city of Surabaya in the east — to attempt to find out.

31

PLASTIFICATION

Tomorrow, more than ever, our life will be "disposable."
— *Sales Management* marketing magazine, 1959

BEFORE ATTEMPTING TO understand the trashing of Indonesia, it was worth trying to grasp something more fundamental. This was plastic itself. It didn't always exist. So where exactly did it come from? Why is there so much of it? And how foreseeable were the problems of its disposal?

In 1942, an American chemical economist and historian named Williams Haynes ventured an outlandish prediction. In the uncertain early months of the Second World War, just weeks after the Japanese attack on Pearl Harbor, Haynes insisted that a scarcely known substance — something referred to as "plastic" for its ability to shape-shift into all types of sizes and forms — would have "more effect on the lives of our great-grandchildren than Hitler or Mussolini."[1]

Plastic today may be ubiquitous. And in the ongoing green transition, it has never been more indispensable to the fossil fuel economy: As fewer cars require gasoline, as fewer ships are powered by heavy diesel fuels, as less electricity is supplied by natural gas, consumption — and the production rhythms of plastic it intensifies — increasingly represents the future of petrochemical profit. It has been estimated that plastics, currently driving 12 percent of global oil demand, will be responsible by 2030 for one-third of it. By 2050, they will constitute half.[2]

And yet the origins of all of this—a six-hundred-billion-dollar product that a great chunk of humanity handles on an hourly basis—were humble and, indeed, almost accidental. Synthetics like plastic started out as a marginal subset of the fuel industry. They were only first mass-produced in the 1940s, during desperate wartime attempts to convert the chemical byproducts of coal and petroleum into something of usable value on the battlefield.

Plastic, in other words, doesn't just turn into waste. In many respects it starts out as waste.

True enough, the general public's introduction to plastic predated the 1940s and Williams Haynes's brazen prediction. A showcase of "colorful synthetics" had been unveiled at the London International Exhibition in 1862, winning its patentee, a Birmingham metallurgist named Alexander Parkes, a bronze medal for "excellence of product," while later, in 1878, a pamphlet published by the Celluloid Manufacturing Company boasted that "it will no longer be necessary to ransack the earth in pursuit of substances which are constantly growing scarcer."[3]

It nevertheless took decades for mass manufacture to begin. The turning point was the Second World War. As the political scientist Adam Hanieh has argued, in the 1930s parallel chemical revolutions were taking place on both sides of the Atlantic to find synthetic substitutes for the natural resources required to wage any future conflict. In Europe, it was the Germans who managed the greatest success. Lacking access to the colonies that kept other European powers supplied with raw materials, Berlin ordered the chemical arm of I. G. Farben—successfully "Nazifying" its board in 1937—to source cotton and rubber substitutes out of Germany's own ample supplies of coal. Through a process known as "cracking," German laboratories were able to convert coal-based waste into a range of synthetic substances — polyvinyl chloride, nylon, Teflon — with which troops could be outfitted and armed.

The situation in the United States was not altogether different, though across the Atlantic it was petroleum, not coal, that would serve as feedstock for the host of emerging materials. Cut off from the rubber plantations of Ceylon, India, and Malaysia in the wake of the Japanese attack on Pearl Harbor, and already facing a shortage of raw materials ranging from metal to wood to cotton — crippling enough that by 1942 the first national speed limit was introduced in the United States to preserve automobile tires — Washington handed a billion dollars to the private sector in 1943 to begin pumping synthetics out en masse, an endeavor one historian has called the "chemical equivalent of the Manhattan project."[4]

Synthetics like TNT and asphalt were soon a significant element of the US war machine. So was plastic. In May 1944, board members of the Society of the Plastics Industry were brought to Washington to help outfit the troops in preparation for D-Day.[5] By July, US soldiers were marching across France to the tune of bugles made of plastic, arranging their hair with plastic combs, sleeping in tents made of plastic canvas, and flying airplanes that had been dispatched across the Atlantic packaged in plastic Saran Wrap to shield them from salt spray. That August, after the liberation of Paris, the US military sketched out plans to seize the Germans' industrial blueprints, an initiative closely coordinated by US oil executives, who oversaw the stealth removal of five million pages of technical documents out of Third Reich laboratories and the dissolution of I. G. Farben — the biggest potential postwar competitor to American plastic manufacturers — into three less formidable entities.[6]

Over four years of conflict, US synthetic production quadrupled.[7] It was ultimately the outbreak of peace that posed the greatest challenge to the country's erupting petrochemical sector. How to sustain such outputs now that the war was over? The production rhythms of plastic had been scaled to wage the largest conflict in human history. What now?

No commercial petrochemical industry had existed in the United States prior to 1945. The explosion of plastic in the second half of the twentieth century is the story of how an industrial byproduct catapulted into the exigencies of conflict was redeployed to the home front and retrofitted into a material of everyday use. For plastic's wartime producers, here was nothing if not an opportunity to re-landscape an entire society, to turn all that had only ever been encountered within the realm of nature into a synthetic facsimile through a chemical process that may well have been alchemy to the general public. "Virtually nothing was made of plastic," one executive would recall, speculating in the months after the Japanese surrender, "and anything could be."[8]

As early as 1943, the US chemical giant DuPont had been preparing for a synthetic overhaul of American life with a corporate division devoted to developing plastic prototypes of existing household products.[9] After 1945, the process began in earnest, across the sector. US oil companies began restructuring themselves into "vertically integrated corporations" that dealt not just in hydrocarbons but in their derivative byproducts too. Huge new plants — behemoth contraptions of vats and pipes that increased tenfold in size on average over the next two decades — went up in Texas and Louisiana. "There is good news in the work of American laboratories," ran advertisements for new rotary phones and steering wheels made of Celanese plastic. "It is...a promise for the future."[10] Or as the president of the Society of the Plastics Industry boasted in April 1946 at the first annual National Plastics Exposition in New York, which welcomed twenty thousand spectators on its opening morning: "Nothing can stop plastic."[11]

It wouldn't take long for the synthetic materials that had helped liberate Europe to be found all over the United States. By the 1950s, they had drastically infiltrated American society. Cotton and wool textiles were giving way to nylon fibers. Soaps were being replaced by detergents. Natural fertilizers were being substituted with chemical ones, pumice stones with microbeads, silk fishing lines with fluorocarbon ones. "The artificial has become so commonplace that the natural begins to seem contrived," observed the historian Daniel Boorstin.[12]

Nothing matched the inroads of plastic. If few Americans had heard of plastic in the 1930s, less than two decades later it would be difficult for them to go a day without purchasing some form of it. By the mid-1950s, Americans were picking up dry-cleaned shirts packaged in clear polyethylene bags, purchasing vegetables sealed in polystyrene trays wrapped in celluloid, and clothing their newborns in disposable diapers. A material that had helped ensure the victory of democracy over fascism offered an even more elusive promise: By democratizing consumption, plastics would create a world of mobility and freedom its benefactors liked to pit favorably against the regimented outputs of the new geopolitical threat of Soviet collectivism. "You will have a greater chance to be yourself than any people in the history of civilization," *House Beautiful* magazine informed its readers in a 1953 issue devoted to the benefits of plastic.[13]

Postwar plastic production within the United States was closely tied to the emergence of a petroleum-based American empire abroad. It's a crucial subplot within the story of postwar US hegemony, one in which US dominion over greater and greater chunks of the world worked in lockstep with its plasticization of them. The rivaling methods of synthetic production that had preceded and marked the Second World War — that of US petroleum versus German coal — was a competition that would be won only a decade after Berlin's capitulation with Marshall Plan funding, a tenth of which went to weaning Europeans off coal by boosting their petroleum sectors. By 1955, Western Europe was refining five times more petroleum than it had at war's end, the bulk of it imported from the United States. It marked an important shift, one in which the United States increasingly held the spigot to European energy flows, and those energy flows in turn generated the material basis of society itself.

Even before the postwar decolonization movement had begun, one advantage of synthetics was inarguable: They could reduce dependence on overseas territories by chemically replicating that which otherwise required exploitative systems of labor and transportation to harvest and extract. By the 1960s, whole manufacturing processes could be started and completed in petrochemical plants at a pittance of the cost of traditional methods of production. As with the history of the hazardous waste trade, the problem would ultimately be where to put those synthetics once they had been recklessly overproduced—and in this, in its eventual conversion of the Global South into a repository for oversupply, the US-led synthetics revolution was to function as both a continuation and an inversion of earlier forms of colonialism. The lands that had once offered wood, latex, and cotton to European capitals were to be flipped into dumping grounds for the synthetic proxies of those same materials—the plastic, rubber, and nylon that were to prove the material hallmarks of a US-led world.

Plastic did not just provide a new material basis for postwar life. It offered up a superabundance of it. Spices, gold, fur pelts—such materials had driven centuries of explorers across oceans. But from the outset plastic was something else. A market did not exist; it had to be invented and imposed. The conundrum that petrochemical executives faced in the postwar era was the opposite of traditional problems of supply and demand. Because the production of plastic was derivative of another process—*the production of energy itself*—its supply was not correspondent to conventional fluctuations in demand. So long as Americans drove their cars, heated their homes, and kept their food cold, the chemical building blocks of plastic were getting continuously generated regardless. America's increasing reliance on petroleum in the second half of the twentieth century was closely trailed by its saturation in plastic. And, already by the mid-1950s, one thing was apparent. There were not enough uses for it all.

One industry solution? Make up new uses. Plastic would be enlisted "not so much to serve social needs as to invent them," claimed one industry insider.[14] "Rather than manufacturing known products by a known method for a known market," recommended a handbook published by the Hooker Chemical Company, "the research department is free to develop any product that looks promising. If there is not a market for it, the sales department group seeks to create one."[15]

The other solution to plastic's overabundance? Accelerating those patterns of consumption. "We must cut down the time lag in expanding consumption to absorb this production," the head of J. Walter Thompson, a New York advertising agency, explained to his rank and file.[16]

What if, instead of selling one plastic fork, you could sell... two? By the late 1950s, a new world beckoned: that of single-use plastic, a product that was nothing less than garbage upon purchase. "The future of plastics is in the trash can," Lloyd Stouffer, editor of *Modern Packaging*, claimed at a Society of the Plastics Industry meeting in New York City in 1956. "It [is] time for the plastics industry to stop thinking about 'reuse' packages and concentrate on single use. For the package that is used once and thrown away, like a tin can or a paper carton, represents not a one-shot market for a few thousand units, but an everyday recurring market measured by the billions of units."[17] Yet one more advantage of plastic presented itself. The ability not merely to replace wood and glass products but to replace them *multiple* times throughout the course of any single day at what advertising executives liked to call "toss-away prices."

Plastic was widely hailed as a miracle, the generator of a "flood of new products...transforming...the American way of life," or so marveled *Time* magazine.[18] It was no less than "ubiquity made visible," concluded French philosopher Roland Barthes after his visit to a synthetics exhibition in Paris in 1957. Plastic, for Barthes, presented "the first magical material that consents to be prosaic," the end-all conquest of chemical concoctions over materials to be found in the natural world. Plastic was simultaneously everything yet nothing, ephemeral yet eternal, disposable yet indispensable, flimsy yet indestructible. "The hierarchy of substances is abolished," Barthes concluded. "A single one replaces them all: the whole world *can* be plasticized."[19]

The balance sheets soon spoke for themselves. In the 1950s, the annual growth rate of the petrochemical industry was doubling that of the roaring US economy.[20] By 1960, plastic production had surpassed that of aluminum. By 1972, 22.4 billion tons of synthetics had been pumped out, leading the industry's in-house trade journal, *Modern Plastics*, to boast that by decade's end, plastic might even outpace steel as the most ubiquitous material on Earth, ushering in the "beginning of the synthetic age."[21] The prediction proved incorrect. Plastic production overtook steel in 1979, a year earlier than its own manufacturers projected.[22]

At the same time, critiques were mounting. Was any of this good? Natural? Necessary? We "divorced ourselves from the materials of the earth, the rock, the wood, the iron ore," claimed novelist Norman Mailer, who, writing in the wake of the Cuban Missile Crisis, found that the idea of a *potential* nuclear apocalypse occluded the *actual* disaster unfolding in plain sight: Humanity had relinquished its freedom to a "creeping plasticism" that laid bare bourgeois emptiness. "We looked to new materials which were cooked in vats, long complex derivatives of urine which we called plastic," lamented Mailer.[23] Several years later, in *The Greening of America*, an analysis of 1960s counterculture, Charles Reich insisted, "Our life activities have become plastic, vicarious, and false to our genuine needs, activities fabricated by others and forced upon us." The corporate American state had weaned its citizenry off natural pleasures, claimed Reich, and turned them into addicts of the deadening convenience of the plastic straw.[24]

More cutting was the analysis of an American biologist named Barry Commoner. In *The Closing Circle: Nature, Man, and Technology*, a 1971 account of postwar American production, Commoner singled out plastic as a scourge masquerading as a solution. True, the public had finally begun to "contemplate the fate of the billions of pounds of plastic already produced" and the incomprehensible quantities of waste it all generated. But for Commoner, this was the wrong way to understand the phenomenon. For the petrochemical industry, plastic's fate as waste was hardly a problem at all: It was precisely how the profit model had been designed. The very point of producing plastic, argued *The Closing Circle*, was to mass-manufacture a material so uniform, regimented, and worthless that you didn't think twice about throwing it away. "If you asked a craftsman to make you a special pair of candlesticks he would be delighted; if you asked for two million pairs he would be appalled," wrote Commoner. "Yet if you asked a plastics molder for one pair of candlesticks he would be appalled, but delighted if you asked for two million pairs."[25]

The deeper problem with plastic, according to Commoner, was not just that it happened to destroy the environment; so did coal. The true threat of plastic was that it was an ecological blight of an unprecedented, even alien, caliber, the material equivalent of introducing an unknown disease into a secluded gene pool. Liberal environmentalists of the late 1980s would aspire to turn waste into a utility, citing the way that trees absorbed the nutrients of their own falling leaves, and insist that with enough market incentive, our garbage could naturally cycle back into rhythms of production. Commoner struck down such ideas two decades before their arrival with his insistence that plastic could never naturally embed itself into patterns of production and disposal because it was inherently *unnatural*. Its advent marked an irreparable shift in humankind's relationship with the environment. By substituting wood and cotton and iron with a laboratory concoction, humanity had sundered its relationship with Earth's self-sustaining material rhythms. Plastic presented the world with the first "intrusion into an ecosystem of a substance wholly foreign to it," a tear in the "ecological fabric that has, for millions of years, sustained the planet's life."[26]

By 1970, the plastics industry had become alive to its mounting reputational crisis. "Let's look into the future and see what kinds of problems may exist in 1980 if plastics production continues to grow at its predicted rate in this country and nothing is done to achieve recyclability or disposability," Hugh Connolly, deputy director of the Bureau of Waste Management, challenged the Society of Plastics Engineers that year.[27]

What were those problems? The biggest one was space. Over the course of the 1950s and '60s, plastic came to crowd Western landfills. Accounting for 8 percent of total waste weight, it took up almost a third of landfill volume. The environmental legislation of the mid-1970s only deepened this problem by making landfills both prohibitively expensive and fewer in number, with the United States possessing half the number of landfills in 1976 than it had in 1956, and an additional two-thirds scheduled to disappear by 1986.[28] To minimize plastic's disposal footprint, the solution pushed by the petrochemical industry was to burn it. By the late 1980s, hundreds of trash incinerators were operational or under construction across the United States.[29]

At the same time, the industry began directing blame elsewhere. In a series of marketing campaigns every bit as calculated as those that had force-fed plastic onto the public in the first place, petrochemical and bottling companies shifted the burden of environmental responsibility from those who mass-manufactured synthetics to those who had been lured into purchasing them. "People, not containers, are responsible," claimed the National Soft Drink Association in 1967 with respect to the growing ecological crisis.[30] "The problem of garbage," argued Sidney Gross, editor of *Modern Plastics,* stemmed from "our civilization, our exploding population, our lifestyle, our technology."[31] "People start pollution," ran a 1971 television advertisement produced by Keep America Beautiful, an environmental nonprofit heavily funded by the soda industry. "People can stop it."[32]

The extent of the devastation was nevertheless becoming hard to hide. In 1966, marine biologists studying the Pearl and Hermes Atoll northwest of Hawaii discovered something that stunned them: synthetic particles lodged in the stomachs of newly hatched Laysan albatross chicks. Six years later, in 1972, scientists at Massachusetts's Woods Hole Oceanographic Institution uncovered something more remarkable still in the Long Island Sound off New York City: slivers of plastic drifting in the water's tidal currents, invisible to the naked eye. By the mid-1970s, tiny particles of plastic — what would be coined "microplastics" in 2004 by marine biologist Richard Thompson — were found in the stomachs of fish. In 1997, an American yachtsman named Charles Moore was sailing through the Northeast Pacific when he began noticing, every five minutes, over the course of seven days, a piece of plastic bobbing by his boat. Two years later, he returned to the area with an enormous net. "That was the real aha moment, when we pulled up that net," Moore told me. "I thought, *Wow, this is a much bigger problem than I could have ever imagined.*" For every pound of plankton Moore caught, there were six pounds of plastic. Three times the size of France, the Pacific garbage patch was determined to be one of at least five major oceanic gyres absorbing trash from the world's currents, a total mass of plastic that, Moore told me, citing research he has published in a series of academic journals, would require "at least seventy thousand years for humanity to remove from the seas."[33]

And so it continued. By 2006, so-called plastiglomerates — amalgamations of volcanic rock, seashells, and coral fused together into a molten goo by melted plastic — were being discovered along the shores of Hawaii. By 2009, plastic was found in samples of thawed-out ice from Antarctica. Four years later, it was found floating in outer space. Five years after that it was discovered at the most distant possible extreme, when a plastic bag was caught on camera floating 36,000 feet below sea level, in the lowest trough of the Mariana Trench, the deepest place anywhere on Earth. By 2019, microplastics were found near the summit of Mount Everest. Three years after that, they were observed to have infiltrated their creators, flowing through the blood of three in four humans, and leading researchers to speculate that the average person — the person reading this book, for example — likely consumes a credit card's worth of plastic every week.[34]

The problem with plastic was not just that it never truly went away, regardless of whether you incinerated or landfilled it. The problem was that the immediate benefits plastic had offered billions of consumers across the world had — since its mass rollout at trade fairs and expos in the 1940s — been coming at the belated expense of Earth itself, amounting to a kind of Faustian bargain humanity had made with its planetary host, trading short-term commercial convenience for long-term environmental cataclysm. Plastic had been marketed to the public as a material that would slip from sight as soon as you, the consumer, had decided you were done with it. The reality was that plastic was a ticking time bomb, a delayed ecological fuse in which all the plastic ever consumed was not disappearing but disintegrating into an infinite number of unfathomably tiny pieces and contaminants. Our realization of the true and terrifying extent of plastic's ubiquity came decades after we had *already* put it *everywhere*.

Many of these revelations were yet to come. But even by the late 1980s, the extent of the problem was stark enough. "The image of plastics is deteriorating at an alarming rate," Larry Thomas, president of the Society of the Plastics Industry, warned in an industry memo in 1989. "We are approaching a point of no return."[35]

32

THE GREATEST MIRACLE YET

> The plastics industry is waking up to the fact that consumers are very frustrated with packaging that goes straight from the grocery bag to their trash can.
> — John Ruston, analyst with the Environmental Defense Fund, 1989[1]

IN THE LATE 1980S, the petrochemical industry hatched a new solution to the deepening crisis of plastic's disposal. The answer was to tack on yet one more miracle to the ever-expanding list of plastic's mesmerizing qualities. Discarding plastic, contended its producers in a new marketing pivot, was easy. You didn't need to throw it into a landfill. You didn't need to burn it.

You just needed to recycle it.

Recycling per se is no myth. It is possible to turn an old issue of the *New York Times* into a new issue of the *New York Times*. It is possible to turn an old aluminum can of Dr Pepper into a new aluminum can of Dr Pepper. The copper extracted from electronic waste sent to Ghana and the steel sourced from ships dispatched to Turkey for dismantling do end up in new electronic products and new steel structures.

But the idea that the majority of *plastics* could ever be effectively recycled was to prove just another gamey stroke of industry advertising, a deliberate conflation of genuine examples of circular economies and the downgrading life cycle of plastic itself. And it's a particularly cunning one, because many of its purchasers have become accustomed to thinking that plastic rose to such prominence precisely because it can be recycled. The truth, though, is the opposite: For the first forty years of its mass production, few of plastic's producers ever mentioned "recyclability." Only when reputational disaster struck in the 1980s did they begin doing so. And then, as now, it has been a disingenuous claim.

For it has never been possible to efficiently convert the overwhelming majority of old plastic into new plastic. The process just doesn't work.

The most obvious difficulty pertains to the material. "Plastic" is a broad term for what is in fact thousands of different combinations of synthetic polymers that differ mainly in chemical complexion and quantity of additives. Wastepaper, cardboard, scrap steel — these materials can be gathered from every corner of the world and, broadly speaking, get pulped or melted down together. The same is not true of plastic. One recycling expert has likened plastics to cheeses: Much as it is unfeasible to melt mozzarella down and expect to produce parmesan, so with plastic is it impossible to shred and reduce polyethylene in order to get polystyrene or polypropylene or polyvinyl chloride, even as all these different types of plastic invariably get tossed into the same recycling bins.[2] It was a problem evident as early as 1969 when, in a collection of studies funded by Esso and Chevron, along with the American Petroleum Institute, the petrochemical industry lamented the fact that the miracle of plastic stemmed from its peculiar chemical makeup.

"[I]t is ironic that the very molecular structure that has made [plastic] so popular creates certain disposal problems," conceded Thomas Becnel, a Dow Chemical executive. Plastic didn't naturally break down in landfills. It couldn't be re-smelted. It just kept accumulating. "[T]he problem," Becnel concluded, "is merely moved from one place to another."[3]

Then there are the economics of recycling. There has scarcely ever been profit in it, namely because manufacturing *new* plastic has always been cheaper than attempting to resurrect *old* plastic. "It is always possible that scientists and engineers will learn to recycle or dispose of wastes at a profit, but that does not seem likely to happen soon on a broad basis," the American Chemistry Council claimed in a 1969 report entitled *Cleaning Our Environment*. Or so worried one executive in a speech delivered to the petrochemical industry in 1974, when the oil crisis had made the production of new plastic increasingly expensive: "There is serious doubt that [recycling plastic] can ever be made viable on an economic basis."[4]

There is a related problem here. Even if old plastic could be turned into new plastic at a profit, it is not a process that can be replicated countless times — as it can with, say, steel. After two or three uses, plastic eventually wears down beyond any ability to serve as future feedstock, meaning that recycling never prevents final disposal; it merely delays it. As the Vinyl Institute, a plastics lobby, conceded in 1986: "[R]ecycling cannot be considered a permanent solid waste solution, as it merely prolongs the time until an item is disposed of."[5]

The final reason why recycling offers no feasible solution to the problem of plastic's disposal? It is increasingly being revealed to be a poisoning process. Consumer plastic contains a variety of unregulated additives — flame retardants, plasticizers, stabilizers — that, were they to be discarded and exported to developing countries in steel drums, would be considered hazardous forms of waste.[6] Indeed, within most of the types of plastic you interact with on any given day, the International Pollutants Elimination Network has concluded that upwards of a hundred toxic contaminants are likely to be chemically ingrained.[7] And while you might be forgiven for thinking that the recycling process — washing and shredding those plastics and melting them down — would eliminate these poisons, recycling has the opposite effect: leaching those toxins out and diffusing them throughout newly created plastic, a process known as "migration." This is all in addition to any residual contaminants still stewing within the plastic containers you toss into recycling bins, and that never completely disappear. "You have soda bottles getting thrown in with bottles of fertilizers and Windex and all sorts of other products," Jan Dell, founder of The Last Beach Cleanup, told me at her house in Southern California. The Last Beach Cleanup is an NGO that aims to shift public awareness of plastic's problems into legislative action. "And these bottles aren't empty, right? There's still some stuff inside. And that stuff is all getting mixed together during the recycling process. The analogy I like to use is pasta sauce in Tupperware. Have you ever tried washing pasta sauce out of Tupperware? It never really comes out, does it? It kind of sticks to the side of the Tupperware no matter how many times you wash it. That's what's happening during the plastic recycling process. Nothing's ever really getting washed out. It's just getting mixed in, mixed in."[8]

The most important point might be this, however: Plastic recycling, even if it were to work, even if it were to be profitable, even if it were to be safe, would still never address the engine driving our global trash crisis. This is our unsustainable level of *production*. There now exists thirty years of evidence to demonstrate that countries claiming to "recycle" more *also* produce more plastic waste. Because there are a limited number of times that it can be resurrected, plastic invariably requires inputs of virgin resin during the manufacturing process, meaning that even the act of recycling plastic is never reducing waste but only guaranteeing more of it. One need only to consider the fact that since the "solution" of plastic recycling was presented to the public, net plastic waste outputs in the US have skyrocketed, up from 60 pounds per person per year in 1980 to 218 pounds per person in 2018.[9]

The petrochemical industry knows all of this. It has known it for over a generation now. But, all the same, recycling emerged as its answer to a trash pandemic of its own making.

By the late 1980s, plastic's manufacturers were spending $50 million a year to promote recycling. They lobbied nearly forty states to put the chasing-arrows symbol on their products. Founded in 1987, the

Institute of Scrap Recycling Industries became an umbrella organization beneath which the interests of the plastics industry could be merged with that of genuinely recyclable materials, such as paper and steel. Arguments were mustered. Plastic was said to make our societies safer because it was sanitary. It was said to be less burdensome than aluminum or wood and easier to transport. And its ecological benefits were assiduously preached.

"I believe you'd agree that environmental protection has become a core value for consumers," claimed Larry Thomas at a dinner for the Society of the Plastics Industry, at the Willard InterContinental Hotel in Washington, DC, in 1992. "Unfortunately, however, consumers do not yet understand how plastics fit with these values. Plastic packaging has become the solid waste scapegoat, and recycling is viewed as the cure-all," he continued. "One critical task is to inform consumers about the personal and environmental benefits of plastics. We call this task 'Outreach.'"[10]

To signal their willingness to help solve the problem they had created, oil and petrochemical companies such as Exxon, DuPont, and Union Carbide entered the recycling business themselves. In a chapter of the plastic recycling saga that's now been all but forgotten, the world's most powerful industry invested millions of dollars into building fourteen recycling facilities across the United States, with the American Plastics Council pledging that by 1995 one in four discarded bottles or containers would enter recycling streams and assuring the public that plastic would be the "most recycled material" in the world by the turn of the millennium.

By 2000, however, only one of those fourteen plants was still working. "Recycling requires a whole process of collecting, grinding, washing, drying, sorting, and pelletizing," rued a Union Carbide manager after the shuttering of its recycling facility in New Jersey in 1996. "After five years of being in the business, we were unable to produce cash flow." A year later, in 1997, the American Plastics Council was confronted with the awkward fact that after years of the petrochemical industry pouring millions into demonstrating that recycling could work, plastic recycling rates in the United States had actually *decreased* over the decade. "The plastics recycling markets didn't take off as we expected them to, and we don't see them getting better any time soon," a Phillips Petroleum representative conceded within a week of the closure of its Oklahoma recycling facility in 1998.[11]

The fact that oil and petrochemical companies had unceremoniously quit their own expensive forays into the recycling world ought to have been as clear a sign as any that in the eyes of its own producers, the circular economies that applied to steel and paper and cardboard could not be efficiently extended to plastic.

In many respects, however, this was the very point of the industry's aborted attempt to clean up the waste crisis of its own making: not to demonstrate that plastic could be effectively recycled, no, but to make a token show of acknowledging that plastic was a problem they were theoretically committed to help solving. For it would turn out that at just the time when it was funding recycling plants, the petrochemical industry was spending *even more money* dismantling legislation aimed at reining in production volumes.

As the petrochemical industry saw it, promoting recycling was a means to an end: making more plastic. "No doubt about it, legislation [restricting future plastic] is the single most important reason why we are looking at recycling," Wayne Pearson, executive director of the Plastics Recycling Foundation, a group whose advocates included Coca-Cola and Pepsi, explained in 1988.[12] "The basic issue is economics," a company director in the petrochemical industry said at a vinyl industry meeting. "Attempt to preempt or influence recycling legislation so it does not disrupt market economics."[13] Or as William Carteaux, president of the Society of the Plastics Industry, insisted: "Legislation and regulation threaten to fundamentally change our business model. We can't continue to fight back just at the reactive stage when things are emotionally charged. We have to take the offensive."[14]

Within just a few years, the results of that offensive were clear.

By 2000, the American Chemistry Council could boast that 195 "onerous bills" had been defeated, 17 laws unfriendly to the plastics industry had been amended or repealed, and, in one state after another, potential bans on plastic bags had been dropped.[15] Plastic was not the "most recycled material" in the world, no, but this didn't slow its conquest of supermarket aisles and refrigerator shelves, to the detriment of genuinely recyclable materials, such as refillable glass bottles, whose use drastically dwindled throughout the 1990s.

33

ONE-MAN MULTINATIONAL

> There's a great future in plastics. Think about it.
> — *The Graduate*, 1967

IN THE BACKGROUND of these developments — a general public that was growing more and more skeptical of plastic's environmental footprint, futile industry forays into the recycling sector that were bleeding millions of dollars — came an extraordinary twist.

In November 1989, the Berlin Wall fell. Huge swathes of the globe that had limited experience of plastic were soon rendered wide-open to its intrusion.

It proved to be a seismic geopolitical moment, yes, one in which the collapse of a rival economic system, Soviet Communism, would allow the most capitalist invention conceivable — single-use plastic packaging used to wrap plastic consumer products — to infiltrate great additional parts of Earth, sometimes new, sometimes used. But it's also a story about individuals, a handful of wily entrepreneurs who entered the post–Cold War world bearing an unusual, almost uncanny set of personal connections.

I had arranged to meet Steve Wong one morning in the hills of eastern Los Angeles County, in his well-to-do city of Diamond Bar, at a concrete strip mall that featured a couple of bao bun joints and a nail salon. Ten minutes late, he emerged from a vanilla Mercedes and hustled over to me exuding apologies. He was a trim man, younger-looking than his sixty-six years, dressed in a black tracksuit that he'd thrown on for that morning's neighborhood soccer league scrimmage.

Wong, almost by way of introduction, informed me that he was no longer welcome in China. Sure, for decades he had been one of Beijing's more consistent suppliers of plastic waste, one of the circuits in the globalized chain that in the 1990s aspired to convert plastic forks discarded in California into new plastic laundry baskets in China. "You can recycle anything if you put your mind to it," Wong told me. "This isn't rocket science."[1] He started pointing to objects on our table — napkins, his coffee cup lid, then the table itself — as if resigning them to the recycling bin by dint of his index finger.

Wong, it turned out, was no mere circuit in the global plastics trade. He was a veritable node of globalization unto himself. In the halcyon days of Western waste flows to China in the mid-2000s, he claims to have been exporting 390,000 tons of plastic every year, or the equivalent of 1,600 cargo containers every *month;* by 2010, he was single-handedly responsible for almost 7 percent of all discarded plastic entering China. Were he a country, Wong would have ranked among the biggest plastic exporters in the world: a Holland, maybe, or a Canada. Better yet, were Wong a beverage company, one that handled 70 million pounds of plastic every month, it's not improbable that he might be listed somewhere on the New York Stock Exchange.

Instead, he's just Steve Wong, a man who plays soccer on weekday mornings in the hills of Southern California and — in the bizarre geo-economics of trash — functions essentially as a one-man multinational.

His was as political a story as that of the globalization of garbage. In the 1940s, Wong's grandfather Sheng Yun Wong was a wealthy landowner in the Chinese port city of Shantou; in addition to a portfolio of farms and trading houses, he founded the construction company responsible for the building of the port's first five-story structure.

"We were a prominent family," Wong, who was born in Shantou in 1957, told me. The Maoists threatened the prosperity of the Wongs with their takeover of the Chinese state in 1949. "It was a time when you could not trust your own children," claimed Wong. "The Communists would take them in, interrogate them — 'How did you make so much money? Where were you keeping it?' — and then use that information against you."

In 1960, Steve Wong's grandfather dug up the gold and antiques and rifles he'd hidden from the Maoists beneath his fields and shepherded his seventeen children — along with his three-year-old grandson, Steve — south to Hong Kong, where he began lending money out to small businesses in exchange for a one percent cut of their monthly profits. Ever wary of Communist encroachment on Hong Kong, Sheng Yun kept rounds of ammunition tucked away in the ceiling boards of his new office, a stone's throw from Victoria Harbour.

"My grandfather was right to be paranoid," Wong told me. "Look at what's happened to Hong Kong today."

Among the businesses Sheng Yun funded was an upstart recycling facility that paid for truckloads of Hong Kong's discarded plastic, sorted and cleaned and shredded it, then sold it to a man by the name of Li Ka-shing, an entrepreneur who specialized in infusing old plastic with eccentric colors to create bouquets of fake flowers — an endeavor that, by the 1980s, had turned Li Ka-shing into a billionaire and, by the 1990s, into Hong Kong's richest man.

Long before all that, however, and to ensure that his loans were being put to careful use at that Hong Kong recycling center in the 1960s, Steve Wong's grandfather lobbied to install Steve Wong's father as its warehouse manager; Wong himself, at age eight, was assigned to the warehouse floor, sorting old plastics after school. "Other kids had sports," he said. "I had plastic."

Wong told me that his life took a turn with the oil crisis of 1973. He recalls sitting on the floor of the sorting yard and overhearing two trash collectors arriving at his father's recycling facility and explaining how they would now be charging double the price for discarded plastic. It was, for Steve, now sixteen, a revelation. Nothing had changed with the plastic itself. It was still the same material. And yet owing to events occurring thousands of miles away, events that few people in Hong Kong could very adequately explain, old soy sauce containers and rice packaging had magically doubled in value overnight. "The traders set the price and made the rules," Wong told me. "And we in the warehouse couldn't do anything about it. It was then that I realized that there was no power in being an importer of plastic. And there was no power in being an exporter of plastic. There was power in being a middleman, a trader, someone who could work both sides of this system."

That day, Wong claims, he became determined to get himself out of Hong Kong. To improve his English, he arranged through a British foreign council program to apply to boarding schools across the United Kingdom. One of them, the Bembridge School on the Isle of Wight, accepted him; in the late summer of 1974, following "the first family meeting I can ever remember," Steve Wong's mother agreed to sell off her jewelry to purchase him a plane ticket to London. Within a semester he had transferred from Bembridge to a trade school in the city of Birmingham, had begun paying his rent by working nights as a Chinese food delivery driver for five pounds a night, and was filling out his Rolodex with contacts.

His first job? Selling wristwatches. For years, Wong essentially lived out of his car, driving up and down the United Kingdom — and later across Western Europe and the United States — pitching Lighton timepieces to retail stores. But his mind was always on plastic; every place Wong visited, he trawled the yellow pages for any potential producer or consumer of plastic — recycling plants, bottlers, factories, farms — that he could purchase in cash, then sell back to his father's contacts in Hong Kong. Nothing was too worthless. Wong scooped up overstock vinyl lawn chairs, factory-reject polyurethane carpet padding, torn industrial fishing nets. "This was before GPS," he told me. "And that's what made it easy. You would drive to a place, show up, and be the only person there — probably the only person who had come there in search of plastic."

And once he got to that recycling plant or bottler or factory or farm, Wong's expertise — a childhood spent sorting polyvinyl chloride from polyethylene terephthalate from polystyrene — did its own fast-talking. "Many people can identify the seven main types of plastic," Wong told me. "But there are really more than a hundred thousand different types of plastic. And I can identify several thousand of them at least — by feel, by weight, by flexibility. No matter who I met, it was clear that I knew what I was talking about. They trusted me. So they agreed to sell to me."

His expertise was one talking point. The other? Few others were buying the stuff in the 1980s. Few others *knew* to buy the stuff. Wong says his first purchase came in 1984: twenty tons of flexible PVC he bought off the Hunt Brothers of Leicester Waste Management for £3,200, which he then sold to importers in Hong Kong, who in turn used it to manufacture the trimmings on faux-designer purses. Just a decade later, Wong was arranging the purchase of that much plastic every few hours, with hundreds of companies, from Las Vegas to Leuven, relying on him as their waste handler; Wong in turn sold to "more than two thousand" importers across southeast China who, he admits, often operated on the border of legality, sometimes rigging customs imports, sometimes paying off inspectors.

By 2000, Wong told me he was making a little more than $10 million a year through the purchase and sale of Western plastic waste. What had begun as a reverse door-to-door sales job — driving from one town and country to the next, not selling things but *buying* them, even out of the bottom of a dumpster when necessary — had turned Wong into an indispensable one-man hinge within the revolving door of consumerism that shipped cheap goods east across the Pacific and dispatched plastic detritus back in the opposite direction. And what began as Sheng Yun Wong's loan to a Hong Kong recycling facility — the result of a rich businessman forced to flee the Communists in 1960 — had become Steve Wong's globe-spanning trash empire, culminating full circle in 2017 with the Chinese Communist Party's insistence that the frenzy of plastic waste importation finally come to an end.

To this day, Wong, a millionaire several times over, remains something of an economic bottom-feeder. I met him in Diamond Bar fresh off a flight from Atlanta, where he was looking to help fund a new center for "advanced recycling," a process that amounts to doing away with any attempt to convert old plastic into new plastic and instead melting it down into liquid fuel; within a few days he would head to Turkey — "I think Turkey can be our next China," he proclaimed, like an explorer who's just spotted an uncharted new continent — then Slovenia, where he'd located a new business outside the capital of Ljubljana that seemed to be "producing a lot of PVC they don't know what to do with."

In California's environmentalist circles, Wong had become a villain of almost legendary dimensions. Every so often in the months after I met him, a network of contacts would forward me advertisements that Wong had listed on eBay for old plastic of one type or another; one, for a consignment of "defective PVC air mattresses," included the hashtags #polyvinylchloride and #sell. Naturally enough, a man who drove a Mercedes through the chiseled suburbs of Los Angeles didn't need to be auctioning such things on eBay. But then Steve Wong became Steve Wong precisely by being the type of man who accumulates defective PVC plastic mattresses in order to hawk them across the internet.

34

PLASTIC CHINA

> The road we have long been traveling is deceptively easy, a smooth superhighway on which we progress with great speed, but at its end lies disaster.
> — Rachel Carson, *Silent Spring*, 1962

THOUGH MAYBE LESS eccentric than Steve Wong, dozens of similar traders emerged at the same time, middlemen who stepped into a peculiar economic no-man's-land in the 1990s: helping the US petrochemical industry divert the spiraling disposal crisis it had itself generated, while also helping an erupting Chinese manufacturing sector source feedstock for the production of flimsy plastic goods — shower sandals, phone covers, garbage bins — whose consumers tended to be that of China's own rising middle class.

From the start, however, many plastic traders weren't sending their material exclusively to China. Within months of the fall of the Berlin Wall, they had begun seizing the opportunity offered by Eastern Europe: a frontier of states desperate for cash and distracted by other problems. In countries like Bulgaria and Romania, where forms of plastic such as packaging and bottles scarcely existed under half a century of Communist rule, it would often make its first entry in the holds of dump trucks rumbling east out of wealthy nations like the Netherlands.

In 1992 alone, Poland intercepted 1,332 waste shipments from Western Europe; nearly 2,000 would get stopped the next year. "Very often it says [on the customs declaration] that it's a donation," claimed a customs officer that year in the city of Katowice, where Western medical and plastic waste would get dumped throughout the decade in lieu of paying exorbitant landfill fees in the other half of Europe.[1] By 1994, the US Chamber of Commerce had begun leading delegations of "environmental entrepreneurs" into the former Communist bloc to scout out possible destinations for discarded American plastic. Meanwhile, farther to the east, in Russia, a privatized derivative of the former Soviet military—a group calling itself the Chetek Corporation, which in the mayhem of the 1990s had secured a "fleet of private jets and limousines, offices in eight cities in several republics and a luxurious retreat outside Moscow"—would begin frequenting international trade expos offering to bring Western trash and toxins to the wildernesses of Russia and, for fees ranging from $300 to $1,200 per metric ton, use "peaceful nuclear explosives" ignited deep underground to vaporize them into "glass silicate."[2]

But by the mid-1990s, China had surpassed Eastern Europe as the greatest destination of Western refuse. And there was to prove an unmistakable difference between shipping plastic waste to China and trucking it to states like Poland: In China, most old plastic wasn't attempting to mask its identity. It wasn't getting smuggled past border guards. It wasn't getting covertly dumped along highways. A lot of it was getting used.

China, unlike Eastern Europe, had entered the post–Cold War era with a giant manufacturing sector not merely intact but supremely positioned to emerge as the world's factory. Labor was cheap. Regulations were slim. Energy costs were low. Infrastructure and shipping lanes advantageously connected it to North America. A strategic intercoupling of the Chinese and American economies had begun as early as the late 1970s, but by the 1990s a unique arrangement had fully locked the two countries together: China hoarded American debt, eventually sitting atop huge stockpiles of US treasuries, while Americans filled their homes with Chinese-manufactured goods.

And undergirding it all — the indispensable but often unacknowledged connective tissue of this mighty American-Chinese economic bond — was plastic, used and new, billions of pounds of it, getting shipped endlessly back and forth across the Pacific.

It's a common misconception that for thirty years half the world's trash went to China. The reality is weirder: Half the plastic deposited into recycling bins across the world went to China. In other words, it was all the plastic that had been diligently discarded, and whose consumers were convinced that doing so was helping the planet, that was traveling around the world — and often resulting in great ecological carnage.

The statistics tell the story best. In 1997, China's plastic waste imports amounted to $476 million. By 2004, the United States was sending four thousand cargo containers of plastic waste to the ports of Shenzhen and Guangzhou every day. By 2008, discarded plastic had surpassed airplane parts and electronics as the United States' biggest dollar value export to China, meaning, in other words, that while China exported a staggering quantity of manufactured goods to the United States, the United States' own biggest export to China was... the stuff Americans were throwing away. By 2012, annual plastic waste importation volumes in China had reached nearly $8 billion, a fifteen-fold increase over a decade and a half.[3]

From the United States, most of it left from California or Washington. But in a sign of just how much cheaper it had become to ship plastic waste to China instead of attempting to "recycle" it within the United States, increasing quantities were arriving from the East Coast, fourteen thousand miles and half the globe away, via the Panama Canal. For its part, Europe's reliance on China was even greater. By 2005, nearly 90 percent of the European Union's plastic waste was going to China for "recycling."[4]

What happened when all those millions of pounds of old plastic got to Asia? Its fate tended to be unclear even to the Westerners who sent it.

"There's no transparency in China at all. I can't even figure out who the reclaimers are," Patty Moore would claim in 2003. Moore is the founder of a California recycling company that was shipping enormous quantities of PET bottles to Hong Kong, from which many were then rerouted to mainland China.[5] "Someone in China would have a relative in California," Moore later explained to me. "And they would call their relative up and say, 'Hey! I need this stuff. Go find it!'"

The US State Department and American Chemistry Council were only too happy to facilitate an exchange that rid the United States of a mounting liability. In the late 1990s, they began helping California's plastic recyclers locate contacts across southeastern China, arranging and funding business trips and meet-and-greets.

"The American Chemistry Council was saying that all plastic is recyclable," Moore continued telling me. "Now, technically this is true, if you have it all in one place. But in California they started selling a lot of mixed plastic to China. And you would then travel to China and see it. The plastic began at one facility, the one with the connection to the original recycler in California. They would search out their type of plastic — say, polyethylene. And they would train their farmers to recognize that stuff. Because it was farmers doing the sorting. They lived in dorms and would sometimes spend half that day planting crops before they began sorting plastic. They would pull it all out. Then they would have a pile left over full of, say, styrene and mixed resin as well as some plastic from old electronics. The next facility would then pull out what they wanted before sending it on to the next facility. And they would take their plastic out — say, red Solo cups. As the plastic shipment got condensed, different products became more predominant. But by the end there were the leftovers. And this is what got dumped. It could be anywhere from 2 to 40 percent of what had originally been sent from the United States."[6]

Even 2 percent would mean a hundred thousand pounds of Western plastic waste getting dumped in the Chinese countryside every year for thirty years. Almost as soon as the importation began, it was clear to Beijing that it carried huge environmental risk. There were, yes, the extremely polluting and carbon-intensive methods by which millions of tons of foreign plastic waste had to be washed, shredded, and melted, a process that across southeast China can be directly linked to brutal spikes in regional air pollution starting in the mid-1990s.[7] But there was also the filth to be found within the trash itself. For good reason has worthless plastic traditionally been known in Western "recycling" circles as "Chinese plastic." Flimsy, cheap, often caked in grime and grease—there was typically little that could be done with it in the United States other than burn or landfill it, or load it into vessels that were heading to Chinese ports stacked with otherwise empty containers.[8]

China did complain. "Ironically, it is the United States that has always been claiming it is concerned about human rights and environmental protection," Beijing insisted in 1996. "If the US government is at all concerned about human rights, it should do something to stop the dirty business. That is the basic demand of respecting human rights, of international convention and human morality."[9] The ethics of garbage exportation even got enlisted as a Chinese counterargument against US accusations of technological theft. There was no equivalence, Beijing contended, between Chinese entrepreneurs producing bootleg DVDs and American entrepreneurs shipping cargo containers of trash to Chinese ports. What the United States was doing was worse.[10] After several hundred tons of US household garbage—including diapers "creating a stench so bad"—were found in a dumpsite outside Beijing, Xing Demao, director of the Bureau for Inspection for Shandong Province, asked: "Why do some countries strictly control export of their advanced technology and equipment but openly permit the export of harmful waste they produced?"[11] Later, Shanghai officials fumed that "the whole world is using our country as a garbage dump," adding, "We're getting saturated with rotten fruit, disposable diapers, dead fish, rancid meat and everything else other nations don't want. Our message to the rest of the world is 'Don't dump on us' and we intend to take strong punitive action against anyone who violates our laws."[12]

But much like the Chinese-American economic coupling more broadly, the problems appeared to be superseded by the benefits. By the early 2000s, American plastic waste had become a crucial pillar of the Chinese manufacturing sector. Its importation and handling put as many as five million people to work, with whole communities of rural China reliant on sorting and washing old Western soda bottles and takeaway containers.[13] The fate of the town of Lianjiao in the province of Guangdong — where, after 1992, more than a thousand factories were built to process scrap, a recycling cluster estimated to have employed anywhere between fifteen thousand and one hundred thousand migrant workers — is instructive. In 2007, London-based Sky News released a story about how household plastic waste from the United Kingdom was overwhelming Lianjiao. When it arrived, one local explained that "a lot of waste is dumped and burned at open-air sites, when it should have been delivered to qualified processing factories and supervised by local environment authorities."[14] Within a month of Sky's report, bureaucrats from Guangzhou ordered the closing of Lianjiao's recycling facilities. For two weeks the city's electricity and water were shut off; roadblocks were set up to prevent outside vehicles from entering; more than a thousand officials conducted inspections. By month's end, claimed Chinese media, a recycling cluster that had received two million tons of foreign plastic waste and registered an income of 830 million Chinese yuan in 2005 — greater than the GDP of many prefectures of Guangdong Province — had been shuttered. All the same, imports were reported to have resumed shortly after the inspectors departed.[15]

Shipments to China continued to be highly contaminated. Up to 30 percent could be unusable municipal trash, according to the Institute of Scrap Recycling Industries, capable only of being burned or dumped.[16] What made the system worth continuing was nevertheless its sheer scale. For every three million pounds of plastic waste that China imported, at least two million pounds could be washed, shredded, and used as feedstock for new plastic—a process that, even when it did seem to work, required vast inputs of energy and water and released innumerable microplastics and toxins into local ecosystems.

The result? Plastic's environmental desecration in the late 1980s had significantly receded from Western view by the late 1990s.

"Exports to China became the mirage behind which the lie of plastic recycling could hide," Jim Puckett, head of the Basel Action Network, which aims to continue the legislative work of the Basel Convention, told me. "I remember going to Asia in the early 1990s and being horrified at what I saw. But the plastic trade was just not being looked into very much at the time. We did not know what we know now—just how toxic all of this is."[17]

But it was also in China where the narrative of plastic's easy recyclability would fall to ruin. In 2012, the 18th National Congress of the Communist Party of China announced a new campaign promoting "a green society" and "a beautiful China."[18] It was one of the opening salvos in what the legal scholar Ying Xia has called Beijing's pivot from a "pollute first, clean up later" model of national progress to an "ecological civilization" one, a transition that swapped out "construction" and "development" for "governmental quality" and "citizen protection."[19]

Two years later, China announced its intention to "fight a war on pollution with the same determination as it fought poverty."[20] Within four years, it had planted over three hundred million hectares of forest. It had taken twenty million fossil fuel cars off the roads. It had closed tens of thousands of factories.[21] And not by coincidence was *Plastic China* released in the last year of this epic environmental push, a documentary depicting two Chinese families who make their living sorting *yang laji*—"foreign waste"—and who come to serve as personifications of what Beijing's role as the world's plastic bin was doing to China more broadly. Alleged to have shocked the Chinese Communist Party, *Plastic China* was in fact allowed to circulate by covert order of the

CCP, which proceeded to shut down the booming plastics importation it had grimly exposed.

On the face of it, Beijing stopped accepting plastic waste in 2018 because of the pollution such imports had herded into China. But as the China scholar Joshua Goldstein has argued, the ban was only one element within a broader, systemic shift in the Communist Party's historic relationship with trash. China's import ban came in the wake of years of domestic policies aimed at eradicating the informal class of garbage merchants who roamed from countryside to city and proved to be one more problematic population — itinerant, polluting, informal — the Communist Party felt compelled to control.[22]

And, as importantly, the ban came in the wake of years of Chinese leadership attempting to get ordinary Chinese citizens to think more carefully about how they sorted their waste. "Garbage classification is related to the people's living environment and the economical use of resources," Xi Jinping would stress on trips to recycling centers across China. "It is also an important embodiment of the level of civic-mindedness."[23]

Not unlike the way Japan in the 1980s was a huge importer of Western wastepaper and old cardboard, only to evolve into one of the world's greatest exporters of those same materials, the long-term strategy in Beijing appears to be one in which China muscles its way — through legislative action and reallocation of industrial might — into no longer being reliant on plastic waste from the rest of the world; China will marshal a swelling middle class into separating and discarding its own garbage outputs with supreme efficiency, thereby accumulating enough old plastic of its own to serve as feedstock for new plastic.

There's also a simpler explanation. China no longer needed huge amounts of foreign plastic waste to source feedstock for domestic production. It could manufacture more and more virgin resin on its own. In 2000, China had a negligible petrochemical industry, amounting to no more than 10 percent of global capacity. By 2017, the year China announced its ban on plastic waste importation, those facilities had expanded to more than a third of global capacity. In the seven years after that, Beijing constructed more petrochemical plants than all those currently operating in Europe, Japan, and South Korea — combined.[24] And over the next decade, it's expected to grow larger still, with a third of forthcoming global petrochemical growth set to occur in China, far and away the world's greatest share. Beijing, in other words, doesn't need to import millions of pounds of foreign plastic waste to manufacture new plastic.[25] It can make what it needs from scratch — at cheaper cost.

The ramifications of China's plastic waste ban proved instant and immense. One of the crucial differences between the used plastics trade and that of electronics or scrap steel is that the former was, from the outset, almost entirely upheld by a single importing nation. Overnight, that destination ceased to exist. Millions of pounds of plastic placed inside recycling bins around the world every year needed to head to a new distant location in order for the fiction upholding plastic's production — that it was not wrecking the planet — to be sustained. And what followed, as one industry leader described it to me, was a "rat race" to locate that new destination.

35

MAD SCRAMBLE

> Do they [China] care about the global environment or only their own environment because we are land-filling perfectly good materials now because of the actions that they're taking.
> — Adina Renee Adler, senior director at the Institute of Scrap Recycling Industries, 2018[1]

THERE WAS AN opportunity here to finally address the plastic problem at its source: a globalized, hydrocarbon-based energy regime that imposed synthetics upon the world at the expense of genuinely recyclable materials such as glass and paper for no other reason than to boost already-gargantuan profits. But after 2018, that isn't what happened. In the wake of China's import ban, an easier solution presented itself to petrochemical companies and plastic traders: to continue making more plastic than ever before while redirecting the inevitable pollution blight from China to more desperate countries.

In the late 2010s, nations across much of the Global South were confronted with a choice not so different from that of the earlier iterations of the international waste trade in the late 1980s: to protect their environments or — potentially — boost their economies by entering the madhouse business of waste importation. For here's one thing that Western plastic exporters occasionally offered that domestic waste management systems across developing states rarely could: *relatively* well-sorted plastic, much of it of higher quality than could be collected locally, for little or no cost.

Within two years of China's ban, US plastic waste exports to Central America had doubled.[2] Much of it was reaching Mexico or Honduras from California, a state that in 2000 mandated 50 percent "diversion" rates on its discarded plastic — in other words, half of the plastic placed into recycling bins in California was legally required to leave California, turning a state often associated with environmental due diligence into one of the world's foremost contributors to waste exportation.[3] No longer able to ship plastic to China, brokers in Los Angeles directed it south instead in what marked a bleak return to the Cold War era and the city's sordid history of treating Central American states as repositories for its detritus. "The exploitation is worse now than it ever was then," Ricardo Navarro, a Salvadoran environmentalist who in 1993 led a campaign to prevent tires from New Orleans from being shipped to San Salvador and burnt as fuel, told me.[4]

Across the Atlantic, it was Eastern European countries that reprised their roles from the late Cold War era, when, in their desperation for hard currency, they had opened themselves up to trash imports from the West. Tens of thousands of pounds of German garbage would get hustled into Poland, leading one minister in Warsaw to slam the hypocrisy of Polish taxpayers funding the waste management of "one of the main initiators of the current climate policy of the EU."[5] Farther to the east, weather reports from the Romanian capital of Bucharest would begin to include the state of smog in the city's outskirts, much of it derived from the burning of plastic that had been placed in recycling bins in France and Germany, transported east by ship to the Black Sea, barged inland via the Danube–Black Sea Canal, then used as fuel in scores of cement factories that had been snapped up for a pittance by French and German multinationals during the post-Communist privatization drive of the 1990s; these French- and German-owned cement companies outside Bucharest got *paid* by Western European waste brokers to incinerate French and German plastic waste and wreck Romanian air quality. "The rest of Europe thinks of us as a second-rate nation," state prosecutor Teodor Niță, who has made his reputation cracking down on illegal waste traffickers in Romania, told me in his office in Constanța, home of the Black Sea's biggest port.[6]

At that same time, on the other cusp of the Black Sea, Turkey—a country whose First Lady had, you will recall, christened it a "zero waste" nation only one year earlier—had become the recipient of 225,000 tons of foreign plastic in 2018 alone. Within two years, that amount had almost quadrupled to some 800,000 tons—imports allegedly facilitating the production of new synthetics, yet entering a country so hopelessly overwhelmed with its *own* plastic waste that scores of recycling facility owners had resorted to setting their stockpiles on fire by night in lieu of any other disposal option.[7]

"Half the recycling yards here were burning our Turkish plastic because they had no place to put it," Fatih Doğan, a plastic importer based in the port city of İzmir, told me. "And the other half kept bringing in more and more of it from Europe."[8] When, in 2021, owing to pressure from environmentalists, Turkey announced an end to foreign polyethylene plastic imports, the ban lasted just eight days. More ruthless pressure from the petrochemical industry resulted in plastic imports resuming—and, unsurprisingly, exploding.[9] Baran Bozoğlu, the head of Turkey's Chamber of Environmental Engineers, described the paradox best: "It's like we have flour and water and, instead of making our own bread, we import bread from abroad! Does that make any sense to you?"[10]

To the south, in Africa, plastic waste imports from the Global North quadrupled in the wake of China's ban.[11] Sub-Saharan nations that had only ever figured at the margins of the toxins trade of the 1980s—

Senegal, Ethiopia, Kenya — entered the 2020s finding themselves in the crosshairs of the plastics one.[12]

"They knew we had a huge plastic problem that we could not solve," Kevin Mtai, a Kenyan climate justice activist, told me. "But because the United States and Europe have more power than we have in Africa, they were able to tell us, 'If you're not going to take our waste material, we are not going to give you a good market with which to trade with us.'"[13]

Plastic hadn't just stopped entering China. Many of China's plastic importers had also left their native land. They were encouraged to migrate by Chinese state-affiliated scrap associations, which loaded Chinese social media sites with nonsense claims that countries like the Philippines imposed "no licensing controls" on what kind of plastic waste could be imported; those same associations paid for Chinese plastic traders to take weeklong tours of Southeast Asian nations to scout out potential warehouses to shred and melt old Western plastic.[14] *Come to the Philippines to process your plastic scrap!*

By late 2018, a diaspora of Chinese trash traders had fanned out across Southeast Asia — and beyond. One afternoon in Dar es Salaam, Tanzania's largest city, I took a taxi north of the metropolis, down a one-lane highway that buzzed with *bajaji* — tuk-tuks — and motorcycles encrusted by sea salt whipping off the adjacent Indian Ocean. After half an hour my car exited the highway, toward an industrial zone full of garment factories and a place that, or so I was told by an Indian plastics trader in downtown Dar es Salaam, "the Chinese now run."

After some wandering, I found my way to a heavy iron gate flanked by a couple of dagger-wielding Maasai tribesmen, to whom the recent Chinese arrivals had begun outsourcing their security. The men agreed to let me inside. I entered the plastics sorting yard, which sat in the shadow of a five-story mountain of empty water bottles resembling a steep, crystalline ziggurat. Circling its base, dozens of Tanzanian women were processing its contents, stripping paper labels and removing bottle caps. Labor at the yard, I later learned, was divided by gender.

Women segregated plastic; half-naked men worked the greasy, roaring machinery that washed and shredded it into transportable flakes.

Nearby, inside an unadorned office, I met a Chinese woman who introduced herself — and was only known to the hundred or so Tanzanians who worked for her — as "Madame." She'd relocated from the province of Guangdong two years earlier. I asked her how business was going in Tanzania. She looked at me uncertainly, then explained in faltering English that the drier months in Africa were better for business. In the drier months, Africans drank more water, consuming and discarding more plastic bottles, meaning greater hauls of plastic flake to be exported; she made a motion with her hands that resembled the act of pushing something away. And while an earlier generation of Chinese plastic-processing colonies in Tanzania had shipped East Africa's plastic back to their homeland, the plastic collected now was destined for a new frontier of pollution havens: Ukraine and Turkey and Thailand. The Chinese were doing to other countries what half of the world had done to China for thirty years: locating places where labor was cheap and environmental regulations tended to be recommendations.

Still, the greatest upheaval after 2018 would overtake Southeast Asia. Plastic exports to Thailand increased twentyfold in the wake of China's ban. To Vietnam they doubled. To the Philippines they tripled.[15] "I will not allow that kind of shit," threatened that country's President Rodrigo Duterte — a former prosecutor who had run on a plank of confronting "Western bullying" — in 2019 in reference to a filthy shipment of Canadian garbage that had in fact reached the Port of Manila years earlier. "I'll give a warning to Canada, maybe next week, that they better pull that thing out — or I will set sail [and return the rubbish]," continued Duterte, beating his press conference table with his hand. "I cannot understand why they're making us a dump site."[16]

The Philippines threatening Canada over cargo containers full of rotting diapers — a strange new world had arrived, one that laid bare for all to see the problems layered into the narrative of plastic recycling. And even as all these countries across Central America and Africa and

Southeast Asia had emerged as new receptacles for the plastic waste that had once gone to China, in none of them did the numbers even come close to making sense. Let's take Malaysia, where more than seven hundred new trash importers sprung into existence in the wake of China's ban, and which in early 2018 had outstripped China as the biggest recipient of US plastic waste on the planet. It will come as little surprise that many of these new import operations were illegal, opened without proper permits or government authorization; as in Tanzania, the majority were being pitched up by the very Chinese plastic traders the CCP had just shut down back in their native land.

But they still had Rolodexes full of Western exporters. And to purchase a ton of US plastic waste, even if much of it was contaminated, still offered *possible* opportunity: Were you able to pay local labor cents an hour to sort that plastic, and were your energy bills low, and were your pollution regulations lax, and were your inspectors able to be bought off, you could source enough uniform amounts of plastic, shred it, and sell that flake to a distant plastic processor for a sliver of profit. To work, the process had to be repeated at great scale, which it often was. By the end of 2018, the hundreds of Chinese facilities that had popped up in Malaysia had imported—on paper, at least; the true numbers were certainly higher—835,000 tons of plastic waste. This importation was in addition to the 2.4 million tons of plastic Malaysians themselves discarded that year.

The total annual capacity of Malaysia's recycling sectors at the time? Some 515,000 tons.[17]

In other words, in 2018 alone, nearly 3 million tons of domestic and imported plastic waste piling up in Malaysia had no place to go to be "recycled." No option existed other than to landfill it, incinerate it, or dump it into the sea.[18] "We don't need your waste because our own waste is enough to give us problems," declared the country's prime minister in May 2019, even as the Chinese waste brokers who had relocated to his country a year earlier continued hustling in more and more bales of Western plastic garbage. That year, Port Klang, the second biggest harbor in Southeast Asia, which is to be found at the mouth of the fourth-most-plastic-infested river in the world, earned the dubious distinction of becoming the world's single greatest recipient port of foreign plastic waste, shipments of which included PET bottles from Australia that Malaysian customs inspectors found to be breeding grounds for maggots.[19] "We came to the realization that the only real way to stop the import of foreign waste would be to close all our ports entirely," the country's former environmental minister, Yeo Bee Yin, told me.[20]

Here's the irony of the tens of millions of pounds of plastic waste that have been sent south over the last thirty years: This was precisely the pollution externalization that the international legislation of the late 1980s and early 1990s had been drafted to prevent.

Recall for a moment the farcical journey of the *Khian Sea*, which roamed the world's oceans for more than two years scouring for a place to dump thirty thousand pounds of Philadelphia's incinerated garbage. At a time when the Western waste crisis was a far greater object of public conversation than it is today, the *Khian Sea* had become shorthand for what not to do. Sending the burnt remnants of your citizens' trash to foreign countries was, it was clear in the late 1980s, unforgivable.

"There's a great incentive for us to ship our garbage overseas," John Edward Porter, a congressman from Illinois, warned in a 1989 government subcommittee prompted by the *Khian Sea*'s voyage. "And unfortunately, with the environmental standards we have here at home, the incentive to ship it overseas where there aren't standards, where one can bribe local officials, is very, very high."[21]

Yet fast-forward a few years and, instead of receiving the charred output of a Philadelphia trash incinerator, the developing world was on the receiving end of a far greater spectrum of plastic pollution output. Entire container loads of plastic waste were arriving. The toxins and contaminants from their burning were infiltrating local food supplies.

The resultant ash was getting dumped across fields. The microplastics were entering river systems and, after that, the ocean. And the scale of the problem only kept compounding. By the 2000s, it wasn't a rogue barge out of Philadelphia, aimlessly wandering the seas for years on end, that was shifting a single city's consumption footprint from north to south. It was dozens and dozens of ships every day, moving thousands of pounds of plastic waste every hour, from one hemisphere to the other.

36

A TRASH CHIEF

> The Dutch are gone from Modjokuto now, their estate and factory system shaken by the depression and shattered by the war and revolution. What remains is a peasantry very used both to money and to foreign goods, tremendous underemployment both rural and urban, and an overcomplex economic system in which the Chinese minority controls the main streams of trade.
>
> — Clifford Geertz, *The Religion of Java*, 1960 [1]

SHIPLOADS OF DIRTY Western plastic waste were also arriving in Indonesia. And how exactly they were getting there was unclear. In the immediate wake of China's ban, Indonesia's plastic waste imports spiked, as they did throughout most of Southeast Asia, more than doubling by 2019.[2] But in 2020, following the example of states like Vietnam, Indonesia announced that it would no longer be permitting the importation of foreign plastic waste.

Yet the trash towns of Java didn't disappear. Quite the opposite, actually: In the years after 2020, photos showed them to be piling higher and higher with American and European garbage.

How?

It's a phenomenon best understood by briefly stepping back to examine Indonesia writ large, a country that should never have been in any position to emerge as one of the world's foremost recipients of discarded plastic. For, not unlike the case of Malaysia, the nation's six thousand inhabited islands can barely handle their own garbage output. Since the 1980s, the great metropolis of Jakarta has grown into the largest city in Southeast Asia, its population expanding by seven million people in the span of a generation and becoming so densely packed that by the 1990s it began — and continues — to sink at the rate of approximately one foot per year.

At the same time, its citizens were getting richer, capable of purchasing more things. By the 2000s Indonesia was producing 2 to 4 percent more solid waste each year, even as nearly half the country's urban population had zero access to any form of waste collection services whatsoever.[3] As for US multinationals, whose earnings at the time depended heavily on expanding markets across the Global South, Indonesia had become a byword for future profit.

"When I think of Indonesia — a country on the Equator with 180 million people, a median age of 18, and a Moslem ban on alcohol — I feel I know what Heaven looks like," gushed the president of Coca-Cola in 1991.[4]

"The sun never sets on McDonald's, whether we're serving customers in the world's great metropolitan centers or near the picturesque rice fields carved into the landscape of the Indonesia island of Bali," claimed a 1997 annual report by the McDonald's Corporation.[5]

More Indonesians being sold more plastic and producing more trash in a country where nearly half the urbanized population had no access to proper waste management — Indonesia was a recipe for disaster, and representative of a broader trend. By the 2000s, many developing countries had become the objects of a kind of pincer attack, as the recipients of both new plastic products pushed by Western multinationals and — what would soon become clear — torrents of old foreign plastic exported by Western waste brokers.

To this day, Indonesia remains disastrous at waste management. It possesses a mere 437 landfills, or approximately 1 per every 620,000 citizens; only a quarter are "sanitary," according to the World Bank.[6] And the question in Jakarta, as throughout much of Southeast Asia, isn't whether plastic gets properly "recycled" or not. It's whether it ends up in any kind of bin whatsoever. By 2010, as much plastic in Indonesia was believed to be entering recycling centers as was entering the sea. A country that had negligible experience with plastic before the 1980s had, by 2015, become the third largest contributor to oceanic plastic pollution on Earth, exceeded only by China and India. Of the twenty most plastic-infested rivers in the world, four flowed through Indonesia, dumping the equivalent of thirty-four million plastic water bottles into the ocean every year. The world's most polluted river, the Citarum, coiled just beyond Jakarta, a putrefying slosh of sludge in which fecal coliform bacteria levels exceeded mandatory limits by a factor of five thousand.[7]

Average Indonesians were being asked to make sacrifices to clean up their country's horrifying environmental record. In 2020, malls and markets in Jakarta banned single-use plastic bags, with fines for consumers running the equivalent of $1,800, more than double the average monthly salary.[8] And on the other side of Java, in the port city of Surabaya, plastic litter had become so ubiquitous that passengers were being directed to pay for public transportation with five empty PET bottles in lieu of bus tickets.[9]

But by this point it should come as little surprise that this plastic — the plastic being purchased and tossed away within Indonesia — was really only half the story.

In 1992, the same year the Basel Convention seemed poised to finally put an end to hazardous waste transfers between the Global North and the Global South, one of Greenpeace's field operatives, Annie Leonard, journeyed to Java after hearing rumors of discarded US plastic making its way to the island.

"Where do all our 'recycled' plastics go?" Leonard wondered. In one of the earliest revelations of its kind, Leonard meticulously traced the journey of single-use plastics discarded in Seattle — the city that claimed to have the highest recycling rates anywhere in the United States — to the Indonesian capital of Jakarta, where, in a "crowded, unventilated room" within a "vast slum" on the metropolis's outskirts, she was shocked by what she saw: hundreds of Javanese women who had developed breathing problems and skin rashes as a result of sorting and shredding Seattle's plastic waste.

And it got worse. One importer Annie Leonard met in Jakarta told her that 40 percent of the plastics he received from the United States were so contaminated, his only option was to bury the material.[10] "I had a sense that, with the enormous increase of waste in the West and the plasticization of everything, this was probably only the beginning of what was to come," Leonard later told me.[11] As for Europe, the fate of the plastic it sent to Java at the time was no less problematic. "People in Germany are told that their plastic wastes are being recycled in an orderly fashion, but their so-called recycling system is turning Indonesia into a rubbish tip," insisted a hazardous waste expert from Hamburg.[12]

That was in 1992, a year in which thirty-five million pounds of Western plastic waste winding up in Indonesia was still remarkable enough to warrant outcries from local newspapers and politicians and, indeed, the few who realized the scale of the crisis it portended.

Yet it was only a fraction of what was to come.

Twenty-five years later, Indonesia was receiving tens of millions of pounds of plastic from the United States and Europe *every month*. The figures were to make those that horrified Annie Leonard seem puny by comparison. By 2019, the greatest exporter of plastic to Indonesia was its old ruler, the Netherlands, which was single-handedly dispatching more than *165 million pounds* of plastic to its former colony every year.[13] The United States wasn't far behind. One of the largest exporters

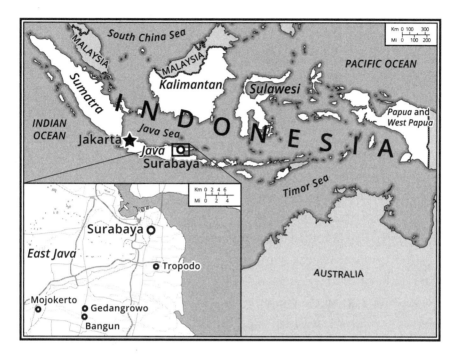

of plastic to Indonesia anywhere in the world was Arizona, a state comprised of great desert stretches beneath which its residents' plastic waste might easily be landfilled. Diverting it thousands of miles away to Indonesia came with an advantage, though. Arizona got to claim that all that plastic waste was being recycled, even as its true fate was often to get dumped along a hillside — or, as I was to discover deep in the highlands of Java, something far worse.

The most important thing to understand about Indonesia's trash villages is that they should not exist. Since 2020, Indonesia has "stopped" importing most forms of post-consumer plastic waste, such as shopping bags, food packaging, and water bottles, from Western countries. And, according to Indonesia's own laws, if a cargo container is discovered in one of the country's ports containing plastic with more than 2 percent contaminants — that is, nonrecyclable plastic or food or some other form of waste — it must be sent back to the country that exported it.[14]

So here's the crucial question: How, in the wake of such legislation, do mounds of foreign garbage continue to get photographed all across Java?

The reason Indonesia's trash towns exist partly owes to a mistake—stemming from ignorance or laziness or dysfunction—that gets committed thousands of miles away, millions of times over, likely by many of the people reading this book. How well do you *really* sort your trash when you discard it? Do you separate plastics from paper? And when you separate the paper, how thoroughly do you separate it? Do you take care to remove *all* the pieces of plastic from that paper—a plastic coupon stapled onto a brown bag, a plastic fork tucked into a napkin?

That is one answer to why acres of Western trash are to be found in the middle of Indonesia's islands today: Citizens in countries like Belgium or Italy are not placing their plastic in the correct bins. They are placing their plastic waste in paper bins.

In the United States the situation is slightly different, though the result is much the same. Unlike Europe, most of the United States does not have two separate waste streams. All "recyclable" material is discarded together. From your home or nearest blue bin, it travels to what's known as a materials recovery facility system where, in theory, paper gets sorted from plastic along a mechanized assembly line. In reality, the scanners used to separate plastic from paper often fail to distinguish between the two, placing large amounts of paper-like plastic—blue-and-white plastic Amazon envelopes, for instance—into paper streams.

Regardless of *why* it happens, Indonesia's trash towns exist because, in Western countries themselves, thousands of tons of old plastic keep going where they should not be going. It is getting labeled "paper for recycling," baled, then shipped to secondhand paper importers, often in developing nations. To landfill a ton of mixed paper in California can cost upward of $150. To send it to Indonesia costs no money at all. In fact, a waste broker in Los Angeles *gets* money, approximately $200, to sell it to an importer in Indonesia.

A TRASH CHIEF

For the last thirty years Indonesia has been importing staggering quantities of this old paper. And on the face of things, it has been doing so for admirable reasons. Indonesia is one of the world's biggest paper producers, and old paper really can be recycled and turned into new paper. Approximately eighty paper mills sprinkled across Indonesia are reliant on making new paper by sourcing more than three billion pounds of paper discarded each year in the richer countries of the world.[15] Italian newspapers, German take-out menus, Korean calendars, American junk mail — dozens of cargo containers of them enter the ports of Java every day. Indonesia's paper producers — one of the country's signature industries, registering upwards of $7 billion a year in sales — could not exist without them. The alternative would be cutting down more trees in a country that, in 2012, surpassed Brazil as the fastest-forest-clearing nation in the world.[16]

Importing lots of paper and receiving some plastic instead — it may seem like the trash world's equivalent of a rounding error, the cost of doing business in the haphazard world of globalized recycling. Yet its consequences are colossal. For a generation now, for every ten pounds of old paper that is unloaded in Indonesia, *at least* one pound of plastic waste has arrived alongside it. The result? Under the guise of being recyclable paper that will help preserve what remains of Java's primeval forests, millions of pounds of dirty foreign plastic must be dealt with in Indonesia every month — so much plastic waste, in fact, that ancient forests have had to be chopped down to accommodate its arrival.

Too voluminous to be landfilled, too worthless to even attempt to recycle, the foreign plastic waste that ends up in Indonesia is ultimately burnt as "fuel" in several hundred tofu and cracker factories dotting the archipelago. It's an environmental wrecking ball whose ramifications are becoming more and more incontestable after having occurred each and every day for the last twenty years: Torching plastic Doritos bags to bake one of the staples of your national cuisine is unbelievably dangerous. And it would also be hard to think of a more searing example of the steamrolling dominance of globalized plastic: so plentiful, it requires the razing of forests in far-off Indonesia, so difficult to get rid of once it arrives there, the only solution is to incinerate it.

That, at any rate, was the situation before Beijing informed the world that China would no longer be receiving half its old plastic. By the time I arrived in Indonesia, in late 2022, the problem had become considerably worse. Over the previous four years, customs authorities in the port of Surabaya told me, a curious new trend was afoot: cargo containers from countries like the United States that are imported as "recyclable paper" and, when crowbarred open, are revealed to be a *majority* plastic waste. Paper and plastic were reaching Indonesia in virtually interchangeable ratios. "From what we can see, it's usually about 40 percent paper," Iman Nurifa, one of the inspectors, told me.[17] And far more shipments were arriving too. In 2021, Indonesia imported twice as much "secondhand paper" as it had in 2020.[18]

Reporting conducted by Indonesia's own environmental watchdogs offered more striking figures. According to Nexus3, an Indonesian NGO that works to safeguard the public against environmental threats, up to 70 percent of the paper that has been exported to Java since 2018 has consisted of plastic trash.[19]

Daru Setyorini, a river biologist based in East Java who has spent decades studying Indonesia's plastic problem, told me, "It's getting harder and harder to tell where it's all coming from."[20] The plastic waste trade had increasingly come to resemble that of globalized finance, Setyorini said. Plastic waste may have been getting shipped from California or the United Kingdom, but more and more it was getting routed through the Marshall Islands or Singapore to obfuscate its national identity. Export companies changed their names by the month. They operated out of post office boxes. They looped transactions through stacks of shell firms incorporated in tax havens.

It was a dizzying problem to untangle. Yet whatever statistic was cited, and whoever was sending the plastic, the truth was difficult to deny. Someone was lying about what they were exporting to

Indonesia. And Indonesia's efforts to learn from the rest of Southeast Asia and attempt to eradicate flows of foreign plastic waste were being brazenly flouted.

Barring the plastic waste from arriving was challenging, Iman Nurifa, the customs agent, told me, because many segments of Indonesian society had a great interest in sustaining its importation. Java's paper industry is not just massive. It is also politically connected. In Indonesia, as in many other places, the relationship between environmental ministries, sustainability initiatives, and business is a revolving door. It's not uncommon for environmental ministers in Jakarta to leave their posts to become paper industry consultants in Surabaya, or for government ministers to be large shareholders in the companies they've been tasked with monitoring.

The clearest evidence of the power wielded by paper mills in Indonesia? How secondhand paper gets to them. Consider for a moment that in the ports of Java today, one "status" among a choice of three is given to all incoming shipments. A "red" status is assigned to those shipments — construction materials, for instance, or medical supplies — that require the greatest scrutiny, for they are the most liable to contain drugs or some other form of contraband. Red shipments are routinely checked. Next is "yellow" status, reserved for those shipments that need to be checked sporadically — "approximately one percent of the time," according to Iman Nurifa — for their shippers are known to occasionally violate import restrictions. Finally, there is "green" status, assigned to those that need only to be checked and searched in the fewest possible instances because their shippers have demonstrated themselves to be the least likely to falsify their shipments' true contents.

In Indonesia, secondhand imported paper — an industry that makes little secret of the illegal plastic waste importation it masks — enters the country under this "green" status.

This latest twist in the curious story of Western plastic's arrival in Indonesia is, according to many of the environmentalists I spoke to in

Java, not accidental but deliberate: attempts by Western waste brokers to stash worthless trash — used plastic that can no longer be shipped to China, and which has no viable place to go in the United States or Europe — inside shipments of paper bound for Indonesia to be recycled.

And so, here we are: Indonesia's trash towns owe their existence to tremendous quantities of plastic waste getting shipped across the ocean within shipments of used paper. And it's a phenomenon — an accident, a deception, a conspiracy — that has manufactured complex garbage economies running up and down the bony spine of Java.[21]

The two most notorious trash towns of East Java are to be found a little more than an hour south of the city of Surabaya, where in late 1945 Javanese revolutionaries ripped the blue third off the Dutch flag flying atop the Majapahit Hotel, producing the red-and-white flag of the independent Indonesian state. From the coast, I was driven headlong toward a volcano chain whose black cones were swathed in wisps of cotton-candy cloud. Somewhere near Surabaya's outskirts we passed the Grand Heaven, a gaudy nine-story glass building propped atop cream-cheese-colored Ionic columns that looked to have been airlifted into East Java from the Vegas Strip; it's a funeral home. For miles thereafter the country sat in a halfway state of industrialization. Bamboo forests were being chopped down to make way for cement plants. Trucks rumbled along a highway crowded with villagers on bicycles riding out to work the paddy fields. The land began to tilt toward the mountains, which came into clearer view, looming larger and larger. Above hideous cinder-block towns rose columns of acrid smoke, twirling in the wind parallel to dragon-shaped kites flown by local children, a sport said to have originated centuries ago as communal overtures to the gods for an abundant rice harvest.

Gedangrowo exuded something ominous. The village felt like a jungle outpost, a drug den, a rebel encampment. Expressionless, lacking a single street sign, oppressively quiet, it is a place that does its utmost not to make known the reason why it exists — or how it makes its money. A network of dirt lanes led us off the main thoroughfare from Surabaya, taking us along a river called the Gedangpulut, which was spanned by a succession of thin bamboo bridges. After a few moments we took a hairpin turn toward the mountains. On the left bank of the road, a shoulder of trash started to pile up; it grew larger, rising higher and higher over the course of a mile, eventually clearing the roof of the car. Pickup trucks stacked with trash gradually became visible in the distance, jouncing over potholed roads and sluicing streams of sudsy, coffee-colored wastewater out from their beds. Farther on, a lanky blue edifice shot forth from the vegetation: a water tower that might easily have been mistaken for a lookout tower, which it also happened to be. I had the sense that I was being observed upon entering Gedangrowo — as I was later to find out, I was.

The first thing that struck me once I was in Gedangrowo — my first realization that it was no ordinary village — was, half obscured through a thick shroud of foliage, a sunbaked expanse of lusterless gray. It rolled in irregular waves. It was unmoving. It gave off zero smell. It extended for acres. A few steps closer and there could be no mistaking it for what it was: a plantation of plastic, piled at least two feet high at its lowest points and rising as much as two stories in certain areas. A landscape had built itself up around the garbage, in some places absorbing it, in others accommodating it. Trees sprung out of the waste, their roots coiling through a forest floor of wrinkled shopping bags and crumpled soda bottles. Green shrubs winked across the grayed-out plain. Paths cut through the garbage, some lined with bamboo fences. In the distance, approximately half a mile away, forming a backdrop to the scene, a steel aqueduct cut across the sky, feeding water into a huge paper factory known as Pakerin, which sat next to Gedangrowo and hovered over it like a feudal keep.

Pakerin was technically Pabrik Kertas Indonesia. And beyond the fact that it loomed over Gedangrowo, occupying more land than the village itself, it was never easy to forget it existed. Owing to the use of caustic sodas for its pulping operations, at all hours of the day Pakerin exuded an excruciating sulfur stench that caused the eyes and throats of anyone not accustomed to its smell to scratch and burn for hours after leaving Gedangrowo; from Pakerin, the sodas were disposed of in the Gedangpulut River, turning its waters blackjack-gray and encrusting its banks with tuffs of solidified chemical fizz that resembled dried-out sponges and stank of expired milk. Meanwhile, every few hours, from somewhere behind Pakerin's thick concrete walls, an alarm clanging like an air-raid siren sounded, announcing changes in shifts for the factory's several hundred workers, some of whom commuted from Gedangrowo itself.

Then there was the plastic waste that passed forth from Pakerin's gates in the holds of dump trucks. They had been coming every day for the last thirty years. Gedangrowo was full of discarded foreign plastic, and Pakerin was the reason any of it reached Java in the first place: The factory was not just one of Indonesia's leading producers and exporters of new paper but also one of its greatest importers of old paper, most of it acquired at the port in Surabaya via Chinese purchasing agents whose relations with waste brokers from New Zealand to Los Angeles date back to the early 1990s. Old paper was imported at a cut rate, pulped, then used to produce new paper, some of it for export. Only much of that imported old paper was really just plastic, meaning that Pakerin was in the curious position of purchasing material from hundreds, or even thousands, of miles away as much as a quarter of which was useless to the company.

Or apparently useless. In 2007, Indonesia earned the distinction of becoming the first country in the world to turn corporate social responsibility (CSR) into a legal obligation. The premise of CSR is, on the face of things, admirable. It ensures that a factory polluting nearby villages is required to compensate those residents for the damage done to their health and surroundings — a fee that some Indonesians call "the ransom." In practice, however, CSR can effectively amount to a bribe. In Gedangrowo, Pakerin paid two types of "ransoms." The first came every Eid, the Islamic festival that marks the end of Ramadan, when the factory gave four lambs to the village and four kilograms of rice to every villager.

Then there was the other type of "ransom." It got paid every other day. For the terms of Pakerin's CSR also required the factory to hand over to the town, free of charge, all the plastic that had reached the factory in cargo containers full of old paper from Western countries.

The trash field that blanketed Gedangrowo for acre after acre was what had resulted. And walking around the great trash field, one might be forgiven for believing that all the plastic visible had sat there for years, possibly even decades. But most of it turned out to have reached Indonesia *that month*. This was one reason why it emitted no smell; it hadn't had enough time to start rotting. The other reason was that it had all been washed in Pakerin before being dumped in Gedangrowo; it had reached the town, if not exactly clean, at the very least rinsed through with thousands of gallons of factory water. The point of layering all the plastic across the great baking plain on the outskirts of town was precisely that: to cook it in the sun, letting it dry out over the course of a week or two, before selling it to tofu factories.

The strangest thing about Gedangrowo — even stranger than the fact that the men who appeared to be mindlessly sipping tea in a bamboo shack at the distant end of the trash plain were keeping careful watch of my every movement — was the realization I came to while watching truck after truck pull up and dump mountains of soggy, damp garbage across its plain, and then truck after truck arrive and take away piles of crispy, dry garbage: If used paper from the United States or Europe stopped coming to Indonesia for just two weeks, Gedangrowo would go back to resembling any other mountain town in Java. The trash now piled across its plain would eventually all get trucked away; no new plastic would arrive to take its place; the great trash field might, after the passage of a few weeks, turn back into an actual field.

The reality about Gedangrowo — often described in local and foreign newspapers as an unwitting recipient of the developed world's waste — turned out to be more complicated and bizarre: The town was a halfway house for Western waste, a place that had carved out an unusually lucrative role for itself in the chameleon economics of the garbage world, in which waste often manages to increase in value not by changing form or substance but merely by virtue of moving from one location to another and getting relabeled "scrap" or "fuel." Here was a place where plastic that should never have been entering Indonesia to begin with could hide and dry, then get incinerated and vanish, before anyone had much idea that it had arrived in the first place. The police did not come to Gedangrowo. The government pretended it didn't exist. Most environmentalists were scared of entering it. And at each step of the plastic's journey — from the moment it arrived in Gedangrowo in the hold of a dump truck to the time it left in the bed of a pickup truck — the town's residents got incrementally richer.

"Don't take any photos," Teguh Suyono, the trash chief of Gedangrowo, told me on my second visit to the town.[22] A wiry man with a black mustache and darting brown eyes, Suyono had agreed to escort me around the garbage plain in exchange for a box of cakes I had brought with me from Surabaya.

We set out across the great trash field, which was layered with a mesmerizing range of foreign trash. The outskirts of Gedangrowo had essentially been turned into a time capsule of products tossed away weeks earlier and thousands of miles across the world, microevents now embedding themselves into the interior of Java's jungle. There were Styrofoam noodle cups from South Korea whose lids were still peppered with chalky yellow seasoning. Bags of pet food existed in ridiculous preponderance. A brand called Royal Canin from France featured a black-and-white-checkered puppy alongside a loaf of perspiring meat: "*Gastro-intestinale.*" A Canadian brand called Orijen vowed to "nourish as nature intended." There were empty shampoo bottles from Italy and squeezed-out toothpaste tubes from the Netherlands and used containers of lavender-scented laundry detergent from Portugal. Calling codes on plastic phone cards belied their national origins: +49 had reached Java from Germany, +372 from Estonia. A bag of "Island Princess Macadamia Popcorn Crunch" purported to contain "Hawaii's Finest Macadamia Nuts." A stick of deodorant—Weleda for Men 24h Roll-On—had once graced an Australian armpit, or at least an armpit raised somewhere in Australia. And how, I couldn't help but wonder, did the California license plate 6LIT257 end up all the way here in Indonesia?

Scattered throughout the field, clusters of women in conical bamboo hats were raking the garbage with brooms, rotating it every hour or so to expose it to the shifting sun. Garbage for the residents of Gedangrowo had become, among much else, a way of tracking the passage of time. The women of half the town's 150 families tended to the plastic in the morning, the women of the other half handled it in the afternoon. As for the men, they passed both halves of the day shaded beneath the bamboo overhang of their *warung*, a refreshments shack, where they made their own contribution to the trash business: chain-sipping tea, counting trucks, and keeping a lookout for prying outsiders. Dotted around the plain, meanwhile, wooden shelters resembling wigwams had been erected to keep the plastic dry during the monsoon season.

"December is our toughest time," explained Suyono. "It's when the rains hit heaviest and the plastic takes the longest to dry."

The miracle of garbage was that apart from raking and watching, it amounted to very little work at all. True, working with trash bore certain resemblances to harvesting rice. It, too, needed to be laid out and dried in the sun. But rice! Rice needed to be planted, fertilized, watered, reaped, threshed. A rice-farming village would compete with a thousand other villages across Java attempting to earn their living the exact same way. A multitude of factors beyond their control—bad harvests, poor soil, unpredictable climate, international markets—conspired against them. Yet trash! Trash came to a town that had dedicated itself to handling garbage. They got it for free. It never spoiled. It never lagged in supply. They got money to give it away. Its use as fuel never cratered in demand. Only the environmental NGOs in Jakarta — or journalists hailing from the very countries shipping plastic to Indonesia — seemed to want to make it all stop. Hence the only real work required of the men of Gedangrowo: smoking cigarettes, sipping tea, counting trucks, and keeping an eye out for do-gooding activists and reporters. It was about as close to printing free money as one could get.

The first time I visited Gedangrowo, I arrived with a teacher I had hired for a few hundred thousand rupiah out of an English-language high school in downtown Surabaya. I had made no contact with the town's residents before my arrival. Late one morning the teacher and I exited our taxi and, unaware that a hundred eyes were collectively trained on us from the shadowy remove of the *warung*, began wandering the great trash plantation in disbelief. Within moments a large man with sunglasses and a ponytail had been called in. He hopped off a moped and strutted out toward us.

"What are you doing here?" he asked, coming closer. "Who gave you permission to be here?"

"No one," I replied, determined to ignore his arrival.

"Well," he said, in confident, almost officious English. "What kind of passport did you arrive in Indonesia on? A tourist visa? A research visa? If you came as a tourist, and you are doing research, you've lied to our authorities. Our country is probably not like your country. In Indonesia lying to the authorities can lead to trouble."

The knowledge of English, the referencing of visas, the vague threat of a forthcoming problem at some higher level — it was a performance carefully stage-managed to intimidate outsiders.

The high school teacher turned to me with a petrified look in his eyes. "I'm leaving," he said, his voice starting to crack.

And so we left.

The next day I returned to Gedangrowo by myself ("Never," the English teacher replied to my text message asking if he would be willing to join me again) and explained the nature of my visit. I wasn't interested in destroying their business, I told the men at the *warung*. I just wanted to understand how it worked. I brought some cakes to apologize for having walked around the day before without their permission.

"Does anyone in the town ever complain about the trash?" I asked, hoping to gauge their willingness to acknowledge the controversy behind how they made their living. The men shrugged, almost in unison, as if to say: *What is there to complain about?*

"Is there ever too much trash to handle?" They chuckled at the idea.

Trash in Gedangrowo was a great, communal enterprise. Three days a week, approximately a dozen garbage trucks arrived from Pakerin—the men of the *warung* pointed in the direction of the factory's walls, festooned with barbed wiring—filled to the brim with plastic waste. Once it was dumped in the trash field, the plastic sat for ten days, give or take, all the while getting raked and dried by some of their wives in the morning, the rest of their wives in the afternoon. Sometimes the men ferried trash over to their own homes, loading it into bamboo baskets fastened to the backs of their motorcycles, for the sake of making a little extra cash by fishing out recyclable material. Many houses in Gedangrowo featured scrubby dirt yards piled with mountains of garbage, atop which a child or two could often be seen scouring for water or soda bottles, their arms elbow-deep in grungy polyethylene. Once it had been rinsed, of bottles or any other worthy pieces of plastic, the trash was hustled back to the great communal field by motorcycle.

A week or two after Pakerin's plastic arrived, Gedangrowo's residents sold it by the truckload to any one of the hundreds of tofu and cracker factories across Java that used it for fuel. A pickup truck of plastic could typically be exchanged for 500,000 rupiah, or about $30. Approximately 200 pickups piled with dry plastic left Gedangrowo each week. The town divided the profits evenly among its 150 families, with each receiving more or less 8 million rupiah — about $500 — per month, according to Teguh Suyono, the trash chief. Sometimes they agreed to divert profits to public projects. The bamboo bridges spanning the Gedangpulut had been constructed with trash cash. So had the great blue water tower, used to store water safely away from the toxic runoff of Pakerin — and to monitor the roads in and out of town. For more than twenty years the business model had never really changed. Given trash for free, the residents of Gedangrowo got paid cash to pass it on.

And it was hard not to detect a certain symbolism in the transformation: A village that had subsisted on harvesting rice for as long as anyone could remember had shifted to drying plastic waste — and had become a good deal more prosperous by doing so. To the men of the *warung*, there was nothing very odd about any of it. Trash might as well have been firewood. And if they ever got tired of selling plastic, they insisted, they could always go back to harvesting rice. But why would they?

Later that week I visited Teguh Suyono at his home. It sat a two-minute motorcycle ride away from the *warung*, on a dirt street lined with modest brick homes that had been ornamented by many of their trash-rich owners with an architectural indulgence or two. Where windows had once consisted of clear plastic sheeting, many houses in Gedangrowo twinkled with exuberant stained glass. Where concrete pillars would have sufficed, some homes were held aloft by white plaster columns featuring ornate capitals. Terra-cotta roofs were going up across much of the town, sometimes atop protruding American-style porches. Suyono's house, though, possessed no such ostentations. In its yard, a pile of damp plastic rubbish about the height and width of an automobile sat in the shadow of a pair of banana trees. "Are the bananas okay to eat?" I asked Suyono as we approached his door.

"Yes," he replied, then paused, racking his brain to come up with a single bad thing that trash had done to Gedangrowo. "We sometimes find pieces of plastic in our chickens," he finally answered. "That's the problem with raising chickens in a town and not the countryside. They never really learn the difference between garbage and food."

The front of Suyono's house featured a crude concrete courtyard beneath an exposed brick roof. A couple of benches had been hammered together out of planks of splintery wood, opposite a dusty couch whose springs coiled forth from tan upholstery. Along the wall were scratched-out math equations where Suyono's eleven-year-old daughter had been doing her homework in chalk. Next to the front door was a brooder box full of chicken eggs, one of which hatched during the hour I sat with Suyono as he explained his life story.

He was a soft-spoken but self-assured man. His father, like his father before him, had worked the rice fields, and Suyono was old enough to remember when Pakerin first started approaching Gedangrowo with its excess trash. Suyono's earliest memories are of rummaging through it in search of valuables.

This was, in the early days, all that trash offered the town: an incidental source of wealth, something to claw through if you had time to spare. "We realized there were treasures in some of the garbage that got dumped here," Suyono told me. In addition to everything else they tossed into a recycling bin, on occasion Westerners would discard something of possible value. Car keys would be left in the pockets of a tossed pair of pants. Lost in the piles of sandwich wrappers and Big Gulp straws and coffee cup lids were gold teeth, wristwatches, fountain pens, even the occasional handgun. It was a scavenger hunt.

And sometimes there was money. In Gedangrowo today tales circulate of men like Didik Utomo, who years ago found a stash of 15,000 Singapore dollars in a stack of plastic waste. Not that Didik Utomo could make much use of 15,000 Singapore dollars: foreign money, claimed Suyono, was an exhilarating discovery but not a very practical one. Currency exchanges in the port city of Surabaya were rarely willing to swap out dirty Italian lire or crinkled Canadian dollars for Indonesian rupiah. But the discovery of foreign cash could boost a

Gedangrowian's stature among his peers all the same, a manifestation of one man's luck digging through another man's rubbish. Over the years, townspeople specialized in taping halves of torn banknotes together. Other residents became trading houses of a sort, willing to exchange pesos for pounds, none of which actually left Gedangrowo. And even today, it's not the trash the Gedangrowians want to show you once they've agreed not to escort you off their land. It's the cash.

One morning at the *warung*, Suyono told me that it might be a good idea for me to give a few of the men a small donation for the time they took to answer my questions. I agreed, and took out my wallet to get some Indonesian rupiah. "Don't give them rupiah," Suyono told me. "Give them dollars."

"American dollars? But they can't use them."

"They don't want to use them," said Suyono. "They want to display them."

But I had no dollars. At the bottom of my bag was only a ragged wad of Serbian dinars. "That works," Suyono said, then began distributing the bills to the men, who ran their fingers over Nikola Tesla's face and medieval Orthodox monasteries, nodding approvingly.

Teguh Suyono remembers being a teenager when Gedangrowo's relationship with garbage began to change. Two things happened. First, by the time he was a teenager, in the mid-1990s, Western trash was no longer coming in sporadic arrivals. It showed up several times every week, in greater and greater quantities.

Second, Suyono recalls the residents of Gedangrowo coming to the realization that garbage's value didn't stem from a Singapore dollar tacked to your wall or a German's lost gold tooth afforded pride of place on your bedside table. No: The value of garbage, it turned out, was the garbage. "Trash soon became the only thing that mattered here," Suyono explained, claiming that by the late 1990s almost no one in Gedangrowo was harvesting rice anymore. A greater source of prosperity was sitting right in front of their eyes. Enormous heaps of trash had been building up around the town for years, some of which could be sorted and sold to recycling firms in Surabaya. The rest — no matter its quantity or quality — could be burned. And, in the industrializing hinterlands of Java, where what little wood still existed had become expensive owing to a raft of new environmental regulations, there was never a shortage of demand for cheap fuel.

Teguh Suyono was sixteen years old when, in 1998, the residents of Gedangrowo agreed to repurpose a swamp on the outskirts of their town. They spent weeks filling in the marshland with thousands of tons of industrial fly ash donated by Pakerin, shoveling it into the estuary by hand and clogging its channels and gullies with more than five feet of soot. The result was a flat, parking-lot-like expanse that stretched for ten acres and doubled as a magnet for sunlight, a veritable drying rack for the truckloads of damp plastic that by now were arriving every other day from Pakerin. To any outsiders who chanced upon the erstwhile swamp, it proved the strangest of sites: a colossal plain of Western garbage fountaining out from the volcanic slope of a remote mountain jungle. The reality was even weirder: The residents of Gedangrowo had presided over the creation of what was in effect an inside-out oil well, a great repository of carbon piled *atop* the earth that — owing to consumers thousands of miles away who couldn't stop throwing their plastic into paper recycling bins — managed to fill up at the exact same rate it depleted.

Over the years, the work became more organized. Shifts — raking for the women, keeping watch for the men — were split evenly among the town's 150 families, each assigned one half of the day. The Gedangrowians agreed to elect a "trash chief," whose job was to forge relationships with tofu factories across Java and convince them to pay Gedangrowo to take the garbage the town was getting free of charge from Pakerin. When I asked him if too much trash was ever a problem, Suyono shook his head and insisted that this was impossible. "There could never exist too much garbage in Gedangrowo," he claimed. "Our trash now comes from across the world," he added proudly, "Korea, China, America, Canada," checking off his fingers one after the next, as if he had ventured to all those far-off places and fetched it himself.

As for the greatest threat to this bustling trash enterprise, Suyono continued, it was not the environmental NGOs or the Indonesian government or even the intrusive journalists who showed up at his house and asked their obnoxious questions.

No. The problem, Suyono explained, lowering his voice, was Bangun.

37

A TRASH SCION

> The headman is the key to village development (*kunci pembangunan*) and a gatekeeper (*perantara*) between state and village.... It is their patronage that is their power.
> — Hans Antlöv, *Exemplary Centre, Administrative Periphery: Rural Leadership and the New Order in Java*, 1995 [1]

THE VILLAGE OF Bangun was on the opposite side of Pakerin. It was not directly visible from Gedangrowo, but in most respects it was identical. For the last generation, Bangun had also been a receptacle for the plastic waste that made its way to Pakerin. American and European garbage layered its streets, overwhelmed its yards, rose as high as ten feet from the ground, and washed up against the windows of its houses in waves. The Banguners, too, made money by drying that plastic waste, then selling it as "fuel" to tofu and cracker factories scattered across Java. The relationship between Gedangrowo and Bangun was, one might say, symbiotic. They took turns taking in Pakerin's plastic. Gedangrowo got it Monday, Wednesday, and Friday. Bangun got it the other four days of the week.

The only real difference between the two towns? Who controlled trash once it did arrive. In Bangun, trash had never developed into any kind of vehicle of horizontal prosperity. There was no man like Teguh

Suyono whom the community had voted in as trash chief. Plastic was instead the domain of a single family, a powerful dynasty called the Ikhsans, who had lorded it over Bangun for as long as almost anyone could remember. Hadji Ikhsan, a man so named because he was rich enough to have once made the hajj pilgrimage to Mecca, was the town's current mayor; his brother was its village secretary; their father, Mohammed Ikhsan, had been mayor in the 1990s; Mohammed's father had been mayor before that, dating back to the early 1980s. If in Gedangrowo trash had managed to elevate a subsistence farming community to a position of relative prosperity and comfortable living, in Bangun it had come to be the symbol and currency of one family's political muscle. The Ikhsans had used trash to get into office nearly forty years earlier — and they had used trash to stay in power ever since.

Most Indonesians I met in Jakarta or Surabaya had never heard of Gedangrowo. But of Bangun many knew enough. The town had a gangster reputation across much of Java. Most of the rumors proved difficult to verify: the claim, for instance, that many residents of Bangun possessed slingshots capable of knocking camera drones out of the sky. But others were true. Local government officials were routinely blocked from entering Bangun. Environmentalists were personae non gratae. A few months before I visited Bangun, an Indonesian journalist had entered the town and asked some questions about the origin of its garbage, only to be stabbed by a Banguner with a pocketknife; photos I was sent by environmentalists in Surabaya showed a scar the length of half a plastic water bottle running beneath his right shoulder.

After a week spent roaming Gedangrowo, I began directing my taxis to the other side of the Pakerin complex, toward Bangun. For the first few days, I stalked its outskirts, where I managed to arrange a meeting with a Banguner named Abah Yasan, a heavyset, garrulous man who claimed to have been driven — literally and metaphorically — to his town's edge. Two years earlier, Yasan had run for mayor of Bangun against Hadji Ikhsan. It didn't just turn him into the closest thing the town had to a political opposition. It also represented an affront to how things worked in Bangun, a threat to the unique economic fabric that had been woven over the last forty years.

"To understand the Ikhsans, you need to understand the trash," Abah Yasan told me. For it was the Ikhsan family who, according to Yasan, single-handedly managed the relationship between Pakerin and Bangun, acting like a faucet of how much garbage flowed from factory to town. "If Bangun needs more garbage, the Ikhsans can just call Pakerin and ask for it," he continued. And it wasn't only the amount of garbage that arrived in Bangun. More importantly, it was a matter of *who* in Bangun got it. In Abah Yasan's telling, if you voted for the Ikhsans, you got rewarded with Western plastic waste, which you in turn sold to tofu factories yourself. If you didn't vote for the Ikhsans, your fate was to go back to harvesting rice.

Abah Yasan insisted he was cut out of the trash trade for attempting to end this patronage system, an account most Banguners I met confirmed. And, true enough, by the time I met him, Yasan lived an almost hermetic existence on the margins of society, shunned by many of Bangun's citizens, idling away his days sipping tea in a moldering ranch house. As for the Ikhsan family itself, Abah Yasan added as I got up to leave his porch, it was only natural that they reserved for themselves the greatest and most uncontaminated shares of Pakerin's plastic waste.

"You'll see what I mean if you ever meet them," he added, spreading open his arms in a gesture indicating magnificence. "These are rich people. Very rich people."[2]

I eventually tracked down Mohammed Ikhsan — the son of the man who ran Bangun in the 1980s, the father of the man who runs it now, and himself the mayor of Bangun throughout the crucial trash-importation decades of the 1990s and 2000s. Early one Friday afternoon after midday prayer had let out, I found him resting outside an imposing green hangar-like structure that turned out to be the Ikhsan family mosque. It had been constructed, I was later informed, with proceeds from trash, and its colossal proportions seemed to be a statement to all those who found their way into this peculiar little community in the bushy interior of Java.

Plastic hadn't just made Bangun rich. It had also brought it closer to God.

After I spent a few moments explaining what I was doing in Indonesia, Mohammed Ikhsan agreed to speak to me on the condition that I deposit a small donation in his mosque's charity box. I dropped a few thousand rupiah inside, and Mohammed Ikhsan proceeded to lead me down the street toward his house, which was just as out of proportion and out of place in Bangun as his family's great mosque. Thick wrought-iron gates opened onto a cool courtyard bubbling with fountains and shaded by luxuriant palm trees. Inside, in a spacious foyer, baroque glass chandeliers hung from sculpted cathedral ceilings. Servants and hangers-on circled round offering tea and ashtrays. Gold-framed paintings on the walls showed bucolic scenes of rice farmers laboring beneath the emerald mantle of a mountain chain. I resisted the urge to ask Mohammed Ikhsan if the paintings showed what Bangun had looked like before its residents had shifted to laboring beneath mountains of garbage — shiploads of French dog food bags and Australian deodorant sticks that had generated enough cash to erect the preposterous mansion in which I found myself.

Mohammed Ikhsan was a rotund man with gray-green eyes sunken into a sagging face. He had a narrow mouth from which thin white cigarettes moved back and forth. Worry beads slithered over his knuckles as he folded and unfolded his arms in slow, almost viscous motions, rather like a man underwater. "Sit," he told me, gesturing to a puffy brown leather couch.

I had gone to Mohammed Ikhsan's house fully prepared to confront him over the ecological devastation he and his family had presided over for the last thirty years. In Gedangrowo, I had found, it was difficult to know whom exactly to blame for the trash epidemic — apart, of course, from the countries shipping the garbage in the first place. In

Gedangrowo, plastic was communally owned and communally sold. Just about everyone was making a living off it. No one had an interest in making it stop. But two miles away, opposite the Pakerin complex, in the rival trash village of Bangun, where I now found myself, there never appeared to be any real question about who had the biggest stake in perpetuating the importation of plastic waste.

It was Mohammed Ikhsan and his family.

Yet Mohammed Ikhsan proved cannily prepared for my arrival at his house. Here was a man who knew the claims that were made about Bangun and his kin, and had his responses at the ready. As I began to speak, to ask about the garbage massed all around his town, he raised his hand, almost as though he was telling a child to settle down.

Then he began.

Sure, Ikhsan began, foreigners such as myself had come to Bangun and been stunned by what they saw. Trash was everywhere. It was a strange sight to behold. "But you should know that there's not a single piece of plastic in this town that ever gets away from us," Ikhsan said, slowly wagging his finger a foot or less in front of my nose.

"Countries that don't respect their environments are the ones that throw their trash wherever they want. That's not us in Bangun. Go check our rivers. There is no trash in them. There is no garbage in our mountains. We value every piece of plastic here in Bangun. To us trash is money. It's our future. We look after it. We don't lose track of it. You see a lot of trash here. We see money."[3]

For thousands of years, merchant enclaves clustering the great ports of Java, traders hailing from Arabia and China and Armenia and the Netherlands, to say nothing of Java itself, had specialized in shipping all manner of goods — coffee beans, rubber, sugar, rice, tea, spices, paper — in and out of the island to the world beyond. Plastic, in Mohammed Ikhsan's telling, was just the latest thing. Outsiders saw waste. Ikhsan saw a commodity. And he saw the international outrage over the fact that Bangun made money off old water bottles and bags of chips as one more instance of Western countries presuming to know what was best for a place like Indonesia. How was it, Ikhsan asked me, that environmentalists from the very same countries that had spent years shipping trash to the Global South were now so eager to have that trash trade end? Where would this leave a town like Bangun?

"Are we to go back to harvesting rice?" Ikhsan asked me. "Is that what you want us to do?"

All the same, over the next weeks I spent traveling around Indonesia, the long-term damage wrought by the redistribution of foreign plastic in and out of Bangun and Gedangrowo became hard to ignore. For the surrounding countryside of East Java wasn't merely more destitute than either town. It had also been turned into the collateral damage for their curious paths to prosperity.

In March 2019, a team of Indonesian and international researchers traveled to Tropodo, a town home to approximately fifty of the several hundred tofu factories across Java that purchase Gedangrowo's and Bangun's plastic and burn it as fuel. The factories, it should be clear, are not industrial enterprises, not complexes like Pakerin, but minimal structures, rarely exceeding a single room, and virtually all family owned. Touring them, the researchers were quickly struck by the fact that their ovens were not burning the garbage at anything close to the temperature required to break down the scores of toxic additives — dioxins, polychlorinated biphenyls, polybrominated diphenyl ethers, flame-retardant chemicals — that go into the production of today's plastics. Those toxins were instead getting baked into Tropodo's tofu, which was in turn being sold to restaurants and markets across the whole of Java.

"The town was like a war zone, with chimneys and smoke everywhere," Yuyun Ismawati, the Indonesian researcher who collected samples for the study, told me.[4] Toxins were also being bred — literally — into the town's main source of protein. Tropodo's chicken eggs, the scientists concluded, after sending samples to three different laboratories across Europe, were so poisoned, the only comparable levels of toxicity ever recorded anywhere on the continent of Asia were those of "the Bien Hoa site in Vietnam, a former US Army airbase where the soil was contaminated by historic Agent Orange use."[5]

"One of the egg samples came from a house next to a tofu factory," Ismawati continued to explain to me. "I asked the family who lived in the house what they did with the eggs. They said they poached them and gave them to their children before they went to school in the mornings. Other eggs they consumed raw, in an herbal drink they made with honey. They thought it gave them strength."

Internationally, the report caused outrage. Within Indonesia, the official response proved less than concerned. "If the plastic is used as fuel it is not a problem but the pollution should be managed," the country's environmental minister claimed after touring Tropodo for herself. "Chickens are smart," added another government spokesman. "They will not eat something hazardous."[6]

Unsurprisingly, the report did nothing to stop the importation of secondhand paper. But by the time I arrived in Indonesia three years later, one thing *had* changed. It had become more difficult for Gedangrowo and Bangun to source their plastic directly from Pakerin. Indonesian officials, or so claimed Teguh Suyono, had warned the factory's managers that their business with Gedangrowo and Bangun was giving Indonesia a bad reputation in the Western press.

Yet Pakerin was of course just one paper company among many dozens of others operating throughout Java. And in the wake of the attempted ban on Pakerin's garbage entering Gedangrowo and Bangun, nothing was being done to stop those *other* factories from offloading their plastic to any willing takers. The situation in a little parcel of East Java resembled an inversion of the globalized plastic waste trade writ large, in which a single country — China — might have stopped importing plastic, but dozens and dozens of others still remained desperate enough to do so.

The success of Gedangrowo's trash chief, Teguh Suyono, voted in six years before I met him, was his ability to locate more and more sources of plastic waste for his electorate. In the wake of Pakerin's export ban, Suyono told me that he began contacting paper mills all over East Java, which now paid Gedangrowo to accept trucks full of the plastic that arrived within their secondhand paper shipments. Plastic from the Mega Surya Eratama paper factory in Mojokerto? Gedangrowo would take it. Plastic from the Gaya Baru Paperindo paper factory in Malang? Send that to Gedangrowo too. There was PT Star Paper Supply, thirty minutes away in the city of Ngoro, and Buana Megah Paper, an hour away in the city of Pasuruan. Though technically all this was illegal — it is forbidden in Indonesia for waste to be trafficked from factories to communities — Suyono told me that Gedangrowo had devised its own solution to evade Indonesian authorities: Across Java, a network of new, makeshift roads threaded through the mountains and the jungles, connecting Gedangrowo and Bangun to ever-more-distant troves of plastic contraband. Day and night, trucks bounced along them, hustling Dutch milk jugs and American ketchup packets toward the great trash field.

Which is all to say that by the time I arrived in Indonesia, exactly thirty years after the Greenpeace investigator Annie Leonard was appalled by the presence of mounds of plastic waste from Seattle being sorted and processed in a dingy factory outside Jakarta, the situation in the jungly interior of the world's most populous island had become truly headshaking. An outlaw economy had been built on the currency of American and European trash, with Bangun and Gedangrowo having become locked in a mutual scramble to source new troves of the Western plastic that had first brought them such prosperity — and to which they had become economically addicted.

One day Suyono told me the story of a garbage truck that some months earlier had been destined for Gedangrowo. Just as it was pulling into the village, a few residents of Bangun — two of them bearing pistols, according to Suyono — emerged from behind bulwarks of roadside garbage and blocked the only route leading toward the great trash field, directing the driver at gunpoint toward Bangun instead. Several similar incidents occurred in the months that followed, said Suyono.

After thirty years of shaky coexistence, relations between Gedangrowo and Bangun had broken down into a kind of archaic banditry. Only instead of poaching one another's cattle or land, the object of contestation was instead truckloads of... plastic waste. And beneath the pistol-waving bravado, a grotesque, almost inconceivable reality was at play. In a country where trash was so mismanaged that nearly one in five pieces of discarded plastic entered the sea, here were two mountain villages competing at gunpoint for truckloads of *other countries'* garbage.

Over the three weeks I spent in Java, something peculiar began to happen. At first it was the men of Gedangrowo who began updating their WhatsApp profile photos to pictures of me seated at their *warung*. I was confused but flattered—and chalked it up to a local gesture of respect, which it may have been. But soon enough another motive presented itself. For, after a few days spent in Bangun, meeting some of its residents and asking them what they thought of how trash was handled a few miles away over in Gedangrowo, several of them began changing *their* WhatsApp photos to pictures of me, not at a *warung* but in their yards, scribbling notes into my journal. Behind the scores of new profile photos, it was hard not to detect some attempt by residents of each village to make it clear that they had talked to an outsider about the trash trade—and, maybe, to leverage mutual collateral against one another: *We've told someone about you!*

38

BACK TO THE PACIFIC

> By exporting waste, we are sweeping it under another corner of our own carpet.
>
> *— New Scientist,* April 1989

"SNAKE!"[1]

An hour northwest of Bangun and Gedangrowo, ECOTON — Ecological Observation and Wetland Conservation — was an air-conditioned reprieve from the bleak industrial towns of East Java. ECOTON's two-story building, the brick facade of which echoes the babbling of a nearby stream, features chemical laboratories, banks of computers, dorm rooms, and picnic tables where Indonesian students while away their afternoons sipping cups of tea, smoking packs of cigarettes, and filling in Excel documents on laptops. I walked into ECOTON's lobby and beheld its charismatic founder, a ponytailed Indonesian named Prigi Arisandi, dressed in a black-and-gold batik shirt, being handed a Goldman Environmental Prize by Barack Obama; it was a framed photo hung on the wall. And then I heard the Goldman Environmental Prize recipient himself:

"Snake!"

I arrived at ECOTON to find Prigi Arisandi inside his bedroom, two hands wrapped around the tail of a black snake that he and several others were yanking out from behind a desk with what appeared to be all their strength. The scene resembled a game of tug-of-war, the strangest one I had ever witnessed. A handful of Indonesian students seemed to think so too. They had begun to film the exchange from a nearby bed, their phone cameras wobbling between fits of laughter.

"Snakes enter our dorm rooms every few weeks," Arisandi told me after the creature had been dislodged.

"Is it poisonous?"

Arisandi glanced at the snake, now being deposited into a white laundry bag by his colleagues. "Maybe. Want to see our museum?"

On the second floor of an adjacent brick building was what the scientists and conservationists of ECOTON have dubbed their "Museum of Foreign Garbage." In the middle of a classroom, a ten-foot papier-mâché tuna hangs from the ceiling by strands of fishing line; out of its stomach, streaming through a cracked rib cage, a torrent of trash tumbles onto the floor. There are bottles of laundry detergent, bags of Sour Patch Kids, sneakers, cans of Pringles, Amazon envelopes imprinted with recycling triangles. "None of this garbage should have ever come to Indonesia," Arisandi told me. "We purchased it from local paper factories to make a point about what's happening to our country."

"They made you purchase it?"

"Of course," he replied. "They had to purchase it from the United States. They weren't going to just give it to us for free."

Arisandi may be the closest thing Indonesia has to an environmental celebrity. He remains best known for a campaign waged across the 2010s aimed at removing toxic sludge from the Citarum River that flows beyond Jakarta, once considered the most polluted body of water on Earth, owing to textile factories that had been dumping an average of 340,000 tons of wastewater into its currents every day. After several years, Arisandi was successful, managing to raise public awareness about the Citarum by meticulously photographing the waste, traveling its banks by motorcycle, offering "pollution tours" of its waterway, and encouraging locals to become their own "environmental detectives."

Bangun and Gedangrowo, Arisandi told me, are trickier problems. For the problem of the Citarum amounted to one of corporate profits versus local Javanese pride in their own environment. No Indonesian *wanted* the river to be polluted.

Trash in Gedangrowo and Bangun is something else. For the citizens of both towns, plastic has become a lifeline. It has made them richer than they ever were while harvesting rice — and better off than any of the neighboring towns that still rely on the paddies for their income. American and European garbage in the hills of East Java has created a conundrum in many ways representative of the global plastic addiction at large. It is no great secret that plastic will ultimately decimate Gedangrowo and Bangun beyond salvation. It will turn their soil barren. It will chew through the organs of their livestock. It will render their water tables unusable. But until all those things occur, the citizens of both towns are determined to wring every last drop of profit out of plastic that they possibly can.

And the real problem, Arisandi insisted, is not the plastic you see. It's that which you cannot. Late one Friday morning, while the majority of the Banguners were filing into Mohammed Ikhsan's mosque, a couple of members of ECOTON and I drove south of the Pakerin factory, past rice fields flanked by aisles of drying foreign garbage, toward the town's swampy outskirts. Pools of water reflected crests of green mountains in the distance. After a few miles, we located a small opening in the brush. A path led us toward a section of stream where a pipeline emptied Pakerin's industrial discharge — the water that had been used to clean Western plastic and paper — out into the local Gedangpulut River system. From here, the runoff, a steaming off-white gurgle that stank of rotten eggs, then flowed thirty miles north before discharging into the Java Sea. After that came the Pacific Ocean.

Using a test tube tied to the end of a rope, we extracted several samples of water from Pakerin's runoff and brought them back to ECOTON's laboratories. There, local high school students helped us put droplets of the water beneath microscopes. The lenses revealed miniature galaxies composed of fuzzy white backgrounds peppered with innumerable black and green specks.

"Some of these are algae," Muhammad Basyaiban, one of the students, told me, referring to the specks. "And some of them are plastic. You can use this to determine which is which."[2] He handed me a needle.

If you poked the algae with the needle, it registered the incursion; the algae would move, or even split apart. But if you touched the microplastics with the needle, nothing happened. They were immune to its touch, eerily unperturbed and unnerved by the force of outside contact. Neither moving nor breaking, the plastics remained stationary—unfeeling, alien emissaries from the distant side of the Pacific Ocean, a body of water that they would eventually enter and continue saturating, steadily breaking down and multiplying, and multiplying and multiplying again and again, the beginning of a journey that is all but certain to outlive our own on this planet.

CONCLUSION

Whither Waste?

Trash Makers and Trash Takers

> Africa has rejected all forms of external domination. We do not want external domination to come in through the back door in the form of "garbage imperialism."
> — Daniel arap Moi, Kenyan president, 1988[1]

IN MARCH 2022, I spent ten days in a terrifying place called Dandora. Planned and partly financed in 1977 by the World Bank, Dandora was only ever meant to be a stopgap solution to the swelling garbage output of the Kenyan capital of Nairobi. Like many cities in the Global South, Nairobi exploded in population under International Monetary Fund programs that spurred migration from countrysides to cities but rarely gave developing nations the tools to manage the problems such breakneck urbanization exacerbated — among them, how to handle the spiraling waste outputs of more populous and more prosperous societies at just the time when cheap foreign goods began entering their borders en masse.

CONCLUSION

And so, year after year, the garbage at Dandora kept piling up, eventually casting its shadows over the roofs of the slums that surrounded it and expanding outward to the point of acquiring its own gnarled geography. Go to Dandora today and you bear witness to acres upon acres of trash hills congressing into six-story mountain peaks that then cut down into river valleys eroded into existence by the slow drip of water — what locals called the "soup" — oozing forth from compacted food waste. "Even mosquitoes won't breed in this water," Zakariah Njoroge, leader of the Jobless Millionaires, one of the eleven trash-picking gangs of Dandora, told me.[2]

In the northwest corner of Dandora, one of the many Volvo excavators used to routinely redistribute its garbage — to plow out "highways" that can hustle yet more waste into the site — was buried by one of the great trash avalanches of the 1990s, and now sits trapped beneath millions of tons of refuse like a prehistoric mastodon caught midmovement in a tar pit. More striking still are the shoulders of these "highways": In the resulting escarpments of trash, which currently rise on either side as high as seventy feet, it's possible to make out a stratigraphy dating back forty years, a tree-ring-like register of Nairobi's daily detritus over the last generation. The bottommost layers consist of rusty aluminum and ash, before elevating into strata of clothes and cardboard, and then, approximately a quarter of the way up the scarp, in waste teams dating to the late 1980s, *boom!* The first inroads of plastic into the African continent, the trickle that became a deluge. Today, one out of every two plastic bottles discarded in Kenya has been produced by — and has enriched — Coca-Cola.[3] Half the cattle that wander the country's urban areas have been found to possess plastic bags in their stomachs.[4] According to the World Bank, a shocking 70 percent of the plastic discarded in Kenya ends up in a water system of one type or another.[5]

For years, Kenyan politicians — some of whom own the truck companies that get paid to ferry the five million pounds of trash that enter Dandora every day from every neighborhood of Nairobi — have pledged to shut East Africa's biggest dump down and zone a successor site in the scrublands south of their capital, a long-term solution to which the greatest unresolved problem may not even be the preposterous quantities of garbage that must now be trucked dozens of miles to the opposite edge of Nairobi, or even the costs of doing so in a country that can only afford to devote a pittance to waste management. No, the greatest unresolved problem may be what such accumulations of open-air garbage in East Africa have become prone to attracting: a predatory species of sub-Saharan stork known as the marabou, whose population (some hundreds) and wingspans (some ten feet) pose a flight risk to jetliners ascending and landing at Jomo Kenyatta International Airport, which sits near the proposed new destination for Nairobi's waste. Trash attracts birds; birds can take down airplanes; the waste at Dandora therefore must stay put — and, alas, keep on piling up.

Such was the situation Kenya found itself in when, in early 2020, its government began engaging in new rounds of bilateral trade talks with the United States. Tucked within a leaked draft of the trade agreement was a remarkable proposal that had been crafted with the input of the American Chemistry Council, a lobbying wing of the petrochemical industry. Kenya had an unsustainable plastic problem. The United States offered to help solve it by introducing more effective recycling systems to the country. First, however, Kenya would have to agree to reverse its ban on single-use plastic bags — Africa's most successful such ban — and open itself up to greater plastic imports from the United States itself. To become more "efficient" at "recycling" plastic, in other words, Kenya would first need to agree to accept great new quantities of it from the world's richest country.[6]

It's worth noting that the leaked trade agreement didn't advocate Kenya becoming a future hub of just any kind of recycling. Because the then-cratering prices of oil had made it not worthwhile to even attempt to source new plastic from old plastic, and because conventional recycling had been so ably dismantled as a solution to the plastic epidemic, a new strategy had been concocted with which to approach waste-embattled nations like Kenya. The petrochemical industry called it "advanced chemical recycling," though there was nothing very advanced about it, and it could hardly be said to constitute any form of recycling: "Advanced chemical recycling" requires melting plastic waste down into a filthy liquid fuel, a process that has been proven to be more carbon-intensive and damaging to the environment than that of extracting crude oil from the ground, and amounts to a disastrous solution to a problem that has already been largely solved by renewable energy.[7]

Just months earlier, Kaluki Paul Mutuku, a young environmentalist from the onetime British colonial capital of Machakos, had waged a successful campaign to prevent the first ever coal plant in Kenya — to be financed by Chinese capital and feature a smokestack rising higher than any other structure in East Africa — from being constructed in a stretch of mangrove coast opposite Lamu, a UNESCO World Heritage Site along his country's Indian Ocean coast.[8] Now, in the early months of the COVID pandemic, at a time when plastic production had become more vital than ever to the petrochemical industry's future profit margins, a new threat had come to Kenya, camouflaged as an effort to help the country rid itself of a scourge that rich countries themselves had introduced.

"The West has always portrayed Africa as a poor continent — and insisted that solutions to our problems must always come from them," Mutuku told me. "Now Western countries were telling us that they had to teach us how to recycle and that we needed to take in their plastic in order to do so."[9]

Over the next few months, Mutuku and a handful of other environmentalists organized a campaign against plastic importation into Kenya. "Africa is not a dumpster!" they scrawled across posters and circulated around social media. "Big oil industries and global corporations are ever creating the rhetoric of plastic as a necessity in today's world," they wrote in a mission statement, "while also escaping the huge responsibility of managing the generated plastic waste." They called for marches through the streets of Nairobi, forged alliances with NGOs such as the Los Angeles–based Global Children's Campaign and Polluters Out, and pleaded with members of the US Congress, who in turn relayed their concerns to President Donald Trump, whose administration had done more to prop up the plastics industry than any since that of Ronald Reagan.

"While many Americans believe they are recycling their plastic when they sort it at home for collection," read an October 1, 2020, letter to the White House signed by sixty-two members of Congress, "this plastic often ends up as waste in developing countries with poor waste management capabilities and ultimately finds its way into rivers, oceans, and landscapes."[10]

The future of plastic importation into Kenya remains uncertain. But if Mutuku's campaign sounds familiar — an African railing against Western waste exports to his continent — that's because it might just as easily have been waged forty years ago. One need not try to imagine the consternation of an earlier generation of environmental activists who thought the international waste trade had been eradicated by the fury of the 1980s and the legislation of the 1990s; many of them are still alive.

"The fact remains that out of all we did, nothing came of it," Ihekwoaba Chimezie told me. In 1987, Chimezie was a student studying abroad in Pisa, Italy, when he read a local newspaper article about a scheme to ship hundreds of drums of PCB from Milan to his native land of Nigeria, and proceeded to alert his country's press. "Waste and toxic wastes continue to be shipped into Africa and other helpless places in the developing world. I've personally given up hope."[11]

The crime of the waste trade isn't merely that it occurs. It's that after a generation of broad recognition of the inequalities it exacerbates and the injustices it perpetuates, after a generation of legislation against the movement of garbage out of the developed world and a generation of outrage emanating out of the developing world, waste has never stopped traveling from north to south. It's a dereliction that bears certain similarities to international failures to address the climate crisis, though in many respects trash export is a more confounding problem. Atmospheric carbon is an invisible phenomenon. Mounds of foreign garbage rotting in a distant land could hardly present a more incontestable one.

The trash trade has historically sustained itself by purporting to offer *solutions* to environmental quandaries rather than deepening them. Moving waste has been said to offer jobs and cash and resources to emerging nations. It distributes pollution levels more evenly across the globe. It turns useless trash into economic input.

And yet the constant intellectual contortions to justify waste's movement have gone a long way toward avoiding the real problem sitting in plain sight: We must offload our garbage onto poorer countries for the simple reason that we produce far too much of it in the first place—and then we insist on an environment liberated of its consequences.

All too often this injustice has fallen upon scattered individuals—governors in Guatemala, Greenpeace researchers in Maine, chemists in Nigeria, striking shipyard workers in Turkey, marine biologists in Java, activists in Kenya—to expose and attempt to correct. Yet any lasting solution to the waste crisis must start at the point where such a systemic problem begins: requiring waste's perpetuators—tech companies, cruise ship operators, petrochemical conglomerates—to become financially liable for the fate of that which they insist on overproducing. The situation as it currently stands is one that would be all too recognizable to Vance Packard pillorying the American throwaway society six decades ago: one in which all the monetary benefits of generating trash exist to be reaped by corporations and their shareholders, while all the problems are foisted off onto a taxpaying public—or, indeed, a distant country. To solve the waste crisis, and the problems compounded by waste movement, gratuitous production must be turned into a financial burden rather than be allowed to remain a profit boon.

It is not the case that no waste should ever move across borders.

CONCLUSION

Scrap steel and old paper are two examples of truly recyclable detritus that benefit the planet through their circulation into secondhand markets — even if, as this book has attempted to show, various dangers also exist in sending those materials to the Global South. But the waste trade as we know it today began as a parasitic subset of Western industry that exploited global inequalities to save money, not the planet. This remains the motive. Since the 1980s, growing quantities of waste have been diverted from rich countries to poor not because they *can* be safely and effectively recycled but precisely because they *cannot*. It remains easier, cheaper, and more aesthetically attractive to offload the material elsewhere.

Another antidote to today's waste trade is a more earnest public conversation about what it is that we are discarding in the first place. In 1962, Rachel Carson almost single-handedly spawned the postwar environmental movement through her probing of the deadly chemical underbelly of US prosperity. It was an indispensable intervention that never got broadly updated for the multiplying conveniences — smartphones, DVD players, FIJI water bottles — of the twenty-first century, all of which are infused with innumerable but scarcely understood toxic chemicals and contaminants. It's an introspection that would go a long way toward exposing the globalization of many sectors of "recycling" for what they are: a continuation of the hazardous waste trade that spawned so much international outrage in the 1980s and had, or so it seemed, been eradicated in the 1990s.

Nowhere is this problem more alarming than with plastic, the material with which the waste trade reaches a point of unsustainable absurdity and in which timescales — an object taking seconds to consume and requiring tens of thousands of years to go away — are twisted out of all proportion. While the green revolution may reduce our reliance on hydrocarbons as our primary energy source, it's a shift the petrochemical industry has been strategically anticipating for years now. New plans for enrichment are afoot. The fossil future will not be in energy, no, but in consumption, in overpackaging all the plastic stuff you get sold with *more* plastic. And the engines of this transition are already being erected everywhere you look.

Be it along the Gulf Coast of Texas, where Saudi petrocapital and a billion dollars in local tax breaks have conjoined to fund the world's largest plastic plant, capable of manufacturing two million tons of plastic a year and releasing into the atmosphere as much carbon dioxide every year as 430,000 automobiles, or outside Pittsburgh, where Pennsylvania's largest industrial project since the Second World War has already begun pumping out nearly two million more tons of plastic a year, much of it for export, the drive to divest the meteoric profits of the recent shale gas boom in the United States and the natural gas export surge in the Middle East is charting a future of boundless plastic.[12] In North America alone, nearly two hundred new petrochemical plants are currently under construction, an investment totaling some $200 billion — new synthetics for a species that already manages to replicate its weight in new plastic every year.[13] By 2050, it is estimated that humanity will have produced *four* times the *total* amount of plastic it has produced up to the present; half will be single-use.[14]

In spite of the advances of the green transition — though in large part because of them — we are heading toward a future in which plastic ceases to be the commodified byproduct of our global energy regime, and becomes instead *the* fundamental driver of hydrocarbon extraction.

The world of waste export has traditionally been, yes, a hodgepodge of mom-and-pop importers and exporters. But this is changing. Waste and its movement are becoming increasingly financialized. The consulting firm McKinsey & Company publishes regular memos on trash, insisting that the problem of waste entering the ocean is not so much an environmental calamity as a missed financial opportunity; your average ton of uncollected trash is worth $375, claims McKinsey, even as it makes zero mention of plastic's toxicity and concedes that the main problem with plastic today is no different from that of the 1980s or even the 1960s: how to accumulate enough uniform types of plastic in "meaningful volumes" so as to make the "recycling" process "viable."[15]

At the same time, the most powerful industry in human history, fossil fuels, has found enthusiastic allies on Wall Street. Asset managers like Capital Group ("ESG investing that thinks not just in years, but in decades") now hold hundreds of millions of dollars' worth of shares in polymer producers, while banks like HSBC ("Join us on the journey towards a more sustainable future and a global net zero economy") lend billions to their parent companies.[16] You don't just discard plastics. You may be unwittingly invested in their unending production.

Like laundered cash, traveling trash is increasingly camouflaging its national origins by getting rerouted through "waste havens" like Poland, Singapore, Hong Kong, and Thailand.[17] Even notorious financial lairs themselves are emerging as garbage magnets. Dubai is currently constructing the biggest plastic waste incinerator in the Middle East, a billion-dollar enterprise that will burn the equivalent of a thousand dump trucks of trash a day, much of it arriving from neighboring Gulf states. The region that got stupefyingly rich off oil may soon be keeping its lights on with torched water bottles and yogurt cups, a process that is categorically worse for the planet than burning coal, even as it gets euphemized by investment groups like BlackRock as a "circular economy."[18] (If plastic possesses a single environmental benefit, it is that it is sequestered carbon; burning it releases all that carbon into the atmosphere, in addition to a host of toxic additives.) And when it comes to organized crime, whose ties with certain systems of domestic waste management have never been much of a secret, the business of international waste *movement* has never appeared more attractive.

"You have groups getting out of the drug and weapons trade and entering the waste one. The risk is so much lower, the reward so much bigger," Joseph Poux, who oversees Interpol's efforts to track illegal waste shipments, told me. "With drugs, you have to find a supplier and you have to get it to the right people. With waste, there's no shortage — and the supply is free."[19]

CONCLUSION

As the worlds of high finance and organized crime converge around its movement, as more and more of it gets manufactured and more and more of it must go up in flames, there's nevertheless an ongoing effort by its profiteers to portray plastic as an imminently solvable problem. The foremost example is that of The Ocean Cleanup, a nonprofit that has developed robotic "interceptors" capable of trawling trash out of oceans and rivers. While there's significant doubt over how effective such campaigns can be at removing, say, microplastics, in certain respects such speculation is beside the point. Like entrusting the safety of your house to the thieves who just robbed it, the financiers of The Ocean Cleanup — Maersk, Coca-Cola — happen to be the very perpetrators of the problem it alleges to be fixing.[20] The purpose of such initiatives seems to be not so much to remove plastic out of one sea or another as to make plastic despoliation appear like a reversible phenomenon, one liable to eventual solving through NGO do-goodery or scaled technological breakthroughs.

Simply put: Hundreds of billions of dollars go into producing plastic, and billions of dollars go into scrambling for a solution to it.

If there's any one lesson to be drawn from the last seventy years of synthetic output, it's that the petrochemical industry's profit model has always depended on sustaining the fiction of its unproblematic disposal. The longer that *some* salvation from the plastic pandemic appears to be *somewhere* on the horizon — incineration, recycling, advanced recycling, ocean cleanups — the more time the industry has clinched for itself to sustain production levels that climb by the day. And the most illusory solution of all is waste export. As long as plastic keeps getting physically diverted by those who consume it the most, the farther from public concern — and political action — it is likely to remain.

Three years after I first met İzzettin Akman in southeastern Turkey, I contacted him again. I asked him how his orange and lemon trees were doing. In late 2016, as you'll recall, they had narrowly avoided irreversible devastation when a truckload of European garbage was set alight at the edge of Akman's farm, killing off much of the local bee population and saturating his irrigation system with microplastics and contaminants. What now? I asked Akman. Had it happened again? Had anyone ever been held responsible for what occurred?

Nothing has changed, Akman replied. Construction trucks full of British and German trash still pass by his farm several times a month, arriving from the Port of Mersin sixty miles to the south and hustling Europe's waste to a cement plant a few kilometers from his groves, where the bales of foreign plastic waste are incinerated as fuel. Before Akman sees the trucks, he told me, he hears them. They are "louder than a cement truck," letting out a distinct "grind when the driver shifts gears," a harsh clank that reverberates across his fields and ricochets between the trunks of his citrus trees.

Or almost nothing has changed, Akman said. One thing has. He has started fighting back. He won't risk a second mountain of garbage getting clandestinely set alight on the edge of his property again, no.

Today, every time he hears the juddering of the construction truck and the grinding of its gears, which usually comes "every ten days or so," a "little after eight in the morning," when "few other sounds are to be heard," Akman hops into his gray Mitsubishi pickup, speeds down the dirt road leading out of his farm, and follows the garbageman as he jounces toward the cement plant. It has become a ritual. Akman makes a show of staying just a hundred feet or so behind the truck. Sometimes he makes eye contact with the driver through the side mirror to remind him that he's there. After five minutes or so spent pursuing the vehicle, Akman watches — to be "certain beyond doubt" — as it turns into the cement plant, then disappears, its fifteen tons of putrid foreign plastic waste hustled from view.

Akman is a still a farmer, yes, just like many generations of Akmans before him, but he has also become something else, something that only a decade or so ago would have struck him as strange, almost incomprehensible. He has become a lonesome sheriff against a system of globe-spanning waste mismanagement that, owing to inscrutable logic, and driven by rank unfairness, has narrowed in on his ancestral farmland.

"I'll keep following the trucks until they stop coming," Akman told me the last time we spoke. "Or until the world stops sending them."[21]

ACKNOWLEDGMENTS

WASTE WARS COULD not have been started without my agent, William Callahan, or completed without my editor, Vivian Lee; I am immensely grateful to both. I owe special thanks as well to Oxford's St. Antony's College, the Berggruen Institute, and the Whiting Foundation, which provided the libraries and funding for my research and writing. I am also indebted to those individuals who offered help and hospitality throughout my research and travels, many of whom have become good friends along the way: Joe Zigmond, Ian Straus, Asli Osman, Can Sezer, Holly Birkett, Liza Grandia, Marco Puga, Guillermo Muñoz, Daniel del Pinal, Luis Solano, Erwin Garzona, Jim Puckett, Jim Vallette, Annie Leonard, Daniel Mendelsohn, Sarah Murray, Jan Dell, Judith Enck, Sedat Gündoğdu, Muntaka Chasant, Alejandro Arriaza, Nicola Mulinaris, Gerald Anderson, Zakariah Njoroge, Bjorn Beeler, Jonathan Blake, Nils Gilman, Emily Rose Anderson, Nick Mulder, Xenia Kounalaki, Thomas Meaney, Mallory Marshall, Kelsey Brendel, Ratik Asokan, Francisco Asturias, Prigi Arisandi, Kathleen Azali, Daru Setyorini, Yuyun Ismawati, Jorge Cabrera Hidalgo, Eduardo Cofiño, Rosa María Chan, Nathan Fruchter, Roger Goodman, Mark Brenner, Courtney Hodell, Dianna Stirpe, and Hilary McClellen.

Fiskardo, Greece,
August 2024

NOTES

INTRODUCTION: MAYHEM IN MESOPOTAMIA

1. "The Electronic Text Corpus of Sumerian Literature," faculty of Oriental Studies, University of Oxford, accessed March 22, 2023, etcsl.orinst.ox.ac.uk/proverbs/t.6.1.01.html.
2. İzzettin Akman, personal conversation, April 18, 2022.
3. Borzou Daragahi, "Turkey to Launch Groundbreaking Restrictions on Plastic Bags in Fight Against Pollution," *Independent,* December 10, 2018, independent.co.uk/news/world/europe/turkey-plastic-bags-ban-law-supermarkets-apk-pollution-waste-government-a8676221.html.
4. Jörg Taszmann, "Society's Excrement," *Deutsche Welle,* September 12, 2012, www.dw.com/en/uncovering-societys-excrement/a-16435164.
5. United Nations Development Project, "The Zero Waste Project Receives UNDP Turkey's First Global Goals Action Award," news release, March 22, 2021, undp.org/turkiye/press-releases/zero-waste-project-receives-undp-turkeys-first-global-goals-action-award.
6. Daily Sabah with Iha, "Turkish Zero Waste Project Is Not Just Campaign It Is an Emotion," *Daily Sabah,* January 16, 2023, dailysabah.com/turkey/turkish-zero-waste-project-is-not-just-campaign-it-is-an-emotion/news.
7. Diyar Guldogan, "Türkiye Explains Zero-Waste Project to World via Its Missions: Turkish Foreign Minister," *Anadolu Agency,* January 17, 2023, aa.com.tr/en/environment/turkiye-explains-zero-waste-project-to-world-via-its-missions-turkish-foreign-minister/2789838.
8. Laura Parker, "China's Ban on Trash Imports Shifts Waste Crisis to Southeast Asia," *National Geographic,* November 16, 2018, nationalgeographic.com/environment/article/china-ban-plastic-trash-imports-shifts-waste-crisis-southeast-asia-malaysia.
9. While China couldn't completely ban imports of plastic waste due to World Trade Organization free trade agreements, China could restrict imports to more than 0.5 percent contamination, which was effectively a total ban.

NOTES

10 Agnieszka Wądołowska, "Special Unit Created to Fight 'Trash Mafia' That Illegally Burn and Dump Waste in Poland," *Notes from Poland*, August 12, 2020, notesfrompoland.com/2020/08/12/special-unit-created-to-fight-trash-mafia-that-illegally-burn-and-dump-waste-in-poland/; and Katy Lee, "On the French Border, Drowning in a Sea of Trash," *Politico*, February 24, 2020, politico.eu/article/french-border-town-struggle-deluge-illegal-trash-redange/.

11 George Kiernan and Erin Kiely, "How Brexit Is Affecting the UK's Recycling Industry," Let's Recycle It, September 14, 2021, letsrecycleit.eu/news/the-uk-recycling-industry/?utm_source=rss&utm_medium=rss&utm_campaign=the-uk-recycling-industry.

12 George Monbiot, "Britain Through the Looking Glass: My Dead Goldfish Is Now a Registered Waste Disposer," *The Guardian*, December 24, 2021, theguardian.com/commentisfree/2021/dec/24/dead-goldfish-licensed-waste-disposer-system-falling-apart.

13 Greenpeace International, "Investigation Finds Plastic from the UK and German Illegally Dumped in Turkey," news release, May 17, 2021, greenpeace.org/international/press-release/47759/investigation-finds-plastic-from-the-uk-and-germany-illegally-dumped-in-turkey/#:~:text=The%20UK%20exported%20210%2C000%20tonnes,recycled%20is%20actually%20sent%20overseas.

14 Kit Chettel and Wojciech Moskwa, "A Plastic Bag's 2000-Mile Journey Shows the Messy Truth About Recycling," Bloomberg Quicktake, March 29, 2022, YouTube video, 6:16, youtube.com/watch?v=AfOkRpko6eY.

15 Sedat Gündoğdu, personal conversation, May 9, 2022.

16 Alex Hithersay, "Petrochemical Mega-Complex to Be Built in Turkey," *Hydrocarbon Engineering*, September 21, 2018, hydrocarbonengineering.com/petrochemicals/21092018/petrochemical-mega-complex-to-be-built-in-turkey/.

17 Emily Elhacham, Liad Ben-Uri, Jonathan Grozovski, Yinon M. Bar-On, and Ron Milo, "Global Human-Made Mass Exceeds All Living Biomass," *Nature* 588 (2020): 442–44.

18 Quoted in Richard Heinberg, *Afterburn: Society Beyond Fossil Fuels* (New Society Publishers, 2015).

19 John Atlee Kouwenhoven, "Waste Not, Want Not," *Harper's Magazine*, March 1959, https://harpers.org/archive/1959/03/waste-not-have-not/.

20 Elhacham, Ben-Uri, Grozovski, Bar-On, and Milo, "Global Human-Made Mass Exceeds All Living Biomass." Statistic applies to North America and is taken from *The Story of Stuff*, written by Annie Leonard, Louis Fox, and Jonah Sachs, and directed by Louis Fox (Free Range Studios, 2007), 21 min., storyofstuff.org/movies/story-of-stuff/.

21 Diana Ivanova, Konstantin Stadler, Kjartan Steen-Olsen, Richard Wood, Gibran Vita, Arnold Tukker, and Edgar G. Hertwich, "Environmental Impact Assessment of Household Consumption," *Journal of Industrial Ecology* 20, no. 3 (2015): 526–36.

22 "Fact Sheet: Single-Use Plastics," Earth Day, accessed February 21, 2023, earthday.org/fact-sheet-single-use-plastics/.

NOTES

23 Darryl Fears, "There's Literally a Ton of Plastic Garbage for Every Person on Earth," *Washington Post*, July 19, 2017, washingtonpost.com/news/energy-environment/wp/2017/07/19/theres-literally-a-ton-of-plastic-garbage-for-every-person-in-the-world/; and Michael Birnbaum, "There Are 21,000 Pieces of Plastic in the Ocean for Each Person on Earth," *Washington Post*, March 8, 2023, washingtonpost.com/climate-environment/2023/03/08/ocean-plastics-pollution-study/.

24 James Pennington, "Every Minute, One Garbage Truck of Plastic Is Dumped into Our Oceans. This Has to Stop," World Economic Forum, October 27, 2016, weforum.org/agenda/2016/10/every-minute-one-garbage-truck-of-plastic-is-dumped-into-our-oceans/#:~:text=Every%20minute%2C%20one%20garbage%20truck,is%20dumped%20into%20our%20oceans.

25 "Introducing the Interceptor Barricade: The Ocean Cleanup Returns to Guatemala," The Ocean Cleanup, May 16, 2023, theoceancleanup.com/updates/introducing-the-interceptor-barricade-the-ocean-cleanup-returns-to-guatemala/.

26 Uzair Hasan Rizvi, "India Rubbish Mountain to Rise Higher Than Taj Mahal," *Phys.org*, June 4, 2019, phys.org/news/2019-06-india-rubbish-mountain-higher-taj.html.

27 Amia Srinivasan, "What Have We Done to the Whale?" *The New Yorker*, August 17, 2020, https://www.newyorker.com/magazine/2020/08/24/what-have-we-done-to-the-whale.

28 Nell Lewis, "Drones Are Helping to Clean Up the World's Plastic Pollution," *CNN*, June 23, 2021, edition.cnn.com/2021/06/23/europe/ellipsis-drone-plastic-pollution-c2e-spc-intl/index.html.

29 Randeep Ramesh, "Paradise Lost on Maldives' Rubbish Island," *The Guardian*, January 2, 2009, theguardian.com/environment/2009/jan/03/maldives-thilafushi-rubbish-landfill-pollution.

30 Peggy Hollinger and Sam Learner, "How Space Debris Threatens Modern Life," *Financial Times*, June 7, 2022, ig.ft.com/space-debris/.

31 *Waste Export Control: Hearing Before the Subcommittee on Transportation and Hazardous Materials of the Committee on Energy and Commerce*, US House of Representatives, 101st Congress, 1st Session on HR 2525 (US Government Printing Office, 1989), 34.

32 *Waste Atlas: The World's 50 Biggest Dumpsites, 2014 Report* (D-Waste.com, 2014), nswai.org/docs/World%27s%20Fifty%20biggest%20dumpsites,Waste%20Atlas%202014.pdf.

33 Nathan Fruchter, personal conversation, December 7, 2023.

34 Patty Moore, personal conversation, November 15, 2023.

35 Teodor Niță, personal conversation, September 16, 2020.

36 Michael G. Faure and Kévine Kindji, *Environmental Crime Affecting EU Financial Interests, the Economic Recovery, and the EU's Green Deal Objectives* (European Parliament's Committee on Petitions, October 2022), europarl.europa.eu/RegData/etudes/STUD/2022/737869/IPOL_STU(2022)737869_EN.pdf; and Diana Barrowclough, Carolyn Deere Birkbeck, and Julien Christen, *Global Trade in Plastics: Insights from the First Life-Cycle Trade Database* (Research Paper No. 53, United

NOTES

Nations Conference on Trade and Development, December 2020), unctad.org/publication/global-trade-plastics-insights-first-life-cycle-trade-database.

37 Joseph Poux, personal conversation, November 25, 2023.

1. BANANA REPUBLIC

1 Bonar Ludwig Hernández, *Bananas, Ports, and Railroads: A Historiographical Essay on the United Fruit Company in Guatemala, 1901–1944* (n.d.), edisciplinas.usp.br/pluginfile.php/7190911/mod_resource/content/1/Historiography_US_Guatemala.pdf.
2 Erwin Garzona, personal conversation, May 25, 2023.
3 Jerry Schwartz, "Garbage Ash That Nobody Wanted," Associated Press, September 17, 2000, archive.seattletimes.com/archive/?date=20000917&slug=4042847.
4 *Waste Export Control: Hearing Before the Subcommittee on Transportation and Hazardous Materials of the Committee on Energy and Commerce,* US House of Representatives, 101st Congress, 1st Session on HR 2525 (US Government Printing Office, 1989), 207.
5 "Toxic," *San Francisco Examiner,* February 14, 1988, newspapers.com/newspage/461247414/.
6 See, for instance, the story of the Arimany family in Alvaro Gálvez, "Desechos tóxicos causarán daños irreparables en bahia de Amatique," *Prensa Libre,* August 9, 1992, accessed May 16, 2023, in La Biblioteca Nacional de Guatemala "Luis Cardoza y Aragón."
7 "Ex gobernadora sindica de actos ilícitos a Serrano y familiares," *Prensa Libre,* November 5, 1992; and "Tobar Franco pide cese de repression contra ex gobernadora," *Prensa Libre,* December 10, 1992.
8 Carlos Morales Monzón, "Pugna de incriminaciones," *Crónica,* December 29, 1992.
9 Lilian Vásquez de Guzmán, personal conversation, May 26, 2023.

2. THE CHEMICAL CENTURY

1 Rachel Carson, *Silent Spring* (Houghton Mifflin, 1962), chaps. 2, 17.
2 Nava Atlas, "Silent Spring by Rachel Carson (1962): An Environmental Classic," *Literary Ladies Guide,* June 13, 2022, literaryladiesguide.com/book-reviews/silent-spring-by-rachel-carson-1962-an-environmental-classic/.
3 Simone M. Müller, *The Toxic Ship* (Univ. of Washington Press, 2023), 95.
4 Bill D. Moyers and Center for Investigative Reporting, *Global Dumping Ground: The International Traffic in Hazardous Waste* (Seven Locks Press, 1990), 37.
5 Simone M. Müller, "Hidden Externalities: The Globalization of Hazardous Waste," *Business History Review* 93, no. 1 (2019): 55.
6 "Hidden Externalities," 95–96.
7 Moyers and Center for Investigative Reporting, *Global Dumping Ground,* 37.

NOTES

3. CASH FOR TRASH

1. Memo quoted in Michael Weisskopf, "World Bank Official's Irony Backfires," *Washington Post*, February 9, 1992, washingtonpost.com/archive/politics/1992/02/10/world-bank-officials-irony-backfires/4241737a-bd09-4052-a153-df7ff428b859/.
2. Emily Brownell, "Negotiating the New Economic Order of Waste," *Environmental History* 16, no. 2 (2011): 272.
3. Charles E. Davis and James P. Lester, "Hazardous Waste Politics and the Policy Process," in *Dimensions of Hazardous Waste Politics and Policy* (Greenwood Press, 1988), 2–3.
4. Laura A. Strohm, "The Environmental Politics of the International Waste Trade," *Journal of Environment and Development* 2, no. 2 (1993): 133; and James Brooke, "Waste Dumpers Turning to West Africa," *New York Times*, July 17, 1988, nytimes.com/1988/07/17/world/waste-dumpers-turning-to-west-africa.html.
5. Valentina O. Okaru, "The Basel Convention: Controlling the Movement of Hazardous Wastes to Developing Countries," *Fordham Environmental Law Review* 4, no. 2 (2011): 141, ir.lawnet.fordham.edu/cgi/viewcontent.cgi?referer=&httpsredir=1&article=1351&context=elr.
6. Jennifer Clapp, *Toxic Exports: The Transfer of Hazardous Wastes from Rich to Poor Countries* (Cornell Univ. Press, 2001), 23.
7. Okaru, "The Basel Convention," 139.
8. *Waste Export Control: Hearing Before the Subcommittee on Transportation and Hazardous Materials of the Committee on Energy and Commerce*, US House of Representatives, 101st Congress, 1st Session on HR 2525 (US Government Printing Office, 1989), 227.
9. *Waste Export Control*, 1st Session on HR 2525, 213.
10. *Waste Export Control*, 1st Session on HR 2525, 298.
11. *Waste Export Control*, 1st Session on HR 2525, 211.
12. F. James Handley, "Hazardous Waste Exports: A Leak in the System of International Legal Controls," *Environmental Law Reporter* (Environmental Law Institute) 19, no. 4 (1989): 10171.
13. *Waste Export Control*, 1st Session on HR 2525, 300.
14. James Brooke, "Waste Dumpers Turning to West Africa, *New York Times*, July 17, 1988, nytimes.com/1988/07/17/world/waste-dumpers-turning-to-west-africa.html.
15. *Waste Export Control*, 1st Session on HR 2525, 182–83.
16. Quoted in Sylvia F. Liu, "The Koko Incident: Developing International Norms for the Transboundary Movement of Hazardous Waste," *Journal of Natural Resources & Environmental Law* 8, no. 1, article 9 (1992): 126, uknowledge.uky.edu/cgi/viewcontent.cgi?article=1144&context=jnrel.

4. DEBT AND DEVELOPMENT

1. *Waste Export Control: Hearing Before the Subcommittee on Transportation and Hazardous Materials of the Committee on Energy and Commerce*, US House of Representatives, 101st Congress, 1st Session on HR 2525 (US Government Printing Office, 1989), 167.

NOTES

2. Jack Anderson and Dale Van Atta, "How Not to Run a Country," *Washington Post*, March 4, 1990, washingtonpost.com/archive/opinions/1990/03/04/how-not-to-run-a-country/91078c16-08c2-4b9a-bf7b-37d1d0f05c30/.
3. *Waste Export Control*, 1st Session on HR 2525, 238.
4. "Guard Kills Dahomey Minister Found with President's Wife," *New York Times*, June 22, 1975, nytimes.com/1975/06/22/archives/guard-kills-dahomey-minister-found-with-presidents-wife.html.
5. Anderson and Van Atta, "How Not to Run a Country."
6. World Bank, *Benin—Country Economic Memorandum (Vol. 3): The Public Enterprises Sector (English)* (World Bank Publications, 1984), http://documents.worldbank.org/curated/en/175091468208735243/The-public-enterprises-sector.
7. Emily Brownell, "Negotiating the New Economic Order of Waste," *Environmental History* 16, no. 2 (2011): 270.
8. *Waste Export Control*, 1st Session on HR 2525, 161.
9. Emily Brownell, "International Trash and the Politics of Poverty: Conceptualizing the Transnational Waste Trade," in *Nation-States and the Global Environment*, ed. Erika Marie Bsumek, David Kinkela, and Mark Atwood Lawrence (Oxford Univ. Press, 2013), 257.
10. Associated Press, "W. Africa Faces Economic Boon in Toxic Shroud," *DeseretNews*, June 19, 1988, deseret.com/1988/6/19/18769116/w-africa-faces-economic-boon-in-toxic-shroud.
11. *Resource Conservation and Recovery Act Reauthorization: Hearing Before the Subcommittee on Transportation and Hazardous Materials of the Committee on Energy and Commerce*, US House of Representatives, 101st Congress, 2nd Session on HR 3735, HR 3736, and HR 3737 (US Government Printing Office, 1990), 583.
12. Frantz Fanon, *The Wretched of the Earth* (Grove Atlantic, 2007), 4.
13. "Toxic Waste and Mental Decolonialism," *African Concord*, July 5, 1988.
14. Valentina O. Okaru, "The Basel Convention: Controlling the Movement of Hazardous Wastes to Developing Countries," *Fordham Environmental Law Review* 4, no. 2 (2011): 137, ir.lawnet.fordham.edu/cgi/viewcontent.cgi?referer=&httpsredir=1&article=1351&context=elr.
15. Matthew G. Sohm, "'Big Clean,' the 'Death Ship,' and the Hazardous Waste Trade Between West Germany and Turkey, 1987–1988," *Contemporary European History* 33, no. 2 (2022): 459–76.
16. Jim Vallette and Heather Spalding, *The International Trade in Wastes: A Greenpeace Inventory* (Greenpeace USA, 1990), 43.
17. F. James Handley, "Hazardous Waste Exports: A Leak in the System of International Legal Controls," *Environmental Law Reporter* 19, no. 4 (1989): 10171–82.
18. *Resource Conservation and Recovery Act Reauthorization*, 2nd Session on HR 3735, HR 3736, and HR 3737, 72, 296.
19. Herbert Inhaber, *Slaying the NIMBY Dragon* (Routledge, 2018), 196.
20. *Marshall Islands Journal*, September 8, 1989.
21. *Resource Conservation and Recovery Act Reauthorization*, 2nd Session on HR 3735, HR 3736, and HR 3737, 296.

NOTES

5. MERCHANTS OF DISEASE

1 *Waste Export Control: Hearing Before the Subcommittee on Transportation and Hazardous Materials of the Committee on Energy and Commerce,* US House of Representatives, 101st Congress, 1st Session on HR 2525 (US Government Printing Office, 1989), 167.
2 Reuters news reports July 10, 1988, item 254; and July 11, 1988, item 235, in *Waste Export Control,* 1st Session on HR 2525, 189.
3 *Resource Conservation and Recovery Act Reauthorization: Hearing Before the Subcommittee on Transportation and Hazardous Materials of the Committee on Energy and Commerce,* US House of Representatives, 101st Congress, 2nd Session on HR 3735, HR 3736, and HR 3737 (US Government Printing Office, 1990), 587.
4 Blaine Harden, "Africans Turn to Hostages in Battle Against Foreign Waste," *Washington Post,* July 16, 1988.
5 Roland Straub, personal conversation, December 14, 2021.
6 K. A. Gourlay, *World of Waste* (Zed Books, 1992), 12–13.
7 *Waste Export Control,* 168.
8 Javier Salinas Cesareo, "Construirían 25 plantas tratadoras de desechos en Tecámac, Edomex; el edil otorgó los permisos," *Jornada,* March 17, 2004, jornada.com.mx/2004/03/17/038n1est.php?printver=1&fly=.
9 Matthew Sohm, "'Big Clean,' the 'Death Ship,' and the Hazardous Waste Trade Between West Germany and Turkey, 1987–1988," *Contemporary European History* 33, no. 2 (2022): 459–76.
10 "Greenpeace Warns of Toxic Waste Dumping in Central America," United Press International, August 6, 1992, upi.com/Archives/1992/08/06/Greenpeace-warns-of-toxic-dumping-in-Central-America/1629713073600/.
11 For the best account of the Somali allegations, see Eric Herring, Latif Ismail, Thomas Bligh Scott, and Jaap J. Velthuis, "Nuclear Security and Somalia," *Global Security: Health, Science and Policy* 5, no. 1 (2020): 1–16, doi.org/10.1080/23779497.2020.1729220.
12 See Christian Bueger, "Practice, Pirates, and Coast Guards: The Grand Narrative of Somali Piracy," *Third World Quarterly* 34, no. 10 (2013): 1811–27.
13 Bill Lambrecht, personal conversation, October 9, 2023.
14 Sam Oduche, personal conversation, October 15, 2023.

6. GUNS AND GERMS

1 "Sense of House Regarding Family Planning Programs," 145 Cong. Rec. H1510–25 (March 23, 1999), govinfo.gov/content/pkg/CREC-1999-03-23/html/CREC-1999-03-23-pt1-PgH1510-2.htm.
2 Gail Vittori, personal conversation, July 3, 2023.
3 Greg Grandin, *Empire's Workshop: Latin America, the United States, and the Rise of the New Imperlialism* (Holt Paperbacks, 2007).
4 Norman B. Schwartz, "Colonization of Northern Guatemala: The Petén," *Journal of Anthropological Research* 43, no. 2 (1987): 176.

5 Directorate of Intelligence, *Guatemala: Development and Insurgency in the Northern Frontier, An Intelligence Assessment* (Central Intelligence Agency, January 1, 1983), 10, cia.gov/readingroom/docs/DOC_0000720641.pdf.
6 "Desechos tóxicos y radiactivos tiran en territorio guatemalteco," *Prensa Libre*, May 31, 1992.
7 Francisco Asturias, personal conversation, October 9, 2023.
8 Eugenio Gobbato, personal conversation, May 23, 2023.
9 Donald J. Planty, personal conversation, September 25, 2023.
10 Jorge Antonio Serrano Elías, personal conversation, September 18, 2023.
11 Magalí Rey Rosa, personal conversation, May 19, 2023.
12 Mark Brenner, personal conversation, December 14, 2023.
13 Carlos Salazar, personal conversation, December 18, 2023.
14 Bernardo Salas, personal conversation, January 27, 2024.

7. TRASH ASH ODYSSEY

1 *Waste Export Control: Hearing Before the Subcommittee on Transportation and Hazardous Materials of the Committee on Energy and Commerce*, US House of Representatives, 101st Congress, 1st Session on HR 2525 (US Government Printing Office, 1989), 210.
2 Simone M. Müller, *The Toxic Ship* (Univ. of Washington Press, 2023), 81.
3 Bill D. Moyers and Center for Investigative Reporting, *Global Dumping Ground: The International Traffic in Hazardous Waste* (Seven Locks Press, 1990), 24.
4 Moyers and Center for Investigative Reporting, *Global Dumping Ground*, 24.
5 Müller, *The Toxic Ship*, 63.
6 Moyers and Center for Investigative Reporting, *Global Dumping Ground*, 26.
7 "Waste Returns After 16 yrs," *News24*, July 18, 2002, news24.com/news24/waste-returns-after-16-yrs-20020717.
8 "I think it is probably scrap metal in a car somewhere in India," Ken Bruno, a Greenpeace tracker of the *Khian Sea,* would later surmise. "I doubt it exists." See Jerry Schwartz, "The Full Story of the *Khian Sea* and the Gonaives Ash Mountain," Associated Press, September 3, 2000, faculty.webster.edu/corbetre/haiti-archive/msg05049.html.
9 *Waste Export Control*, 1st Session on HR 2525, 2.
10 *Waste Export Control*, 1st Session on HR 2525, 29.
11 *Waste Export Control*, 1st Session on HR 2525, 254.
12 *Waste Export Control*, 1st Session on HR 2525, 370.
13 *Waste Export Control*, 1st Session on HR 2525, 5.
14 *Waste Export Control*, 1st Session on HR 2525, 215.

8. RISING UP

1 François Marie Gabriel André Charles-Ferdinand Roelants du Vivier, personal conversation, December 2021.
2 Jim Vallette and Heather Spalding, *The International Trade in Wastes: A Greenpeace Inventory* (Greenpeace USA, 1990), 29.

NOTES

3. Associated Press, "W. Africa Faces Economic Boon in Toxic Shroud," *DeseretNews*, June 19, 1988, deseret.com/1988/6/19/18769116/w-africa-faces-economic-boon-in-toxic-shroud.
4. James Brooke, "Waste Dumpers Turning to West Africa, *New York Times*, July 17, 1988, nytimes.com/1988/07/17/world/waste-dumpers-turning-to-west-africa.html.
5. Oladele Osibanjo, personal conversation, September 26, 2023.
6. See Steven Bernstein, *The Compromise of Liberal Environmentalism* (Columbia Univ. Press, 2001).
7. David W. Pearce and R. Kerry Turner, *Economics of Natural Resources and the Environment* (Johns Hopkins Univ. Press, 1989).
8. Jim Vallette, personal conversation, December 18, 2021.
9. Annie Leonard, personal conversation, November 16, 2023.
10. *Waste Export Control: Hearing Before the Subcommittee on Transportation and Hazardous Materials of the Committee on Energy and Commerce*, US House of Representatives, 101st Congress, 1st Session on HR 2525 (US Government Printing Office, 1989), 157.
11. *Waste Export Control*, 1st Session on HR 2525, 298.
12. *Greenpeace Waste Trade Update* 2, no. 2 (Greenpeace International Waste Trade Campaign, 1990).

9. AMERICAN EXCEPTIONALISM

1. *Waste Export Control: Hearing Before the Subcommittee on Transportation and Hazardous Materials of the Committee on Energy and Commerce*, US House of Representatives, 101st Congress, 1st Session on HR 2525 (US Government Printing Office, 1989), 112.
2. *Waste Export Control*, 1st Session on HR 2525, 39.
3. *Waste Export Control*, 1st Session on HR 2525, 112.
4. *Waste Export Control*, 1st Session on HR 2525, 112.
5. *Waste Export Control*, 1st Session on HR 2525, 187.

10. THE WASTE TRADE STRIKES BACK

1. Quoted in Harold Crooks, *Giants of Garbage: The Rise of the Global Waste Industry and the Politics of Pollution Control* (Lorimer, 1993), 10.
2. "Naples Garbage Is Mafia Gold," Reuters, January 9, 2008, reuters.com/article/us-italy-waste-mafia-idUSL0830577220080109.
3. For an excellent analysis of those discussions, see C. Jay Ou, "Native Americans and the Monitored Retrievable Storage Plan for Nuclear Wastes: Late Capitalism, Negotiation, and Control," Kroeber Anthropological Society Papers, vol. 92–93 (2006): 128–96, digitalassets.lib.berkeley.edu/anthpubs/ucb/text/kas092_093-006.pdf.
4. Sarah L. Lincoln, "Expensive Shit: Aesthetic Economies of Waste in Postcolonial Africa" (dissertation, Duke University, 2008), hdl.handle.net/10161/696.
5. Jim Vallette, personal conversation, September 1, 2023.
6. Stephen G. Puccini, "The Plastics Problem Part 2: Consumers and Plastic," *Environs*

16, no. 2 (1992): 39, environs.law.ucdavis.edu/archives/16/2/plastics-problem-part-2-consumers-and-plastics.

7   Laura Sullivan, "Plastic Wars: Industry Spent Millions Selling Recycling—To Sell More Plastic," NPR, March 31, 2020, npr.org/2020/03/31/822597631/plastic-wars-three-takeaways-from-the-fight-over-the-future-of-plastics.

8   Jan Dell, personal conversation, February 26, 2024.

11. STATE AND SLUM

1   "'We Must Unite Now or Perish'—President Kwame Nkrumah," *New African,* May 3, 2013, newafricanmagazine.com/3721/.

2   "Ghana Is Free Forever," *BBC World Service,* March 6, 1957, bbc.co.uk/worldservice/focusonafrica/news/story/2007/02/070129_ghana50_independence_speech.shtml.

3   Keith Hartwig, "Digital Waste & Cyber Crime: Examining the Relationship" (research paper for Harvard Kennedy School course IGA 238: Technology, Privacy and the Transnational Nature of the Internet, December 2016), academia.edu/31385539/DIGITAL_WASTE_and_CYBER_CRIME_EXAMINING_THE_RELATIONSHIP.

4   Office of Public Affairs Press Release, "Ghanaian Citizen Extradited in Connection with Prosecution of Africa-Based Cybercrime and Business Email Compromise Conspiracy," U.S. Department of Justice, August 26, 2020, justice.gov/opa/pr/ghanaian-citizen-extradited-connection-prosecution-africa-based-cybercrime-and-business-email.

5   Matthew Barakat, "Annandale Man Paid $500k in Romance Fraud, Part of $42m Scam Headquartered in Ghana," Associated Press, February 17, 2022, wjla.com/news/local/romance-fraud-scam-victim-iflirt-linda-mbimadong-richard-broni-annadale-richard-dorpe-virginia-ghana.

6   The highest ever reported levels of brominated dioxins in eggs were found in chicken eggs from Agbogbloshie, Ghana. See "Highest Level of World's Most Toxic Chemicals Found in African Free-Range Eggs: European E-Waste Dumping a Contributor," Arnika, accessed March 2, 2022, arnika.org/en/news/most-toxic-chemicals-in-african-eggs#:~:text=Researchers%20have%20found%20the%20highest,chicken%20eggs%20in%20Agbogbloshie%2C%20Ghana.

7   Anand Chandrasekhar, "The Battle Against Global E-Waste Dumping Reaches Tipping Point," *SWI swissinfo.ch,* March 27, 2022, swissinfo.ch/eng/society/the-battle-against-global-e-waste-dumping-reaches-tipping-point/47445264#:~:text=%E2%80%9CA%20child%20who%20eats%20just,Digital%20Dumpsites%20released%20last%20year.

8   Nicolas Martin, "E-Waste = Toxic Waste," *Deutsche Welle,* May 13, 2015, www.dw.com/en/report-highlights-growing-global-e-waste-problem/a-18448012.

9   "Do Cows Find Plastic Tasty?" Sid's Farm, accessed February 26, 2022, sidsfarm.com/2023/04/14/do-cows-find-plastic-tasty/.

NOTES

12. TO THE QUAYS OF TEMA

1. Stephen Atta Owusu, "Complaints of Tribalism as Sodom and Gomorrah Is Demolished," *GhanaWeb*, July 13, 2005, ghanaweb.com/GhanaHomePage/features/Complaints-of-Tribalism-as-Sodom-and-Gomorrah-is-Demolished-368127.
2. An EU Commission official estimated that "25–75% of second-hand goods" exported to Africa do not work. See Gert-Jan van der Have, "E-Waste Regulations: Calls for Stronger Enforcement," *Recycling International* 2 (2008): 28. A decade-long study of Agbogbloshie found that "the share of working electronic goods found inside a typical e-waste shipment generally is about 25%." See Kurt Daum, Justin Stoler, and Richard J. Grant, "Toward a More Sustainable Trajectory for E-Waste Policy: A Review of a Decade of E-Waste Research in Accra, Ghana," *International Journal of Environmental Research and Public Health* 14, no. 2 (2017): 135, ncbi.nlm.nih.gov/pmc/articles/PMC5334689/.

13. TREASURE

1. Isaac Kaledzi, "Renewed North Rhine-Westphalia–Ghana Partnership," *Deutsche Welle*, May 23, 2016, www.dw.com/en/north-rhine-westphalia-helps-ghana-fight-consequences-of-e-waste/a-19277409.

14. LOGGING ON

1. Ryszard Kapuściński, *The Shadow of the Sun: My African Life* (Allen Lane, 2001), 23.
2. Samuel Adu-Gyamfi and Richard Oware, "Economy and Health in the Gold Coast, 1902–1957," *African Economic History* 47, no. 2 (2019), muse.jhu.edu/article/746668/pdf.
3. Jonathan Roberts, "Korle and the Mosquito: Histories and Memories of the Antimalaria Campaign in Accra, 1942–5," *Journal of African History* 51, no. 3 (2010): 343–65.
4. Jon Kraus, "The Struggle over Structural Adjustment in Ghana," *Africa Today* 38, no. 4 (1991): 22.
5. Jeffry A. Frieden, *Global Capitalism: Its Rise and Fall in the 20th Century* (W. W. Norton, 2007), 443.
6. Jack Anderson, "How the Aluminum Deal Foiled Ghana," *Washington Post*, December 13, 1980, washingtonpost.com/archive/opinions/1980/12/14/how-the-aluminum-deal-foiled-ghana/f8bc342e-8ecc-48bf-b5e1-456d2e92c8df/.
7. George B. N. Ayittey, "Introduction: The Lost Continent," chap. 1 in *Africa in Chaos* (St. Martin's Press, 1998), found at *New York Times on the Web*, archive.nytimes.com/www.nytimes.com/books/first/a/ayittey-africa.html#:~:text=%22We%20shall%20achieve%20in%20a,Nkrumah%2C%201957%2C%2034; and Daniel Green, "Ghana's 'Adjusted' Democracy," *Review of African Political Economy* 22, no. 66 (1995): 577–85.
8. Aramide Odutayo, "Conditional Development: Ghana Crippled by Structural Adjustment Programmes" (student paper, Western University, 2014), posted online March 1, 2015, at E-International Relations, e-ir.info/2015/03/01/conditional-development-ghana-crippled-by-structural-adjustment-programmes/.

NOTES

9   Gavin M. Hilson, "Structural Adjustment in Ghana: Assessing the Impacts of Mining-Sector Reform," *Africa Today* 51, no. 2 (2004): 53–77.
10 Jason Tockman, *The International Monetary Fund: Funding Deforestation* (American Lands Alliance, 2001), 7–8.
11 Hilson, "Structural Adjustment in Ghana," 66.
12 Hilson, "Structural Adjustment in Ghana," 64–65.
13 *Waste Export Control: Hearing Before the Subcommittee on Transportation and Hazardous Materials of the Committee on Energy and Commerce,* US House of Representatives, 101st Congress, 1st Session on HR 2525 (US Government Printing Office, 1989), 176.
14 Derrick L. Cogburn and Catherine Nyaki Adeya, "Globalization and the Information Economy: Challenges and Opportunities for Africa" (working paper for the African Development Forum '99, October 24–28, 1999, United Nations Conference Centre, United Nations Economic Commission for Africa, Addis Ababa, Ethiopia), 4, archive.unu.edu/africa/papers/cogburn-adeya.pdf.
15 Cogburn and Adeya, "Globalization and the Information Economy," vii, 32–33.
16 Quoted in Ayittey, "Introduction: The Lost Continent," in *Africa in Chaos*. Dr. Kwame Nkrumah, the man who kick-started Ghana's—and Africa's—independence, continually implored his fellow countrymen, "[W]e cannot afford to sit still and be passive onlookers of technological change," found in Kwaku Appiah-Adu and Mahamudu Bawumia, *Key Determinants of National Development* (Routledge, 2016).
17 G. Pascal Zachary, "Black Star: Ghana, Information Technology, and Development in Africa," *First Monday* 9, no. 3 (2004), firstmonday.org/ojs/index.php/fm/article/view/1126/1046#:~:text=%22The%20message%20for%20Ghana%20is,ICT%20Policy%20and%20Plan%20Development.
18 Nii Narku Quaynor, personal conversation, April 13, 2023.
19 World Bank, *World Development Report 1998/99: Knowledge for Development* (World Bank Publications, 1998).
20 Andrei Andries, *Human Development Report 2001: Making New Technologies Work for Human Development* (United Nations Development Programme, 2001), 3, hdr.undp.org/content/human-development-report-2001.
21 *World Telecommunication Developments Report 1998* (International Telecommunication Union, March 1998), itu.int/ITU-D/ict/publications/wtdr_98/wtdr98.pdf.
22 G. Pascal Zachary, "Ghana's Digital Dilemma," *MIT Technology Review,* July 1, 2002, technologyreview.com/2002/07/01/275349/ghanas-digital-dilemma/.
23 Sarah L. Lincoln, "Expensive Shit: Aesthetic Economies of Waste in Postcolonial Africa" (dissertation, Duke University, 2008), hdl.handle.net/10161/696. Cited in Emily Brownell, "International Trash and the Politics of Poverty: Conceptualizing the Transnational Waste Trade," in *Nation-States and the Global Environment*, ed. Erika Marie Bsumek, David Kinkela, and Mark Atwood Lawrence (Oxford Univ. Press, 2013), 258.
24 Zachary, "Black Star."

NOTES

25 Quoted in Charles W. Schmidt, "Unfair E-Waste in Africa," *Environmental Health Perspective* 114, no. 4 (2006): A232–35, ncbi.nlm.nih.gov/pmc/articles/PMC1440802/.
26 "Uganda to Review Ban on Import of Used Electronics," *East African*, March 6, 2010, theeastafrican.co.ke/tea/business/uganda-to-review-ban-on-import-of-used-electronics--1297972.
27 Jenna Burrell, *Invisible Users: Youth in the Internet Cafés of Urban Ghana* (MIT Press, 2012), 163.
28 Schmidt, "Unfair E-Waste in Africa."
29 Anna Boustany and Deborah Kornblut, "Life After William & Mary: How W&M's Donated Laptops Get to Ghana," William & Mary: IT News, July 25, 2019, wm.edu/offices/it/news/ghanalaptops.php?fbclid=IwAR20Lp1oDew_P6I5yq_TjmiGzvG_Ox8T-a54CnQqaUALMfgRBiyY4me8vV4/; Richard Grant, "The 'Urban Mine' in Accra, Ghana," in *Out of Sight, Out of Mind*, ed. Christof Mauch (Rachel Carson Center for Environment and Society, 2016), 23, environmentandsociety.org/sites/default/files/2016_i1.pdf; and Jacklin Kwan, "Your Old Electronics Are Poisoning People at This Toxic Dump in Ghana," *Wired*, November 26, 2020, wired.co.uk/article/ghana-ewaste-dump-electronics.
30 Grant, "The 'Urban Mine' in Accra, Ghana," 23.

15. TECHNOLOGICAL TINKERING

1 Apple Newsroom, "Letter from Tim Cook to Apple Investors," news release, January 2, 2019, apple.com/newsroom/2019/01/letter-from-tim-cook-to-apple-investors/.
2 J. B. MacKinnon, "The L.E.D. Quandary: Why There's No Such Thing as 'Built to Last,'" *The New Yorker*, July 14, 2016, newyorker.com/business/currency/the-l-e-d-quandary-why-theres-no-such-thing-as-built-to-last.
3 Giles Slade, *Made to Break: Technology and Obsolescence in America* (Harvard Univ. Press, 2007).
4 John Harris, "Planned Obsolescence: The Outrage of Our Electronic Waste Mountain," *The Guardian*, April 15, 2020, amp.theguardian.com/technology/2020/apr/15/the-right-to-repair-planned-obsolescence-electronic-waste-mountain.
5 Kim Komando, "How Long Before a Phone Is Outdated? Here's How to Find Your Smartphone's Expiration Date," *USA Today*, October 22, 2023, eu.usatoday.com/story/tech/columnist/komando/2023/10/22/how-to-find-smartphone-expiration-date/71255625007/.
6 Nathan Proctor, personal conversation, August 23, 2023.
7 Nathan Proctor, "Americans Toss 151 Million Phones a Year. What If We Could Repair Them Instead?" WBUR, December 11, 2018, wbur.org/cognoscenti/2018/12/11/right-to-repair-nathan-proctor; "What's the Average Lifespan of Your Electronics?" *Quantum Lifecycle Partners* (blog), July 14, 2021, quantumlifecycle.com/en_CA/blog/whats-the-average-lifespan-of-your-electronics/; Graeme Roberts, "Global Light Vehicle Sales Dip in 2022," JustAuto, May 4, 2023, just-auto.com/news/global-light-vehicle-sales-dip-in-2022/; and Kelly Valencia, "Over 32 Million Bibles Distributed Worldwide in 2021, New Report Shows," *Christian News,* July 20,

2022, premierchristian.news/en/news/article/over-32-million-bibles-distributed-worldwide-in-2021-new-report-shows.

16. THE FLEXIBLE MINE

1. *A New Circular Vision for Electronics: Time for a Global Reboot* (PACE and World Economic Forum, January 2019), www3.weforum.org/docs/WEF_A_New_Circular_Vision_for_Electronics.pdf.
2. Richard Grant and Martin Oteng-Ababio, "The Global Transformation of Materials and the Emergence of Informal Urban Mining in Accra, Ghana," *Africa Today* 62, no. 4 (2016): 4.
3. Katie Campbell and Ken Christensen, "Where Does America's E-Waste End Up? GPS Tracker Tells All," *PBS Newshour*, May 10, 2016, pbs.org/newshour/science/america-e-waste-gps-tracker-tells-all-earthfix.
4. Aniyie Ifeanyichukwu Azuka, "The Influx of Used Electronics into Africa: A Perilous Trend," *Law, Environment and Development Journal* 5, no. 1 (2009): 97, lead-journal.org/content/09090.pdf; and "Trade Summary for European Union 2005," World Integrated Trade Solution, accessed March 21, 2022, wits.worldbank.org/CountryProfile/en/Country/EUN/Year/2005/Summarytext.
5. To produce the same amount of energy as hydrocarbon machines, it's been estimated that green technologies require a tenfold increase in the amount of metals and minerals that we currently unearth. Mark P. Mills, *Mines, Minerals, and "Green" Energy: A Reality Check* (Manhattan Institute, July 9, 2020), manhattan.institute/article/mines-minerals-and-green-energy-a-reality-check; and Guillaume Pitron, *The Rare Metals War*, trans. Bianca Jacobsohn (Scribe, 2020), 159.
6. Jewellord T. Nem Singh, "Geographies in Transition," *Phenomenal World*, June 29, 2022, phenomenalworld.org/analysis/geographies-in-transition/.
7. And those minerals are, according to the UN, "40 to 50 times richer" than ore deposits from "primary mines." In one ton of busted TVs and phones alone, it is possible to extract 131 kilograms of copper, or 15 times the amount of copper to be extracted pound for pound via conventional mining.
8. Henry Sanderson, *Volt Rush: The Winners and Losers in the Race to Go Green* (Oneworld, 2023), 201.
9. Julie Michelle Klinger, *Rare Earth Frontiers: From Terrestrial Subsoils to Lunar Landscapes* (Cornell Univ. Press, 2018), 60–61.
10. Caitlin Yilek, "Jeff Bezos on Future of Spaceflight: 'We Can Move All Heavy Industry and All Polluting Industry off of Earth,'" *CBS News*, July 21, 2021, cbsnews.com/news/jeff-bezos-space-heavy-industry-polluting-industry/.
11. Freyja L. Knapp, "The Birth of the Flexible Mine: Changing Geographies of Mining and the E-Waste Commodity Frontier," *Environment and Planning A* 48, no. 10 (2016): 1889–909.
12. Klinger, *Rare Earth Frontiers*, 55.

NOTES

17. START-UP CESSPOOLS

1. UN Environment Programme, "Nigeria Acts to Fight Growing E-Waste Epidemic," news release, January 5, 2023, unep.org/gef/news-and-stories/press-release/nigeria-acts-fight-growing-e-waste-epidemic.
2. Oladele Osibanjo, personal conversation, September 26, 2023.
3. Amy Yee, "Electronic Marvels Turn into Dangerous Trash in East Africa," *New York Times*, May 12, 2019, nytimes.com/2019/05/12/climate/electronic-marvels-turn-into-dangerous-trash-in-east-africa.html.
4. Annie Leonard, personal conversation, November 18, 2023.
5. Jennifer Joines, "Globalization of E-Waste and the Consequences of Development: A Case Study of China," *Journal of Social Justice* 2 (2012): 9, transformativestudies.org/wp-content/uploads/Globalization-of-E-waste-and-the-Consequence-of-Development.pdf.
6. Ying Xia, "China's Environmental Campaign: How China's 'War on Pollution' Is Transforming the International Trade in Waste," *NYU Journal of International Law and Politics* 51 (2019): 1130, nyujilp.org/wp-content/uploads/2019/09/NYI402.pdf.
7. Aaron Boxerman, "A Deadly Trash Trade Is Poisoning Palestinians in the West Bank," *Times of Israel*, January 10, 2022, timesofisrael.com/a-deadly-trash-trade-is-poisoning-palestinians-in-the-west-bank/.

18. A NEW AGBOGBLOSHIE?

1. Malcolm X, Alex Haley, and Ossie Davis, *The Autobiography of Malcolm X* (Ballantine, 1992), 404–5.
2. Martin Oteng-Ababio, "Electronic Waste Management in Ghana — Issues and Practices," in *Sustainable Development: Authoritative and Leading Edge Content for Environmental Management*, ed. Sime Curkovic (InTechOpen, 2012), intechopen.com/chapters/38097.
3. "A New Agbogbloshie Is Coming — Henry Quartey," *GhanaWebbers*, April 21, 2022, ghanaweb.live/GhanaHomePage/business/A-new-Agbogbloshie-is-coming-Henry-Quartey-1520714.

19. GOING FISHING

1. Quoted in Yepoka Yeebo, *Anansi's Gold: The Man Who Looted the West, Outfoxed Washington, and Swindled the World* (Bloomsbury Publishing, 2023), 72.

21. GLOBAL JUNK HEAP

1. Brandon Carter, "Trump: If You Don't Have Steel, You Don't Have a Country," *Hill*, March 2, 2018, thehill.com/homenews/administration/376408-trump-if-you-dont-have-steel-you-dont-have-a-country/.
2. John Kolesidis and Karolina Tagaris, "Athens Scrap Dealer Defies Taboos in Crisis-Hit Greece," Reuters, February 6, 2013, reuters.com/article/us-greece-scrapdealer-idUSBRE9150MO20130206.

3. Reuters, "Swords to Ploughshares? Soviet Tank Scraps Help Fuel Afghan Building Industry," *NBC News*, July 25, 2012, nbcnews.com/id/wbna48318214.
4. James Glanz, "Arms Equipment Plundered in 2003 Is Surfacing in Iraq," *New York Times*, April 17, 2005, nytimes.com/2005/04/17/world/middleeast/arms-equipment-plundered-in-2003-is-surfacing-in-iraq.html.
5. Alessio Perrone, "'Metal Thieves' Steal 56-Tonne, 23-Metre Rail Bridge in Russia," *Independent*, June 5, 2019, independent.co.uk/news/world/europe/russia-bridge-theft-disappear-arctic-umba-river-rail-a8944971.html.
6. George Iype, "On WTC Debris, These Indian Buildings Stand," *Rediff News*, September 14, 2006, rediff.com/news/2006/sep/14spec.htm.
7. Arno Maierbrugger, "Steel from Historic East German Palace Used to Build Burj Dubai," *Gulf News*, August 10, 2008, gulfnews.com/business/property/steel-from-historic-east-german-palace-used-to-build-burj-dubai-1.124124.
8. Jane Jacobs, *The Economy of Cities* (Vintage, 1970), 110.
9. S. M. Mizanur Rahman, Robert M. Handler, and Audrey L. Mayer, "Life Cycle Assessment of Steel in the Ship Recycling Industry in Bangladesh," *Journal of Cleaner Production* 135 (2016): 963–71.
10. John Seabrook, "American Scrap," *The New Yorker*, January 7, 2008, newyorker.com/magazine/2008/01/14/american-scrap.
11. Vaclav Smil, *How the World Really Works* (Viking, 2022), 92.
12. See Gökdeniz Neşer, Deniz Ünsalan, Nermin Tekoğul, and Frank Stuer-Lauridsen, "The Shipbreaking Industry in Turkey: Environmental, Safety, and Health Issues," *Journal of Cleaner Production* 16, no. 3 (2008): 350–58.
13. Ying Xia, "China's Environmental Campaign: How China's 'War on Pollution' Is Transforming the International Trade in Waste," *NYU Journal of International Law and Politics* 51 (2019): 1108, nyujilp.org/wp-content/uploads/2019/09/NYI402.pdf.
14. "Steel Scrap Availability to Be Challenging as 60 Nations Plan to Ban Trade," *Business Standard*, September 28, 2023, business-standard.com/industry/news/steel-scrap-availability-to-be-challenging-as-60-nations-plan-to-ban-trade-123092801090_1.html.
15. Olumuca Cuoano, Miguel Rodrigo Gonzalo, Frank Farrell, Rainer Remus, Serge Roudier, and Luis Delgado Sancho, *Best Available Techniques (BAT) Reference Document for the Non-Ferrous Metals Industries. Industrial Emissions Directive 2010/75/EU (Integrated Pollution Prevention and Control)*, EUR 28648 EN (Publications Office of the European Union, 2017), 1, publications.jrc.ec.europa.eu/repository/handle/JRC107041.
16. "Scrap Metal Exports to China Up," *Nation*, last modified July 2, 2020, nation.africa/kenya/news/scrap-metal-exports-to-china-up—867014.

22. SHIPPING OUT

1. Quoted in Nicholas Shaxson, *The Finance Curse: How Global Finance Is Making Us All Poorer* (Random House, 2018), 25.
2. R. Scott Frey, "Breaking Ships in the World-System: An Analysis of Two Ship

NOTES

Breaking Capitals, Alang-Sosiya, India, and Chittagong, Bangladesh," *Journal of World-Systems Research* 21, no. 1 (2015): 28.

3 Ana Swanson, "Shipping Contributes Heavily to Climate Change. Are Green Ships the Solution?" *New York Times,* October 30, 2023, nytimes.com/2023/10/30/business/economy/shipping-climate-change-green-fuel.html; and "Shipping at a Turning Point," in "Transport over the Seas," chap. 4 of *World Ocean Review 7* (Maribus gGmbH, 2021), accessed June 10, 2023, worldoceanreview.com/en/wor-7/transport-over-the-seas/shipping-at-a-turning-point/.

4 Martin Cames, Jakob Graichen, Anne Siemons, and Vanessa Cook, *Emission Reduction Targets for International Aviation and Shipping* (Think Tank European Parliament, November 16, 2015), accessed March 22, 2019, europarl.europa.eu/thinktank/en/document/IPOL_STU(2015)569964.

5 Jim Puckett, personal conversation, September 12, 2023.

6 Nikos Mikelis, personal conversation, November 2, 2023.

7 Frey, "Breaking Ships in the World-System," 25–49.

8 S. M. Mizanur Rahman, Robert M. Handler, and Audrey L. Mayer, "Life Cycle Assessment of Steel in the Ship Recycling Industry in Bangladesh," *Journal of Cleaner Production* 135 (2016): 963–71.

23. INTO THE HEART OF ANATOLIA

1 Yaşar Kemal, *Anatolian Tales* (Dodd, Mead, 1969), 54.

2 Costco Travel, "Carnival Inspiration: Onboard the Carnival Inspiration," accessed April 30, 2022, costcotravel.com/Cruises/Carnival-Cruise-Line/Carnival-Inspiration.

3 Carnival Corporations & PLC, "Carnival Corporation Delivers Two Retired Cruise Ships for Responsible Recycling in Turkey," news release, August 3, 2020, carnivalcorp.com/news-releases/news-release-details/carnival-corporation-delivers-two-retired-cruise-ships.

4 Emine Aslı Odman Pérouse, "Call for Evidence in June 2022, by the European Commission on EU-Shipbuilding Regulation-Evaluation," European Commission, June 30, 2022, ec.europa.eu/info/law/better-regulation/have-your-say/initiatives/13377-EU-Ship-Recycling-Regulation-evaluation_en.

5 Mike Schuler, "Seven Shipbreaking Workers Killed in Accidents in Bangladesh, Turkey—NGO," gCaptain, September 20, 2021, gcaptain.com/seven-shipbreaking-workers-killed-in-accidents-in-bangladesh-turkey-ngo/.

6 Carnival Corporations & PLC, "Carnival Corporation Delivers."

24. DEADLY BUSINESS

1 Nikos Kavadias, *The Collected Poems of Nikos Kavadias,* trans. Gail Holst-Warhaft (Cosmos Publishing, 2006).

2 "Ship-Breaking.com: Information on Ship Demolition, #8–11, from January 1 to December 31, 2007," Robin des Bois, January 24, 2008, robindesbois.org/wp-content/uploads/2014/10/shipbreaking-2007.pdf.

3 Akhilesh Tewari et al., "The Effect of Ship Scrapping Industry and Its Associated

NOTES

Wastes on the Biomass Production and Biodiversity of Biota in In Situ Condition at Alang," *Marine Pollution Bulletin*, 42, no. 6 (2001): 462–69, researchgate.net/profile/Subir-Mandal-2/publication/11873363_The_Effect_of_Ship_Scrapping_Industry_and_its_Associated_Wastes_on_the_Biomass_Production_and_Biodiversity_of_Biota_in_in_situ_Condition_at_Alang/links/5a94d7030f7e9ba42970dada/The-Effect-of-Ship-Scrapping-Industry-and-its-Associated-Wastes-on-the-Biomass-Production-and-Biodiversity-of-Biota-in-in-situ-Condition-at-Alang.pdf.

4 Geetanjoy Sahu, *Challenges for the Implementation of Workers' Rights in Hazardous Industries: A Critical Analysis of Alang-Sosiya Ship Breaking Yard, Bhavnagar, Gujarat, from 1983–2013* (Tata Institute of Social Sciences, January 2014), 53.

5 Gopal Krishna, personal conversation, October 25, 2023.

6 Nikola Mulinaris, personal conversation, August 31, 2023.

7 NGO Shipbreaking Platform, "Flags of Convenience," accessed June 15, 2023, shipbreakingplatform.org/issues-of-interest/focs/.

8 Aaron Fronda, "The Nature of Shipbreaking Is Casting a Shadow over the Shipping Industry," World Finance, accessed March 20, 2023, worldfinance.com/markets/the-nature-of-shipbreaking-is-casting-a-shadow-over-the-shipping-industry.

9 Bhavya Dore, "A Unique Bric-a-Brac Market Sells Salvaged Goods from Broken Ships," Atlas Obscura, June 10, 2019, atlasobscura.com/articles/alang-shipyard-market.

10 Sahu, *Challenges for the Implementation of Workers' Rights*, 60.

11 Frankie Youd, "Crewless Cargo: The World's First Autonomous Electric Cargo Ship," Ship Technology, February 24, 2022, ship-technology.com/features/crewless-cargo-the-worlds-first-autonomous-electric-cargo-ship/.

12 Syeda Rizwana Hasan, personal conversation, October 11, 2023.

13 Ebe Daems, Gie Goris, and Nikola Mulinaris, "Where Ships Go to Die: Switzerland and the Uncontrolled Dismantling of Ships," *MO* Magazine* and NGO Shipbreaking Platform, 2018, shipbreakingplatform.org/spotlight-swiss-focus/.

14 Staff Correspondent, "KSRM 'Spreading Falsehood' over Worker's Death," *Business Post*, March 25, 2022, businesspostbd.com/back/ksrm-spreading-falsehood-over-workers-death-2022-03-25.

15 Tony George Puthucherril, *From Shipbreaking to Sustainable Ship Recycling: Evolution of a Legal Regime*, vol. 5 (Legal Aspects of Sustainable Development) (Martinus Nijhoff Publishers, 2010), 28.

16 Javeria Younus, "Pakistan: Over 100 Workers at Ship-Breaking Yard Incinerated Without a Trace," Asia Human Rights Commission, November 12, 2016, humanrights.asia/news/ahrc-news/AHRC-ART-066-2016/; and Chris White, "Two Years Since Pakistan's Gadani Ship-Breaking Disaster, Why Are Workers Still Dying?" *South China Morning Post*, October 28, 2018, scmp.com/week-asia/health-environment/article/2170256/two-years-pakistans-gadani-ship-breaking-disaster-why.

17 Holly Birkett, "Dog-Eating Leopard Caught at Alang Breaking Yard After Living in Tanker," *TradeWinds*, May 12, 2020, tradewindsnews.com/shipyards/dog-eating-leopard-caught-at-alang-breaking-yard-after-living-in-tanker/2-1-805175.

18 Kit Chapman, "The Toxic Tide of Ship Breaking," *Chemistry World*, February 21, 2022, chemistryworld.com/features/the-toxic-tide-of-shipbreaking/4015158.article.
19 Cathy Buyck and Marion Solletty, "EU Tackles Dirty Business of Recycling Ships," *Politico*, August 23, 2017, politico.eu/article/eu-efforts-still-needed-to-tackle-the-dirty-business-of-ships-recycling/.
20 International Federation of Human Rights, Greenpeace, and YPSA (Young Power in Social Action), *End of Life Ships: The Human Cost of Breaking Ships* (International Federation for Human Rights, December 2005), 10, refworld.org/reference/country rep/ifhr/2005/en/57303.

25. SCRAP SHEPHERDS

1 Matthew G. Sohm, "'Big Clean,' the 'Death Ship,' and the Hazardous Waste Trade Between West Germany and Turkey, 1987–1988," *Contemporary European History* 33, no. 2 (2022): 459–76.
2 Carnival Corporation & PLC, *Sustainable from Ship to Shore: 2020 Sustainability Report*, accessed April 13, 2023, carnival-sustainability-2023.nyc3.digitalocean spaces.com/assets/content/pdf/2020-Sustainability-Report_Carnival-Corporation-plc.pdf.
3 Konstantinos Galanis, personal conversation, August 23, 2023.
4 Simon Ward, personal conversation, September 20, 2023.
5 Bestenigar Kara, "Younger and Larger Vessels Dismantled at Aliaga," *TR Monitor*, April 26, 2021, trmonitor.net/younger-and-larger-vessels-dismantled-at-aliaga/.
6 Mike Schuler, "Carnival Corporation Confirms Sale of Two Ships for Recycling in Turkey," gCaptain, August 4, 2020, gcaptain.com/carnival-corporation-confirms-sale-of-two-ships-for-recycling-in-turkey/.
7 "MSC Ship Recycling Policy," MSC, accessed June 12, 2024, msc.com/en/sus tainability/msc-ship-recycling-policy.
8 "Responsible Ship Recycling," Maersk, accessed June 24, 2023, maersk.com/sustain ability/our-esg-priorities/responsible-ship-recycling.
9 "Responsible Ship Recycling."

26. SCRAP NATION

1 David J. Lynch, "Turkey Went on a Building Spree as Its Economy Boomed. Now the Frenzy Is Crashing to a Halt," *Washington Post*, September 25, 2018, washington post.com/business/economy/turkeys-strongman-rule-imperils-gains-from-economic-boom/2018/09/25/e85d788c-b056-11e8-9a6a-565d92a3585d_story.html.
2 "Turkish Economy on the Mend," *Al Jazeera*, October 4, 2003, aljazeera.com /news/2003/10/4/turkish-economy-on-the-mend.
3 International Monetary Fund, "Turkey's Economy: A Future Full of Promise, Speech by Anne O. Krueger, First Deputy Managing Director, IMF" (Istanbul Forum, Istanbul, Turkey, May 5, 2005), imf.org/en/News/Articles/2015/09/28/04/53/sp050505.
4 International Monetary Fund, *Turkey: 2007 Article IV Consultation—Staff Report; Public Information Notice on the Executive Board Discussion; and Statement by the*

NOTES

Executive Director for Turkey (IMF Country Report No. 07/362, November 2007), 36, imf.org/external/pubs/ft/scr/2007/cr07362.pdf.

5 Jim Yong Kim, "Lessons from Turkey After 90 Years of Progress," World Bank, October 31, 2013, worldbank.org/en/news/opinion/2013/10/31/lessons-from-turkey-after-90-years-of-progress.

6 Simon Kuper, "Another Five Years with Turkey's Not-So-Strong Strongman," *Financial Times,* June 22, 2023, ft.com/content/57d990c1-e32c-4bd6-b6b1-78b27b8cba45.

7 Kubra Chohan, "Turkey Seeks to Be Among Top 10 Countries: Erdoğan," *Anadolu Agency,* April 2, 2018, aa.com.tr/en/economy/turkey-seeks-to-be-among-top-10-countries-erdoğan/1106069.

8 Ian Traynor and Constanze Letsch, "Turkey at a Crossroads as Erdoğan Bulldozes His Way to Lasting Legacy," *The Guardian,* June 2, 2015, theguardian.com/world/2015/jun/02/turkish-election-recep-tayyip-erdogan-legacy.

9 Reuters, "Turkey Opens One of the World's Biggest Suspension Bridges," *Gulf Business,* August 28, 2016, gulfbusiness.com/turkey-opens-one-worlds-biggest-suspension-bridges/.

10 United Nations Office on Drugs and Crime (UNODC), "Unwaste: Trendspotting Alert," Bulletin No. 5, 2, July 2023, https://www.unodc.org/res/environment-climate/asia-pacific/unwaste_html/Unwaste_Trendspotting_Alert_No.5.pdf.

11 Ahmet Buğra Tokmakoğlu, "How Aliağa Became Europe's Dumping Ground for Ships," *Inside Turkey,* October 25, 2021, insideturkey.news/2021/10/25/how-aliaga-became-europes-dumping-ground-for-ships/.

12 Brian Taylor, "As Turkey Goes…," *Recycling Today,* October 2018, recyclingtoday.com/article/turkish-ferrous-scrap-market-report/.

13 Vaclav Smil, *How the World Really Works* (Viking, 2022), 92.

14 Temo Bardzimashvili, "Georgia: For Metal Scavengers, a Scrappy Fight to Survive," *Eurasianet,* April 19, 2013, eurasianet.org/georgia-for-metal-scavengers-a-scrappy-fight-to-survive.

15 Nathan Fruchter, personal conversation, August 17, 2023.

16 Emin Sazak, "The Turkish Construction Industry Has Become an International Brand," interview by *The World Folio,* May 6, 2013, theworldfolio.com/interviews/emin-sazak-president/2176/.

17 "Controversial Tender System Allows Turkish Companies to Dominate World Bank Public Investment List," *Ahval,* December 29, 2018, ahvalnews.com/world-bank/controversial-tender-system-allows-turkish-companies-dominate-world-bank-public.

18 Aykan Erdemir, personal conversation, October 6, 2023.

19 Daniel Dombey and Piotr Zalewski, "Turkey Probe Underlines Links Between Construction and Politics," *Financial Times,* January 1, 2014, ft.com/content/ef8b4e0e-72d5-11e3-b05b-00144feabdc0.

20 Bülent Gültekin, personal conversation, October 20, 2023.

21 "The Wheel of Fortune That Keeps Turkey's Erdoğan in Power," *Ahval,* April 26, 2019, ahvalnews.com/turkish-corruption/wheel-fortune-keeps-turkeys-Erdoğan-power.

NOTES

22 "Turkey: 'Former Erdogan Graft Probe Prosecutors' Flee Country," Organized Crime and Corruption Reporting Project, August 12, 2015, occrp.org/en/news/turkey-former-erdogan-graft-probe-prosecutors-flee-country.
23 Adam Corbett, "Demolition Yard Accident in Turkey Claims Two Lives on Carnival Ship," *TradeWinds,* July 27, 2021, tradewindsnews.com/casualties/demolition-yard-accident-in-turkey-claims-two-lives-on-carnival-ship/2-1-1045197.
24 Mehmet Temurtaş, personal conversation, April 29, 2022.

27. AT EUROPE'S EDGE

1 Pinar Demírcan, "Radioactive Facts in Gaziemir and Authoritarian Negligence," *Yeşil Gazete,* June 9, 2020, yesilgazete.org/radioactive-facts-in-gaziemir-and-authoritarian-negligence/.
2 NGO Shipbreaking Platform, "Ship Recycling Workers' Protest Shakes Aliaga," news release, February 16, 2022, shipbreakingplatform.org/workers-strike-shakes-turkey/; and "Cancer Fears in Ship Breaking Facilities," *AnadoluTürkHaber,* August 25, 2013, anadoluturkhaber.com/TR/Detail/Cancer-Fears-In-Shipbreaking-Facilities/404.
3 Jerola Ziaj, "Zgjerohen hetimet për tragetin me mbetje toksike në Vlorë," *Reporter.al,* February 21, 2020, reporter.al/2020/02/21/zgjerohen-hetimet-per-tragetin-me-mbetje-toksike-ne-vlore/.
4 "Two Workers Killed While Disassembling 'Love Boat' in Turkey," *Hurriyet Daily News,* August 11, 2013, hurriyetdailynews.com/two-workers-killed-while-disassembling-love-boat-in-turkey—52319.
5 Ebe Daems, Gie Goris, and Nikola Mulinaris, "Where Ships Go to Die: Switzerland and the Uncontrolled Dismantling of Ships," *MO* Magazine* and NGO Shipbreaking Platform, 2018, shipbreakingplatform.org/spotlight-swiss-focus/.
6 Emine Aslı Odman Pérouse, "Call for Evidence in June 2022, by the European Commission on EU-Shipbuilding Regulation-Evaluation," European Commission, June 30, 2022, ec.europa.eu/info/law/better-regulation/have-your-say/initiatives/13377-EU-Ship-Recycling-Regulation-evaluation_en.
7 "Regulation (EU) No. 1257/2013 of the European Parliament and of the Council of 20 November 2013 on Ship Recycling and Amending Regulation (EC) No. 1013/2006 and Directive 2009/16/EC," *Official Journal of the European Union* 330, no. 1 (2013), eur-lex.europa.eu/legal-content/EN/TXT/PDF/?uri=CELEX:32013R1257&qid=1690963037577.
8 Orhan Demirgil, personal conversation, May 16, 2022.
9 "Striking Ship-Breaking Workers Hold a Mass Rally in Aliağa, Turkey," *World Socialist Web Site,* February 21, 2022, wsws.org/en/articles/2022/02/21/turk-f21.html.
10 Gökhan Çoban, personal conversation, May 15, 2022.

28. GREEKS BEARING GIFTS

1 Quoted in Robert Garland, *Athens Burning: The Persian Invasion of Greece and the Evacuation of Attica* (Johns Hopkins Univ. Press, 2017), 120.
2 Ingvild Jenssen, personal conversation, July 12, 2023.

NOTES

3. Oliver Telling and Richard Milne, "The Shipping Rivals Plotting Divergent Courses on Global Trade," *Financial Times*, June 23, 2023, ft.com/content/c29a373e-4c31-4f48-ac7c-221f1d0652a7.
4. Shipowner, personal conversation, October 6, 2023. The shipowner asked not to be mentioned by name.

31. PLASTIFICATION

1. Quoted in Nena Baker, *The Body Toxic: How the Hazardous Chemistry of Everyday Things Threatens Our Health and Well-Being* (Farrar, Straus and Giroux, 2008), 38.
2. International Energy Agency, *The Future of Petrochemicals* (IEA, October 2018), iea.org/reports/the-future-of-petrochemicals.
3. Quoted in Rebecca Altman, "Five Myths About Plastics," *Washington Post*, January 14, 2022, washingtonpost.com/outlook/2022/01/14/five-myths-plastics/.
4. John Kenly Smith Jr., "World War II and the Transformation of the American Chemical Industry," in *Science, Technology and the Military*, ed. E. Mendelsohn, M. R. Smith, and P. Weingart (Kluwer Academic Publishers, 1988), 307–22.
5. Heather Rogers, *Gone Tomorrow: The Hidden Life of Garbage* (New Press, 2013), 120.
6. Adam Hanieh, "Petrochemical Empire," *New Left Review* 130 (2021).
7. Robert D. Friedel, *A Culture of Improvement: Technology and the Western Millennium* (MIT Press, 1997), 487.
8. Jeffrey Meikle, *American Plastic: A Cultural History* (Rutgers Univ. Press, 1997), 180.
9. Susan Freinkel, "A Brief History of Plastic's Conquest of the World," *Scientific American*, May 29, 2011, scientificamerican.com/article/a-brief-history-of-plastic-world-conquest/.
10. 1943 Bell Telephone System "Rainbow in the Sky" advertisement, *Fortune*, 1943.
11. "Plastic — From Wonder Material to Waste," *Newtrients* (blog), University College Cork, Ireland, December 4, 2017, ucc.ie/en/newtrients/blog/plastic---from-wonder-material-to-waste.html.
12. Meikle, *American Plastic*, 278.
13. Freinkel, "A Brief History of Plastic's Conquest of the World."
14. Barry Commoner, *The Closing Circle: Nature, Man, and Technology* (Alfred A. Knopf, 1971).
15. Kenneth Geiser, *Materials Matter: Toward a Sustainable Materials Policy* (MIT Press, 2001), xi.
16. Quoted in Vance Packard, *The Waste Makers* (IG Publishing, 2011), 21.
17. Max Liboiron, *Pollution Is Colonialism* (Duke Univ. Press, 2021).
18. "New Products: Prometheus Unbound," *Time*, September 19, 1960, time.com/archive/6622821/new-products-prometheus-unbound/.
19. Roland Barthes, "Plastic," in *Mythologies* (Éditions du Seuil, 1957).
20. Smith, "World War II and the Transformation of the American Chemical Industry."
21. Quoted in Duncan Green, *From Poverty to Power* (Practical Action, 2008), 259.
22. Baker, *The Body Toxic*, 39.

NOTES

23. Norman Mailer, "The Big Bite," *Esquire*, April and May 1963; reprinted in Norman Mailer, *The Presidential Papers* (Berkley Medallion, 1970), 159.
24. Charles A. Reich, *The Greening of America* (Bantam, 1971).
25. Commoner, *The Closing Circle*.
26. Commoner, *The Closing Circle*.
27. Hugh H. Connolly, *Plastic Wastes in the Coming Decade* (US Environmental Protection Agency, 1971), 2–3.
28. Rogers, *Gone Tomorrow*, 156.
29. Stephen G. Puccini, "The Plastics Problem Part 1: Plastic Packaging and the Solid-Waste Crisis," *Environs* 16, no. 2 (1992): 21, environs.law.ucdavis.edu/archives/16/2/plastics-problem-part-1-plastic-packaging-and-solid-waste-crisis.
30. *American Soft Drink Journal* (July 1967): 34, quoted by Bartow Jerome Elmore in "Citizen Coke: An Environmental and Political History of the Coca-Cola Company" (PhD dissertation, University of Virginia, 2012), 223.
31. Quoted in Donovan Hahn, "The Great Escape: The Bath Toys That Swam the Pacific," *The Guardian*, February 12, 2012, theguardian.com/environment/2012/feb/12/great-escape-bath-toys-pacific.
32. Heather Rogers, "Garbage Capitalism's Green Commerce," in *Socialist Register 2007: Coming to Terms with Nature*, ed. Leo Panitch and Colin Leys (Merlin Press, Monthly Review Press, and Fernwood Publishing, 2007), 234, socialistregister.com/index.php/srv/article/view/5854/2750.
33. Charles Moore, personal conversation, February 13, 2024.
34. Margaret Osborne, "Microplastics Detected in Human Blood in New Study," *Smithsonian*, March 28, 2022, smithsonianmag.com/smart-news/microplastics-detected-in-human-blood-180979826/.
35. Jack Anderson and Dale Van Atta, "Plastics Deteriorate in Image Only," *Washington Post*, January 29, 1990, washingtonpost.com/archive/lifestyle/1990/01/30/plastics-deteriorate-in-image-only/4cf03c5e-aca8-4585-8dda-f95300b51ec6/.

32. THE GREATEST MIRACLE YET

1. Malcolm Gladwell, "High Cost of Raw Materials Makes Re-Use Attractive: Demand for Recycled Plastic Exceeding Supply," *Los Angeles Times*, May 9, 1989, latimes.com/archives/la-xpm-1989-05-09-fi-2945-story.html.
2. Susan Shain, "Can Plastic Recycling Ever Really Work?" *New York Times*, September 1, 2023, nytimes.com/2023/09/01/headway/plastic-recycling-california-law.html.
3. US EPA Solid Waste Management Office, *First National Conference on Packing Wastes: Proceedings, September 22–24, 1969* (US Government Printing Office, 1971), 85, 87.
4. Laura Sullivan, "How Big Oil Misled the Public into Believing Plastic Would Be Recycled," NPR, September 11, 2020, npr.org/2020/09/11/897692090/how-big-oil-misled-the-public-into-believing-plastic-would-be-recycled.
5. Municipal Solid Waste Task Force, *The Solid Waste Dilemma: An Agenda for Action* (Office of Solid Waste, US Environmental Protection Agency, February 1989), nepis

NOTES

.epa.gov/Exe/ZyNET.exe/1000199O.TXT?ZyActionD=ZyDocument&Client =EPA&Index=1986+Thru+1990&Docs=&Query=&Time=&EndTime=&Search Method=1&TocRestrict=n&Toc=&TocEntry=&QField=&QFieldYear=&QField Month=&QFieldDay=&IntQFieldOp=0&ExtQFieldOp=0&XmlQuery= File=D%3A%5Czyfiles%5CIndex%20Data%5C86thru90%5CTxt %5C00000003%5C1000199O.txt&User=ANONYMOUS&Password=anonymous &SortMethod=h%7C-&MaximumDocuments=1&FuzzyDegree=0&Image Quality=r75g8/r75g8/x150y150g16/i425&Display=hpfr&DefSeekPage=x&Search Back=ZyActionL&Back=ZyActionS&BackDesc=Results%20page&MaximumPages =1&ZyEntry=1&SeekPage=x&ZyPURL. See also Vinyl Institute, "Solid Waste Fact Sheet: Draft," July 18, 1986, cdn.toxicdocs.org/6w/6wr0N7GOdVw85VaozkQqZp 3M9/6wr0N7GOdVw85VaozkQqZp3M9.pdf.

6 Sharon Lerner, "Waste Only: How the Plastics Industry Is Fighting to Keep Polluting the World," *The Intercept*, July 20, 2019, theintercept.com/2019/07/20/plastics-industry-plastic-recycling/.

7 Jindrich Petrlik, Yuyun Ismawati, Joseph DiGangi, Prigi Arisandi, Lee Bell, and Björn Beeler, *Plastic Waste Poisons Indonesia's Food Chain* (IPEN, Nexus3, Arnika, and Ecoton, November 2019), ipen.org/sites/default/files/documents/indonesia-egg-report-v1_9-web.pdf.

8 Jan Dell, personal conversation, February 3, 2023.

9 Sean Mowbray, "Americans Rank First in Plastic Waste Contribution," *Discover*, March 25, 2023, discovermagazine.com/environment/americans-rank-first-in-plastic-waste-contribution.

10 Larry Thomas, "Dinner Remarks" (delivered to the Society of the Plastics Industry, Tripartite Conference, Willard Intercontinental Hotel, Washington, DC, September 2, 1992).

11 Sarah S. Smith, "Phillips Joins Resin Makers' Recycling Exodus," *Plastic News*, September 28, 1998, plasticsnews.com/article/19980928/NEWS/309289998/phillips-joins-resin-makers-recycling-exodus.

12 Tik Root, "Inside the Long War to Protect Plastic," *Center for Public Integrity*, May 16, 2019, publicintegrity.org/environment/pollution/pushing-plastic/inside-the-long-war-to-protect-plastic/.

13 F. E. Krause, "PVC Recycling—An Overview" (presented at B. F. Goodrich Vinyl Industry Tripartite Meeting, Washington, DC, September 3–4, 1992), toxicdocs.org/d/91wxG1YnjQ8KjOnZ3jE9wLxg7?lightbox=1.

14 Alice Delemare Tangpuori, George Harding-Rolls, Nusa Urbancic, and Ximena Purita Banegas Zallio, *Talking Trash: The Corporate Playbook of False Solutions to the Plastic Crisis* (Changing Markets Foundation, September 2020), 84, talking-trash.com/wp-content/uploads/2020/09/TalkingTrash_FullReport.pdf.

15 Thaddeus Herrick, "Ads' Aim Is to Fix Bad Chemistry," *Wall Street Journal*, October 8, 2003, wsj.com/articles/SB106557967477807900.

NOTES

33. ONE-MAN MULTINATIONAL

1 Steve Wong, personal conversation, January 27, 2023.

34. PLASTIC CHINA

1 Steve Coll, "Free Market Intensifies Waste Problem," *Washington Post,* March 22, 1994, washingtonpost.com/archive/politics/1994/03/23/free-market-intensifies-waste-problem/4b4bbb21-084a-4d08-8316-7657b5a7e46d/.

2 *Basel Convention on the Control of Transboundary Movements of Hazardous Wastes and Their Disposal: Hearing Before the Committee on Foreign Relations,* US Senate, 102nd Congress, 2nd Session, March 12, 1992 (US Government Printing Office, 1992), 37.

3 Costas Velis, *Global Recycling Markets: Plastic Waste. A Story for One Player — China* (report prepared by FUELogy on behalf of International Solid Waste Association, Globalisation and Waste Management Task Force, September 2014), greenpeace.org/static/planet4-eastasia-stateless/2019/11/27d1dd21-27d1dd21-tfgwm_report_grm_plastic_china_lr.pdf; and David Carrig, "The US Used to Ship 4,000 Recyclable Containers a Day to China. Where Will the Banned Trash Go Now?" *USA Today,* June 21, 2018, eu.usatoday.com/story/news/nation-now/2018/06/21/china-ban-plastic-waste-recycling/721879002/.

4 Cheryl Katz, "Piling Up: How China's Ban on Imported Waste Has Stalled Global Recycling," *Yale Environment 360,* March 7, 2019, e360.yale.edu/features/piling-up-how-chinas-ban-on-importing-waste-has-stalled-global-recycling.

5 Dan Rademacher, "Plastic Not So Fantastic," *Ecology Center* (blog), November 15, 2003, ecologycenter.org/terrainmagazine/winter-2003/plastic-not-so-fantastic/.

6 Patty Moore, personal conversation, November 15, 2023.

7 Kerstin Unfried and Feicheng Wang, "Importing Air Pollution? Evidence from China's Plastic Waste Imports" (discussion paper, Institute of Labor Economics, April 2022), docs.iza.org/dp15218.pdf.

8 The transportation costs were astonishingly cheap. According to some estimates, in 2013, shipping a container from Los Angeles to Shenzhen cost about US$300, while shipping freight from Shenzhen to Los Angeles cost more than US$2,400, approximately the same shipping cost from Los Angeles to Chicago. See Ying Xia, "China's Environmental Campaign: How China's 'War on Pollution' Is Transforming the International Trade in Waste," *NYU Journal of International Law and Politics* 51 (2019): 1111, nyujilp.org/wp-content/uploads/2019/09/NYI402.pdf.

9 Mark Bradford, "The United States, China & the Basel Convention on the Transboundary Movements of Hazardous Wastes and Their Disposal," *Fordham Environmental Law Journal* 8, no. 2 (1997): 305–49, ir.lawnet.fordham.edu/elr/vol8/iss2/3.

10 Seth Faison, "Conspiracy Theories: China's Garbage War," *New York Times,* June 9, 1996, nytimes.com/1996/06/09/weekinreview/conspiracy-theories-china-s-garbage-war.html.

11 "China Accuses U.S. of Dumping Illegal Waste / Formal Protest Alleges Violation of Treaty," *SFGATE,* June 1, 1996, sfgate.com/news/article/china-accuses-u-s-of

NOTES

-dumping-illegal-waste-2980460.php. See also David Naguib Pellow, *Resisting Global Toxics: Transnational Movements for Environmental Justice* (MIT Press, 2007), 102.

12 "Businessman Arrested for Smuggling Garbage into China!" *Weekly World News*, March 11, 1997, books.google.gr/books?id=0OwDAAAAMBAJ&pg=PA9&lpg=PA9&dq=%22We%E2%80%99re+getting+saturated+with+rotten+fruit,+disposable+diapers%22&source=bl&ots=3nVXtQRB_A&sig=ACfU3U2CIZwfkdp9LBbFMe7UTpFslewxDA&hl=en&sa=X&ved=2ahUKEwjjhtbqtJaFAxUkRvEDHTccAbsQ6AF6BAgKEAM#v=onepage&q=%22We%E2%80%99re%20getting%20saturated%20with%20rotten%20fruit%2C%20disposable%20diapers%22&f=false.

13 Aaron Mak, "Why Does Half of the World's Used Plastic End Up in China?" *Slate*, June 21, 2018, slate.com/technology/2018/06/why-china-import-half-world-used-plastic.html.

14 Andy Peri, Green Sangha, to DTSC (Green Chemistry Initiative), November 15, 2007, public letter, 4, dtsc.ca.gov/wp-content/uploads/sites/31/2018/01/Peri_Green_Sangha_Letter.pdf.

15 Xia, "China's Environmental Campaign," 1139–40.

16 Jessica Heiges and Kate O'Neill, "A Recycling Reckoning: How Operation National Sword Catalyzed a Transition in the U.S. Plastics Recycling System," *Journal of Cleaner Production* 378 (2022): 134367, sciencedirect.com/science/article/pii/S0959652622039397.

17 Jim Puckett, personal conversation, September 12, 2023.

18 Aya Yoshida, "China's Ban of Imported Recyclable Waste and Its Impact on the Waste Plastic Recycling Industry in China and Taiwan," *Journal of Material Cycles and Waste Management* 24 (2022): 74–82, link.springer.com/article/10.1007/s10163-021-01297-2.

19 Xia, "China's Environmental Campaign," 1175.

20 "Li Says China Will Declare War on Pollution as Smog Spreads," *Bloomberg News*, March 5, 2014, bloomberg.com/news/articles/2014-03-05/china-to-declare-war-on-pollution-as-smog-spreads-across-country?embedded-checkout=true.

21 Xia, "China's Environmental Campaign," 1103.

22 Joshua Goldstein, *Remains of the Everyday: A Century of Recycling in Beijing* (Univ. of California Press, 2020), 232–33.

23 "Xi Stresses Habit of Garbage Sorting," *Global Times*, June 3, 2019, globaltimes.cn/content/1152898.shtml.

24 Keith Bradsher, "More Semiconductors, Less Housing: China's New Economic Plan," *New York Times*, November 6, 2023, nytimes.com/2023/11/06/business/china-economy-property-crisis.html.

25 "China to Contribute 28% of Global Petrochemical Capacity Additions by 2030," *Middle East Business*, accessed April 20, 2023, middleeast-business.com/china-to-contribute-28-of-global-petrochemical-capacity-additions-by-2030/.

NOTES

35. MAD SCRAMBLE

1. Michael Taylor, "Southeast Asian Plastic Recyclers Hope to Clean Up After China's Ban," Reuters, January 16, 2018, jp.reuters.com/article/us-asia-environment-waste-plastic/southeast-asian-plastic-recyclers-hope-to-clean-up-after-china-ban-idUSKBN1F504K/.
2. Joe Parkin Daniels, "Latin America Urges US to Reduce Plastic Waste Exports to Region," *The Guardian,* December 24, 2021, theguardian.com/environment/2021/dec/24/latin-america-urges-us-to-reduce-plastic-waste-exports-to-region.
3. Patti Verbanas, "Yard Waste Programs Chop Diversion Rates Diversion Rate with Yard Waste Program Waste Collection," Waste 360, October 1, 1996, waste360.com/yard-waste/yard-waste-programs-chop-diversion-rates-diversion-rate-with-yard-waste-progrma-waste-collection.
4. Ricardo Navarro, personal conversation, July 20, 2023.
5. Alicja Ptak, "Poland Files Complaint to EU Against Germany over Illegal Export of Waste," *Notes from Poland,* July 26, 2023, notesfrompoland.com/2023/07/26/poland-files-complaint-to-eu-against-germany-over-illegal-export-of-waste/.
6. Teodor Niță, personal conversation, September 16, 2020.
7. Louis Gore-Langton, "Liar, Liar, Plastic on Fire: Turkish Recyclers Discovered Arsoning Facilities to Destroy Waste," Packaging Insights, January 10, 2022, packaginginsights.com/news/liar-liar-plastic-on-fire-turkish-recyclers-discovered-arsoning-facilities-to-destroy-waste.html.
8. Fatih Doğan, personal conversation, April 21, 2022.
9. "Turkey Repeals Plastic Import Ban," Waste Management World, July 12, 2021, waste-management-world.com/artikel/turkey-repeals-plastic-import-ban/#:~:text=Recycling%20Turkey%20repeals%20plastic%20import%20ban&text=Turkey%20repealed%20a%20recently%20instated%20ban%20on%20foreign%20polyethylene%20plastic%20imports.&text=Turkey%20revoked%20its%20ban%20on,by%20the%20local%20plastic%20industry.
10. Erik McCormick, Bennett Murray, Carmela Fonbuena, Leonie Kijewski, Gökçe Saraçoğlu, Jamie Fullerton, Alastair Gee, and Charlotte Simmonds, "Where Does Your Plastic Go? Global Investigation Reveals America's Dirty Secret," *The Guardian,* June 17, 2019, theguardian.com/us-news/2019/jun/17/recycled-plastic-america-global-crisis.
11. Hiroko Tabuchi, Michael Corkery, and Carlos Mureithi, "Big Oil Is in Trouble. Its Plan: Flood Africa with Plastic," *New York Times,* August 30, 2020, nytimes.com/2020/08/30/climate/oil-kenya-africa-plastics-trade.html; "COMMON Project: Italian Waste Traffic in Tunisia, Containers Are Still in Sousse," ENI CBCMED, June 11, 2021, enicbcmed.eu/common-project-italian-waste-traffic-tunisia-containers-are-still-sousse; and Thodoris Chondrogiannis, "Illegal Transport of Waste from Greece to Liberia," Govwatch, February 4, 2020, govwatch.gr/en/finds/illegal-transport-of-waste-from-greece-to-liberia/.
12. Inès Magoum, "SÉNÉGAL: La douane saisit 25 tonnes de déchets plastiques en provenance d'Allemagne," *Afrik 21,* May 24, 2021, afrik21.africa/senegal-la-douane-saisit-25-tonnes-de-dechets-plastiques-en-provenance-dallemagne/.

NOTES

13   Kevin Mtai, personal conversation, October 18, 2023.
14   Joshua Goldstein, *Remains of the Everyday* (Univ. of California Press, 2020), 232–33.
15   Hillary Leung, "Southeast Asia Doesn't Want to Be the World's Dumping Ground. Here's How Some Countries Are Pushing Back," *Time,* June 3, 2019, time.com/5598032/southeast-asia-plastic-waste-malaysia-philippines/.
16   Merrit Kennedy, "'Sordid Chapter' Ends as Philippines Sends Back Canada's Trash," NPR, May 31, 2019, npr.org/2019/05/31/728611992/sordid-chapter-ends-as-philippines-sends-back-canada-s-trash.
17   *The Truth Behind Trash: The Scale and Impact of the International Trade in Plastic Waste* (Environmental Investigation Agency, September 2021), eia-international.org/wp-content/uploads/EIA-The-Truth-Behind-Trash-FINAL.pdf.
18   *The Truth Behind Trash.*
19   Leung, "Southeast Asia Doesn't Want to Be the World's Dumping Ground."
20   Yeo Bee Yin, personal conversation, July 20, 2021.
21   *Waste Export Control: Hearing Before the Subcommittee on Transportation and Hazardous Materials of the Committee on Energy and Commerce,* US House of Representatives, 101st Congress, 1st Session on HR 2525 (US Government Printing Office, 1989), 34.

36. A TRASH CHIEF

1   Clifford Geertz, *The Religion of Java* (Univ. of Chicago Press, 1960), 2.
2   Linna Amanda, "Is Indonesia on Track to Becoming a Wasteland?" *Project Planet,* July 19, 2020, projectplanetid.com/post/is-indonesia-on-track-to-becoming-a-wasteland.
3   Aretha Aprilia, Tetsuo Tezuka, and Gert Spaargaren, "Household Solid Waste Management in Jakarta, Indonesia: A Socio-Economic Evaluation," in *Waste Management — An Integrated Vision,* ed. Luis Fernando Marmolejo Rebellon (InTechOpen, 2012), intechopen.com/chapters/40536; and World Bank, *Plastic Waste Discharges from Rivers and Coastlines in Indonesia,* East Asia and Pacific Region: Marine Plastics Series (World Bank Publications, May 2021), 1.
4   Quoted in Roger Cohen, "Coke's World View — A Special Report; For Coke, World Is Its Oyster," *New York Times,* November 21, 1991, nytimes.com/1991/11/21/business/coke-s-world-view-a-special-report-for-coke-world-is-its-oyster.html.
5   World Bank, *What a Waste: Solid Waste Management in Asia* (World Bank Publications, May 1999), 13.
6   World Bank, *Plastic Waste Discharges from Rivers and Coastlines in Indonesia,* 27.
7   Andrea Carrubba, "Rotten River: Life on One of the World's Most Polluted Waterways — Photo Essay," *The Guardian,* November 2, 2020, theguardian.com/global-development/2020/nov/02/rotten-river-life-on-one-of-the-worlds-most-polluted-waterways-photo-essay.
8   "Indonesia's Capital Bans Single-Use Plastic Bags from Markets and Malls," Reuters, January 7, 2020, reuters.com/article/us-indonesia-environment-plastic-idUSKBN1Z612H.
9   "No Money for Your Bus Ride? Pay with Plastic Instead," *Euronews,* October 25, 2018,

euronews.com/2018/10/25/no-money-for-your-bus-ride-pay-with-plastic-instead#:~:text=In%20Indonesia%2C%20city%20buses%20accept,cups%2C%20depending%20on%20their%20size.

10   Ann Leonard, "Where Do All Our 'Recycled' Plastics Go? A Greenpeace Investigation," Greenpeace, September 4, 1992, scorcher.org/~jym/greenpeace/where-recycled-plastics-go.html.
11   Annie Leonard, personal conversation, November 18, 2023.
12   Michael Richardson, *International Herald Tribune*, "The Trashing of Indonesia," *New York Times*, November 24, 1993, nytimes.com/1993/11/24/business/worldbusiness/IHT-the-trashing-of-indonesia.html.
13   Plastic Soup Foundation, *A Neocolonial Plastics Scandal* (Plastic Soup Foundation, 2021), plasticsoupfoundation.org/wp-content/uploads/2022/09/PSF220915_A-neocolonial-plastics-scandal.pdf.
14   "Indonesia Plans New Rules to Curb Jump in Imports of Plastic Waste," Reuters, July 27, 2019, reuters.com/article/us-indonesia-environment-plastic-idUSKCN1UL1E4.
15   Jindrich Petrlik, Yuyun Ismawati, Joseph DiGangi, Prigi Arisandi, Lee Bell, and Björn Beeler, *Plastic Waste Poisons Indonesia's Food Chain* (IPEN, Nexus3, Arnika, and Ecoton, November 2019), 6, ipen.org/sites/default/files/documents/indonesia-egg-report-v1_9-web.pdf.
16   Melissa Hellman, "Indonesia Now Has the Highest Rate of Deforestation in the World," *Time*, July 1, 2014, time.com/2944030/indonesia-now-has-the-highest-rate-of-deforestation-in-the-world/.
17   Iman Nurifa, personal conversation, September 22, 2022
18   See "Annual International Trade Statistics by Country (HS)," Trend Economy, accessed October 22, 2022, trendeconomy.com/data/h2/Indonesia/4707.
19   Petrlik, Ismawati, DiGangi, Arisandi, Bell, and Beeler, *Plastic Waste Poisons Indonesia's Food Chain*.
20   Daru Setyorini, personal conversation, September 24, 2022.
21   In the 1950s, the waste chronicler Vance Packard lamented the state of affairs whereby a mounting number of Americans — the cashier operating the supermarket register, the stockbroker pitching shares of Coca-Cola — had stakes in the perpetuation of a petrochemical economy that was despoiling their communities with unnecessary garbage. In Java today, it's hard not to witness something similar: a broad spectrum of Indonesians — politicians, customs agents, paper-factory managers, mayors, truck drivers, waste pickers, tofu manufacturers — with stakes in sustaining a subterfuge plastics economy that happens to be injecting titanic amounts of poison into their food and river systems.
22   Teguh Suyono, personal conversation, September 13, 2022.

37. A TRASH SCION

1   Hans Antlöv, *Exemplary Centre, Administrative Periphery: Rural Leadership and the New Order in Java* (Curzon Press, 1995).
2   Abah Yasan, personal conversation, September 10, 2022.
3   Mohammed Ikhsan, personal conversation, September 19, 2022.

NOTES

4 Yuyun Ismawati, personal conversation, March 1, 2023.
5 Jindrich Petrlik, Yuyun Ismawati, Joseph DiGangi, Prigi Arisandi, Lee Bell, and Björn Beeler, *Plastic Waste Flooding Indonesia Leads to Toxic Chemical Contamination of the Food Chain* (IPEN, Arnika, Nexus3, and Ecoton, December 2019), 4, ipen.org/sites/default/files/documents/indonesia-egg-report-long-v1_2web-en.pdf.
6 Richard C. Paddock, "Indonesia Lets Plastic Burning Continue Despite Warning on Toxins," *New York Times*, December 19, 2019, nytimes.com/2019/12/19/world/asia/indonesia-dioxin-plastic-tofu.html.

38. BACK TO THE PACIFIC

1 Prigi Arisandi, personal conversation, September 22, 2022.
2 Muhammad Basyaiban, personal conversation, September 22, 2022.

CONCLUSION: WHITHER WASTE?

1 *Waste Export Control: Hearing Before the Subcommittee on Transportation and Hazardous Materials of the Committee on Energy and Commerce*, US House of Representatives, 101st Congress, 1st Session on HR 2525 (US Government Printing Office, 1989), 180.
2 Zakariah Njoroge, personal conversation, March 2, 2022.
3 "One in Every Two Plastic Bottles Waste in Kenya Is from Coca-Cola," Clean Up Kenya, August 29, 2021, cleanupkenya.org/one-in-every-two-plastic-bottles-waste-in-kenya-is-from-coca-cola/.
4 "Kenya Plastic Bag Ban: Three Fruit Sellers Arrested in Nairobi," BBC.com, February 18, 2020, bbc.com/news/world-africa-51546277.
5 Alice Delemare Tangpuori, George Harding-Rolls, Nusa Urbancic, and Ximena Purita Banegas Zallio, *Talking Trash: The Corporate Playbook of False Solutions to the Plastic Crisis* (Changing Markets Foundation, September 2020), 136, talking-trash.com/wp-content/uploads/2020/09/TalkingTrash_FullReport.pdf.
6 Hiroko Tabuchi, Michael Corkery, and Carlos Mureithi, "Big Oil Is in Trouble. Its Plan: Flood Africa with Plastic," *New York Times*, August 30, 2020, nytimes.com/2020/08/30/climate/oil-kenya-africa-plastics-trade.html.
7 Judith Lewis Mernit, "As Plastics Keep Piling Up, Can 'Advanced' Recycling Cut the Waste?" *Yale Environment 360*, June 1, 2023, e360.yale.edu/features/advanced-plastics-recycling-pyrolysis.
8 Shi Yi, "Kenyan Coal Project Shows Why Chinese Investors Need to Take Environmental Risks Seriously," *Dialogue Earth*, March 9, 2021, dialogue.earth/en/energy/lamu-kenyan-coal-project-chinese-investors-take-environmental-risks-seriously/.
9 Kaluki Paul Mutuku, personal conversation, October 10, 2023.
10 Tom Udall, Alan Lowenthal, Jeffrey A. Merkley, Steve Cohen, Earl Blumenauer, and fifty-seven other US Congress members, in a letter sent to the President of the United States, October 1, 2020, agoa.info/images/documents/15799/bicameral-letter-to-potus-us-kenya-free-trade-agreement-and-plastic-pollution-10-1-201.pdf.

NOTES

11 Ihekwoaba Chimezie, personal conversation, October 11, 2023.
12 Oliver Milman, "World's Largest Plastics Plant Rings Alarm Bells on Texas Coast," *The Guardian*, December 26, 2017, theguardian.com/environment/2017/dec/26/worlds-largest-plastics-plant-rings-alarm-bells-on-texas-coast; Reid Frazier, "Shell's Ethane Cracker, a Mammoth Plastics Plant near Pittsburgh, Begins Operations," *StateImpact Pennsylvania*, November 15, 2022, stateimpact.npr.org/pennsylvania/2022/11/15/shells-ethane-cracker-a-mammoth-plastics-plant-near-pittsburgh-begins-operations/; and Beth Gardiner, "Amid Hopes and Fears, a Plastics Boom in Appalachia Is on Hold," *Yale Environment 360*, April 13, 2020, e360.yale.edu/features/plans-to-make-appalachia-a-plastics-hub-face-growing-hurdles.
13 *The Truth Behind Trash: The Scale and Impact of the International Trade in Plastic Waste* (Environmental Investigation Agency, September 2021), eia-international.org/wp-content/uploads/EIA-The-Truth-Behind-Trash-FINAL.pdf.
14 Jindrich Petrlik, Yuyun Ismawati, Joseph DiGangi, Prigi Arisandi, Lee Bell, and Björn Beeler, *Plastic Waste Poisons Indonesia's Food Chain* (IPEN, Nexus3, Arnika, and Ecoton, November 2019), ipen.org/sites/default/files/documents/indonesia-egg-report-v1_9-web.pdf.
15 Hauke Engel, Martin Stuchtey, and Helga Vanthournout, "Managing Waste in Emerging Markets," McKinsey Sustainability, February 17, 2016, mckinsey.com/capabilities/sustainability/our-insights/managing-waste-in-emerging-markets.
16 See "ESG Perspectives: Must-Read ESG Insights and Much More," Capital Group, accessed August 14, 2023, capitalgroup.com/intermediaries/dk/en/investments/esg.html; "Sustainability Hub," HSBC UK, accessed August 14, 2023, hsbc.co.uk/sustainability/; and Dominic Charles and Laurent Kimman, *Plastic Waste Makers Index* (Minderoo Foundation, 2023), 12, cdn.minderoo.org/content/uploads/2023/02/04205527/Plastic-Waste-Makers-Index-2023.pdf.
17 Leonie Cater and Louise Guillot, "To Ban or Not to Ban: Fixing the EU's Global Plastic Waste Mess," *Politico*, October 19, 2023, politico.eu/article/ban-fix-eu-pollution-plastic-waste-myanmar/.
18 "Dubai's USD 1bn Waste-to-Energy Facility," Ramboll, accessed July 20, 2023, ramboll.com/en-us/projects/energy/dubai-s-usd-1bn-waste-to-energy-facility.
19 Joseph Poux, personal conversation, November 25, 2023.
20 See "Thanks to Our Partners We Can Clean the Oceans," The Ocean Cleanup, accessed October 10, 2023, theoceancleanup.com/partners/.
21 İzzettin Akman, personal conversation, February 16, 2023.

INDEX

Note: Italic page numbers refer to illustrations.

Abidjan, Ivory Coast, 95, 104
Aboabo, Ghana, *105,* 170–71
Accra, Ghana. *See also* Agbogbloshie
   (market slum of Accra)
  Abeka Road in, 115
  Ablekuma Central District Hospital
    and, 157
  as British capital, 126
  internet access and, 134, 135
  Kofi Annan Centre of Excellence, 134
  Kwame Nkrumah Circle, 167–68
  map of, *105*
  monuments to independence, 103–4,
    107, 127
  new iPhone economy in, 168–69
  press coverage of bulldozing of
    Agbogbloshie, 157
  ring roads of, 104, 167
  street bazaars of, 114, 167–68
Adana, Turkey, 3–6, *5,* 12, 13
Addis Ababa, Ethiopia, 131
Admiralty Pacific Group, 55–57
Afghanistan, 179
African countries. *See also* West Africa;
  and specific countries
  colonial legacy of, 132, 150
  developmental aid for, 135
  developmental project for, 60, 131–33
  hazardous waste shipped to, 59–60, 86

  incentives for accepting waste, 44
  plastic waste shipped to, 9, 285–86, 287
  as repository of excess capital, 134–35
  scrap metal trade and, 182
  toxic waste dumping in, 54
  wars of, 60
African Development Forum, 131–32
Agbogbloshie (market slum of Accra)
  aluminum in, 111
  appliances in, 109–11, 113, 116
  bola fires of, 118–19, 120, 121, 122, 124,
    125, 145, 155, 157–58
  browser boys of, 106, 159, 160–66,
    168–74, 184
  copper in, 111, 115, 118–19, 120, 121
  dismantlers in, 110–11, 115, 116, 117–18,
    142, 154, 159
  electronics in, 104, *105*–6, 108–11, 112,
    113, 115, 117–18, 121–22, 124, 127,
    136, 141, 145–46, 148–49, 150, 151,
    169
  as flexible mine, 148–49
  government's bulldozing of, 156–58
  Guinea Fowl War and, 130, 142
  juxtaposition with monuments of
    Accra, 107
  Korle Lagoon and, *105,* 107–8, 111, 117,
    120, 121, 125–26, 130, 142, 158, 159,
    160, 164

INDEX

Agbogbloshie (market slum of Accra) (cont.)
  migration to, 142, 145, 154
  Old Fadama, 105, 108, 111, 116, 118, 120, 121, 122, 123–24, 126, 130, 145, 157, 158, 160, 162
  Onion Market, 105, 108, 120, 123, 130
  tribal hierarchy in, 117–18, 130, 156
Ahmed, Brahim Ould Cheibani Ould Cheikh, 45–46
A. H. Robins Inc., 65, 66
Aikpé, Michel, 48
Akan tribespeople, 129
Akin, Fatih, 7
Akman, İzzettin, 3–6, 8, 11–13, 19, 98, 336–38
Akosombo Dam, Ghana, 105, 127–28
Akufo-Addo, Nana, 157
Akwamu tribespeople, 126
Alang Ship Breaking Yard, India, 189, 200, 205
Ali, Sugule, 62
Aliağa, Turkey, 207–15, 218, 224–25, 226, 227–35, 237, 241
Aliağa Ship Breaking Yard, Turkey, 189, 192–94, 196, 197, 228–29
Allotey, Francis, 135
Alpi, Ilaria, 61–62
aluminum, 111, 127
American Chemistry Council, 263, 267, 277, 329
American Iron and Steel Institute, 181
American Petroleum Institute, 262
American Plastics Council, 265, 66
Amin, Idi, 60
Angola, 60, 62
Apple, 24–25
appliances, 109–11, 113–14, 116, 118–19
Arana, Roberto, 71–73, 74
Árbenz, Jacobo, 65
Argentina, 61
Arisandi, Prigi, 322–24
Arslan, Turan, 196
asbestos, 43, 53, 95–96, 194, 201
Astor, 224
Asturias, Francisco, 71–72
Atatürk, Kemal, 219

Athens, Greece, 177–78
Australia, 129, 151–52, 289
Awal, Mala, 117, 123
Awal, Mohammed
  ambitions of, 123
  on bulldozing of Agbogbloshie, 157
  burning work of, 117, 119–20, 121, 122, 123, 128, 155, 158, 160, 161
  cell phones of, 117, 119
  daily wages of, 121, 145, 154
  dismantling work of, 116, 117–20, 150, 154–56
  family village in Tamale, 122–23, 126, 128, 129, 130, 156
  health of, 124
  house of, 117, 159, 160
  interviews with journalists, 122
  literacy and, 164
  wire transfers and, 155–56, 167

Bahamas, 90, 186–87
Bal, Veli, 196
Ban Amendment, 90
bananas, 30–32, 49, 51, 64–65
Bangladesh, 202, 204, 205, 214, 218
Bangladesh Environmental Lawyers Association, 204
Bangun, Java, 312, 313–21, 324
Bank of Ghana, 129
Barros, Filinto, 84
Barthes, Roland, 255
Basel Action Network, 187, 280
Basel Convention on the Control of Transboundary Movements of Hazardous Wastes and Their Disposal, 87–88, 90–91, 93–96, 135, 187–88, 280, 293
Basyaiban, Muhammad, 325
Becnel, Thomas, 262
Belarussiya, 230
Belgium, 9, 53, 296
Belize, 34
Benin, Republic of, 47–48, 49, 50, 131, 132
Berlin Wall, fall of, 21, 87, 268, 274
Bermuda, 77
Bezos, Jeff, 148

INDEX

Bidid, Ilyas, 196
Bikini Atoll, 57
Blacken, John, 84
black flag nations, 200
BlackRock, 335
Blade Maritime Recycling, 233
BMS Gemi Söküm, 234
Boorstin, Daniel, 253
Bozoğlu, Baran, 285
Bradford, Anu, 213
Brazil, 90, 147
Brenner, Mark, 73
Brexit, 10
Brownell, Emily, 50–51
browser boys, 106, 159, 160–66, 168–74, 184
Brussels effect, 213
Bucharest, Romania, 284–85
Buchert, Matthias, 106–7
Bulgaria, 274
Bureau of Waste Management, 257
Burke, Bill, 193
burner boys, *bola* fires of, 118–19, 120, 121, 122, 124, 125, 145, 155, 157–58, 159, 164
Büyüktemiz, Mehmet, 61

Campaign for the Right to Repair, 140
Canada
    electronics recycling in, 113, 114
    Ghana's mineral resources and, 129
    global waste trade and, 22, 91
    hazardous waste received from US, 45
    plastic waste shipped from, 10, 287
capasta. See copper
Capital Group, 335
carbon emissions, 144–45, 181, 186
Caribbean Conservation Association, 78
Caribbean Sea, 18, 202
CARICOM, 89–90
Carnival Cruise Line, 193, 196, 210–11, 212, 215, 232
Carnival Inspiration, 192–97, 202, 213–14, 215, 232, 234
Carnival Jubilee, 215
Carson, Rachel, 36–38, 42, 65, 98, 333
Carteaux, William, 266

Çavuşoğlu, Mevlüt, 8
Celluloid Manufacturing Company, 250
Central America
    hazardous waste shipped to, 61, 66, 74
    Martina Langley's field work in, 65
    plastic waste shipped to, 284, 287
Ceyhan Mega Petrochemical Industry Zone, 13
chemicals
    disposal of, 38–40, 45, 65
    proliferation of, 36–37
    selling of, 39–41
Cherokee Nation, Oklahoma, 79
Chetek Corporation, 275
Chile, 145
Chimezie, Ihekwoaba, 331
China
    ban on plastic waste, 281, 288, 291, 298
    petrochemical industry in, 281–82
    plastic waste shipped to, 8–9, 10, 182, 269, 271–73, 274, 275–82, 283, 286
    rare earth element mining in, 145, 147
    scrap metal trade in, 182
    shipbreaking and, 189
China Ocean Shipping Company, 238
Chinese Communist Party, 8–9
Chios, 237–40
Chiquita Brands International, 31
Chittagong Ship Breaking Yard, Bangladesh, 189, 198
climate change. *See also* environmental damage
    electronic waste and, 144, 155
    evolution from extractivist economy and, 180
    failures in addressing, 332
    Greenpeace on, 96
    Marshall Islands and, 56
    shipping industry and, 240
    trash disposal and, 23–24
cobalt, mining of, 144–45, 150–51
Çoban, Gökhan, 235–36
Cocolí, Guatemala, 32–35, *33*
Colbert, Charles, 39–41, 42, 43, 45, 52, 92
Colbert, Jack, 39–41, 42, 43, 45, 52, 92

INDEX

Cold War
  borders in post–Cold War era, 91
  consumerist culture and, 17, 85
  embargoes of, 240
  global waste trade in post–Cold War era, 20, 23, 52, 53–54
  Guatemala and, 66, 68
  hazardous waste shipments and, 58
  Turkey and, 209
Columbus, Christopher, 237
Commoner, Barry, 256–57
Comoros, 200
Congo, Democratic Republic of, 58, 59, 145, 150–51
Connolly, Hugh, 257
Consumer Reports, 97
contraceptive devices, US export of, 65–66
Conyers, John, Jr., 81–82
copper
  in Agbogbloshie, 111, 115, 118–19, 120, 121, 124, 146, 150, 158, 160, 164
  mining of, 144
  in South America, 52
copyrights, patents compared to, 139–41
corporate social responsibility (CSR), 302–3
Cortés, Hernán, 69
Coşan, Mustafa, 196
Costa Rica, 77, 89
Costa Victoria, 224
cruise ships, 99, 192–93, 196, 210, 224, 232
Cuban Missile Crisis, 256
Cultivators of the West, 79
Czechoslovakia, 86

Dagomba tribespeople, 118, 123, 130, 156, 159
Dalí, Salvador, 61
Dalkon Shield products, 65–66
Dar es Salaam, Tanzania, 151, 286–87, 288
DDT, 37, 38, 39, 40, 42
de Grazia, Victoria, 17
de Guzmán, Lilian Vásquez, 34–35
Dell, Jan, 98, 264
Dell Computers, 136
Delphin, 230

Demir, Yılmaz, 195
Demirgil, Orhan, 233–34
Denmark, 53
developing countries
  capital for industrialization of, 51, 54, 55
  decolonization movement and, 54, 85, 103, 253–54
  economies of, 50–51, 52, 53
  environmental regulations and, 49, 50, 52, 95
  hazardous waste shipped to, 45
  industrialization of, 95
  infrastructure needs of, 50
  labor rights in, 95
  plastic waste trade and, 289
  recyclable materials and, 97–98
  structural adjustment reforms for, 52
  toxic waste shipped to, 54, 81–82, 84, 85
  trash shipped to, 44
  urbanization of, 142
  waste defined as resource for, 96, 97–98
dioxins, 98, 106, 127
dismantlers, in Agbogbloshie, 110–11, 115, 117–18
Doğan, Fatih, 285
Dominican Republic, 77
Douglas, William O., 37
Doymaz, Cemal, 196
drug trade, 208
Dubai, United Arab Emirates, 180, 335
DuPont, 252
Duterte, Rodrigo, 287
Du Vivier, François Marie Gabriel André Charles Ferdinand Roclants, 83–84
Dzidonu, Clement, 133

East Africa, 183
East Asia, 52
Eastern Bloc, 53–54
Eastern Europe, 132, 274, 275
East Germany, 53–54
ecological imperialism, 48
ECOTON (Ecological Observation and Wetland Conservation), 322–23
ECOWAS (Economic Community of West African States), 90
Ege Çelik shipping yard, 214

INDEX

electronics
  in Agbogbloshie, 104, 105–6, 108–11, 112, 113, 115, 117–18, 121–22, 124, 127, 136, 141, 145–46, 148–49, 150, 154–55
  burner boys and, 118–19, 120, 121, 124
  chemical solvents from, 121
  copper extraction and, 119, 120, 121, 150
  lifespan of, 140
  raw materials of, 115, 128, 146–47
  recycling of, 99, 109, 121–22, 136, 144
  secondhand electronics trade, 134–36, 137, 141, 142, 144, 152–53, 163–64, 169, 174
  in Tema, 113–14
  toxicity of, 98, 122, 124, 127
electronic waste trade
  European Union and, 144
  in Ghana, 109, 112–13, 118, 121–22, 125, 128, 131, 143, 144, 150–51, 156
  growth of, 146–47
  in Israel, 152–53
  reusable materials of, 247
  tech industry expansion and, 184
  United States and, 144
Elizabeth II, queen of England, 104
Ellis, Stephen, 163
El Salvador, 61
environmental damage. *See also* climate change
  from chemicals, 36–41
  "economy versus environment" binary and, 86
  from electronic waste trade, 143, 144
  from global waste trade, 22
  from mining, 147–48
  from plastics, 257–58
Environmental Protection Agency, 40, 77, 79, 81, 89, 92
environmental standards, discrepancies in, 42, 43
Erdemir, Aykan, 222
Erdoğan, Emine, 6–8, 10–11, 209, 285
Erdoğan, Recep Tayyip, 13, 216–18, 222–24
Eritrea, hazardous waste shipped to, 58
Escobar, Pablo, 159–60, 169
Ethiopia, DDT sent to, 40
European Anti-Fraud Office, 26
European Commission, 181, 215
European Communities, 84, 90
European Food Safety Authority, 106
European Union
  bribes for acceptance of waste from, 45–46
  colonialism in waste trade and, 47, 90
  electronic waste trade and, 144
  hazardous waste shipped from, 59
  pharmaceutical waste from, 84
  plastic waste shipped from, 9, 276, 294
  profit incentives for shipping trash from, 44
  scrap metal imported by, 181
  toxic waste dumping within, 94
  trash exported from, 4, 25
  Turkey's membership in, 216, 219
  Turkey's shipyards approved by, 211
exosphere, human-produced debris in, 19
Exxon Valdez, 185

Fanon, Frantz, 54
Federal Environmental Pesticide Control Act of 1972 (FEPCA), 38, 39, 43
Federation of American Scientists, 82
Felicia, 79
fertilizers, 66
Fleming, Dan, 56
flexible mines, 148–49
Ford, Cristina, 61
Ford, Henry, II, 61
fossil fuel industry, 335
Frafra tribespeople, 111, 118, 119–20, 121, 130, 156
France
  colonialism in waste trade and, 47–48, 49
  industrial residue received from West Germany, 45
  nuclear weapons testing of, 88
  plastic waste shipped to, 9
  radioactive waste in Benin and, 49, 131
Freetown, Sierra Leone, 59
Frey, Glenn, 37
Frey, R. Scott, 188–89
Friends of the Earth, on Rachel Carson, 37
Fruchter, Nathan, 25, 220–21

INDEX

Gadani Ship Breaking Yard, Pakistan, 189, 203
gadolinium, 149
Galanis, Konstantinos, 212
Galápagos Islands, 18
Gambia, the, 53, 85
garbage imperialism, 21–22, 24–25
Garzona, Erwin, 32
Ga tribespeople, 125
Gedangrowo, Java, 300–312, 313, 314, 316–21, 324
General Agreement on Tariffs and Trade (GATT), 86
germanium, 146
Germany
   Agbogbloshie field clinic and, 124
   electronics recycling in, 113–14
   plastic manufacturing and, 250–51
   plastic waste shipped from, 9, 294
   scrap metal trade in, 178
Gezer, Musa, 196
Ghana. *See also* Accra, Ghana
   ancestral lands of tribespeople, 108, 129–30, 145, 155, 156, 169
   Christian missionaries in, 171
   cocoa industry of, 127, 128, 129, 132, 133, 170
   communications sector of, 134
   economy of, 127–30, 131, 132, 133–34, 149
   electronic waste trade in, 109, 112–13, 118, 121–22, 125, 128, 131, 143, 144, 150–51, 156
   foreign domination of, 127
   gold-mining sector of, 129, 151, 163
   independence of, 107, 127, 128, 131
   internet access and, 133–34
   juju tradition in, 169–74
   map of, 105
   mining concessions in, 129
   population of, 133, 134
   Portuguese in, 125–26
   secondhand electronics trade in, 134–36, 137, 141, 142, 144, 163–64, 169, 174
   wages in, 154
   yam farming in, 123, 128, 129, 142, 145, 155, 159
Glencore, 25

Global Children's Campaign, 331
Global E-Waste Monitor, 143
Global North
   hazardous waste exported by, 89
   incentives for accepting waste from, 44
   raw materials from Global South and, 51
   secondhand electronics trade and, 141
   value of toxic waste shipped from, 54–55
   waste transfers with Global South, 293
Global South
   bridging digital divide in, 149
   campaign to ban global waste trade, 89
   chemicals shipped to, 38, 40, 41
   commodity exports and, 51–52
   contraceptive devices shipped to, 65–66
   debts of, 51, 52, 53
   development aid for, 54–55, 65–66, 81
   elites of, 36
   emerging new citizenries of, 84–85
   global waste trade and, 21–23, 36, 49, 51–53, 98, 283, 333
   hazardous waste shipped to, 49, 53, 70, 84–85, 231–32
   incentives for accepting waste from Global North, 44
   industrialization moving to, 95
   lack of production of industrial toxic waste, 45
   plastic waste trade in, 53, 254, 283–84, 318
   toxic waste dumping in, 49, 52, 53, 88
   value of toxic waste shipped to, 54–55
   waste management systems of, 52, 53
   waste transfers with Global North, 293
global waste trade. *See also* plastic waste trade
   Caribbean states against, 89–90
   colonialism and, 47–48, 49, 84, 90
   continuation of, 98–99
   as criminal enterprise, 10
   definitional differences within, 25–26
   Eastern European countries and, 284
   environmental damage from, 22
   exploitation in, 47
   Federal Environmental Pesticide Control Act of 1972 and, 38

INDEX

Global South and, 21–23, 36, 49, 51–53, 98, 283, 333
growth of, 21–23, 24
Guatemala and, 32–35, 36
human safety and, 197
interconnections of, 58–59
international regulation in 1990s, 96–97
justifications for, 49, 92, 332
legacies of, 50–51, 53, 54, 98
legislation on, 89, 186
metastasization of, 99
oil crisis of 1970s and, 50–51
paths of, 8–10, 11, 12, 23, 25–26
perpetuation of, 331–32
post–Cold War explosion of, 52, 53–54
post–industrial waste export, 96–97
"recyclable" materials and, 97
revenue of, 55
tracking of, 26, 88–89
trash as object of exchange and, 80
US Congress on, 91–93
waste as commodity and, 86–87, 96–97
Gobbato, Eugenio, 72
Göbekli Tepe, Neolithic mound of, 6
gold, 51–52, 115, 127, 146, 151
Gold Coast, 126–27, 128, 151
Gold Coast Government Railway, 170
Goldstein, Joshua, 281
Gonaïves, Haiti, 78, 79
Goode, Wilson, 79
Grandin, Greg, 68
granulators, for copper extraction, 119
Greece
 financial crisis of 2009–2017, 177–78, 240
 Philadelphia's trash ash shipped to, 79
 plastic waste shipped from, 9
 refugees and, 227
 scrap metal trade in, 177–78
 shipping industry and, 237–40
 Turkey and, 226, 227
green energy transition, 13, 144, 147, 334
Green Line, Israel, 152
Greenpeace
 on Rachel Carson, 37
 on environmental injustice, 96
 on global waste trade, 90, 98

International Waste Trade Project of, 92
 on plastic waste, 293
 on toxic waste dumping, 32, 74–75, 88–89
Greenspan, Alan, 179
Gross, Sidney, 258
Guangzhou, China, 182
Guatemala
 civil war in, 69
 colonialism in, 49
 contraceptive devices shipped to, 65–66
 global waste trade and, 32–35, 64
 the Petén in, 66–75, 67
 Philadelphia's trash ash shipped to, 77
 Río Motagua delta of, 18
 toxic waste dumping in, 32, 66–67, 70–75
 United Fruit Company and, 30–32, 64–65
 US weapons in, 66, 68, 70
Guatemala City, Guatemala, 29–30, 32, 33, 68, 69, 72, 73, 74
Guinea, 59
Guinea-Bissau, 77, 84
Guinea Fowl War, 130, 142
Guiyu, China, 152
Güleç, Ahmet, 196
Gulf of Guinea, 103–4, 105, 126, 169
Gültekin, Bülent, 223
Gündoğdu, Sedat, 12
Guyana, 44

Haiti, 40, 58, 77–78, 91, 92
Hanieh, Adam, 250
Hasan, Syeda Rizwana, 204
Haynes, Williams, 249–50
hazardous waste trade
 bribes for acceptance of, 45–46, 87
 continuation of, 333
 designation of, 43, 96
 dumping in East Africa, 183
 history of, 254
 industrialization of Global South and, 95
 prohibition of exports of, 90, 187, 248
 shipbreaking and, 188–89, 194, 201–3, 206, 224, 231–33

INDEX

hazardous waste trade *(cont.)*
   shipment costs of, 43–44
   shipment to Africa, 44
   shipment to developing countries, 45, 58
   shipment to Global South, 49, 53, 70
   shipment to Guatemala, 70
   shipment to West Africa, 54, 61
Herodotus, 228
holmium, 149
Honduras, 77, 95
Hong Kong, 152, 270–72, 277, 335
Hong Kong Convention, 188, 200
Hooker Chemical Company, 255
Horizon, 232
hospital waste, 43
HSBC, 335
human-made objects, 14, 15, 16
human trafficking, 26
Hussein, Saddam, 179
hydraulic fluids, 43

I. G. Farben, 250–51
Ikhsan, Hadji, 314
Ikhsan, Mohammed, 314–18, 324
incineration plants, 59–61
India
   computer training in, 134
   DDT sent to, 40
   global waste trade and, 90
   landfill east of New Delhi, 18
   plastics in cattle food, 107
   scrap metal trade in, 180
   shipbreaking in, 189, 200, 202, 204, 205–6, 209, 214
Indian Ocean, shipbreaking yards of, 198, 199, 205, 208, 211, 215
individual morality, 24
Indonesia
   ban on plastic waste, 291, 295
   corporate social responsibility and, 302–3
   DDT sent to, 40
   map of, *295*
   nickel mining in, 145
   paper for recycling and, 296–97, 298, 299
   paper mills in, 299

   plastic waste trade in, 246–48, 291, 294, 295, 297, 298–300
   trains in, 245–46
   trash towns of, 246, 248, 291–92, 295, 296, 300
   waste management in, 292–93
industrial residue, 21–22, 45
industrial waste, 44, 45
Ingenthron, Robin, 135
Institute for Applied Ecology, 106–7
Institute of Scrap Recycling Industries, 265, 279
Inter-Allied Malaria Control Group, 126
Inter-American Development Bank, 51
international hierarchies, in trash disposal, 23, 24, 86
International Maritime Organization (IMO), 186–88, 239
International Monetary Fund (IMF), 52, 86, 113, 128–29, 142, 156, 216, 219, 327
International Pollutants Elimination Network, 263
International Ship Recycling Association, 212
internet
   information revolution and, 132–34, 144
   scams of browser boys and, 160–62, 163, 164–66, 168–74, 184
Interpol, 26
Ionian Spirit, 230
Iran, 145
Iraq, 179
iridium, 151
Işıksan shipbreaking yard, 225
İskenderun, Turkey, 177
Ismawati, Yuyun, 318–19
Isparta, Turkey, 210
Israel, 152–53
Istanbul, Turkey, 217
Italy
   hazardous waste shipped from, 59, 61–63, 84, 85, 131, 151
   plastic waste shipped from, 9, 296
   "triangle of death" near Naples, 94
Ivory Coast, 95, 104, 128
İzmir, Turkey, 177, 224–25, 227–28, 285

INDEX

Jacobs, Jane, 180
Jadrolinija, 232
Jakarta, Indonesia, 292, 293, 294
Jamaica, 127
Japan, 57, 152, 181, 199, 281
Java
    ECOTON and, 322–25
    Pakerin in, 301–3, 307, 308, 309, 311, 313–15, 318, 319–20, 324
    paper for recycling and, 297
    plastic waste trade and, 246, 291, 293–96, 300
    trash towns of, 300–312, 313, 314–21, 324
Jenssen, Ingvild, 238
Jheisha, Abu, 153
Jobless Millionaires, 328
Jurong Port, Singapore, 79

Kabua, Amata, 55–57, 248
Kaiser, 127
Kasoa, Ghana, 170
Katowice, Poland, 275
Keep America Beautiful, 258
Kennedy, John F., 37, 127
Kenya, 151, 182, 327–30
Kenyatta, Uhuru, 182
Kepone, 42
Kérékou, Mathieu, 47–48
Kessler syndrome, 19
Khabarovsk, Russia, 180
Khian Sea, 32, 77–80, 91, 92, 289
Kicillof, Isidro, 61
King, Michael, 78
King Kong I, 205
Kissinger, Henry, 51
Klinger, Julie Michelle, 147
Knapp, Freyja L., 148
Koahsiung, Taiwan, 203
Koko, Nigeria, 85, 151
Konkomba tribespeople, 118, 119–20, 121, 129–30, 143, 156
Korle (goddess), 125–26
Korle Lagoon, Accra, *105*, 107–8, 111, 117, 120, 121, 125–26, 130, 142, 158, 159, 160, 164
Kosovo, 178

Kouwenhoven, John Atlee, 16
Krishna, Gopal, 200
Kuito, 231
Künzler, Arnold Andreas, 59–60
Kyrgyzstan, 147

Labban, Mazen, 178–79
Lagos, Nigeria, 104, 151
Laguna Verde Nuclear Power Station, Mexico, 75
Lake Izabal, Guatemala, 33, *33*
Lake Petén Itzá, Guatemala, 67, *67*, 68, 69, 70, 73–74
Lake Volta, Ghana, *105*, 127
Lambrecht, Bill, 62
Langley, Martina, 64–67, 70–71
Las Escobas, Guatemala, 32–35, 64, 70, 94
The Last Beach Cleanup, 98, 264
Latin America, 47
Lebanon, 58
Lebow, Victor, 16
Leonard, Annie, 89, 92, 151–52, 293–94, 320
Les Amis de la Nature, 79
Lesvos, 227
Lianjiao, China, 279
Liberia, 9, 44, 186, 187
Liberty Mutual, 134
light-bulb companies, 138
Li, Ka-shing, 270
Lincoln, Sarah, 134–35
lithium mining, 145
Litvinov, Aleksey, 48
London International Exhibition, 250

Machakos, Kenya, 330
McKinsey & Company, 334
Maersk, 212–13, 336
Mailer, Norman, 256
Malaysia, 9–10, 288–89, 292
Maldives, 18–19
Malta, 186
manatees, 33
Mandela, Nelson, 133
Maoism, 270
Marc Rich/Glencore, 220
Marella Dream, 232

INDEX

Mariana Trench, 259
Marshall Islands, 55–57, 58, 62, 186, 298
Marshall Plan, 17, 253
Maya community, 64, 66–71, 73
MD Alpine, 198–99
Mediterranean Shipping Company, 212
Megastar, 232, 234
Mehmet, Hacı, 225
Mennonites, 68
Mesopotamia, 6
Metas shipping yard, 214
Mexico, 60, 75, 91
microplastics
    effects of plastic burning on, 5
    in fish, 258–59
    measurement of, 98
    on Mount Everest, 259
    plastic recycling and, 280, 325
    in river systems, 290
Mikelis, Nikos, 188
mining economies
    flexible mine and, 148–49
    restructuring of, 144–49, 150
mining tailings, 43
Mirador-Río Azul National Park, Guatemala, 72
Mitchell, Joni, 37
Mngcweni River, South Africa, 59, 131
Montkowski, Marianne, 54
Moore, Charles, 259
Moore, Patty, 25, 277
morality performance, 22
Morocco, 47, 58
Moruroa, atoll of, 90
Mtai, Kevin, 286
Mugabe, Robert, 54
Muhammad, Ali Mahdi, 61
Mulinaris, Nicola, 200–201
Mutuku, Kaluki Paul, 330–31

Nairobi, Kenya, Dandora, 327–29
Nakpayili, Ghana, 129–30
Namibia, 59–60
Nanumba tribespeople, 129–30
National Plastics Exposition, 252
National Soft Drink Association, 258
Nations, James, 64, 66

Native American reservations, US nuclear waste stockpiled in, 94–95
NATO, 209, 226
natural resources, appropriation of, 21, 23
Navarro, Ricardo, 284
Nepal, DDT sent to, 40
Nestlé, 24–25
Netherlands
    global waste trade and, 53
    Gold Coast and, 126
    plastic waste shipped from, 9, 12, 274, 294, 304
    waste trade and, 22
Netherlands Antilles, 44–45, 77
Nevis, 200
New Jersey, 131
Nexus3, 298
Niarchos, Stavros, 239
nickel mining, 145
Nigeria
    Benin compared to, 47
    British control of, 163
    electronic waste market in, 109, 151
    hazardous waste shipped to, 59, 62–63, 85, 131, 331
    hydrocarbon discoveries in, 163
    scamming culture of, 163
    scrap dealers of, 119
Niță, Teodor, 25, 285
Njoroge, Zakariah, 328
Nkrumah, Kwame, 103–4, 113, 132–33
Northern Ireland, 87–88
North Korea, 152
Northrop Grumman, 104
Norway
    beached goose-beaked whale in, 18
    electronics recycling in, 114
    hazardous waste shipped from, 59
    shipping industry and, 203
nuclear waste, 57, 94–95, 228
Nurifa, Iman, 298–99

Obama, Barack, 322
Ocean Cay, Bahamas, 77
The Ocean Cleanup, 336
oceanic plastic pollution, 18, 293, 334
Odaw River, Ghana, 125–26

INDEX

Oduche, Sam, 62–63
oil crisis of 1970s, 50–51
oil prices, boom in, 50–51
Oktyabrskaya, Russia, 179
Omatseye, Sam, 85
Önal, Kamil, 212
Onassis, Aristotle, 239
Onion Market, Agbogbloshie, *105,* 108, 120, 123, 130
Oregon, 55–56
Organisation for Economic Cooperation and Development, 209
Organization of Arab Petroleum Exporting Countries, 50
organized crime networks, 20, 61, 335, 336
Osibanjo, Oladele, 85, 151
Otopan, 231
Ottoman Empire, 210, 218, 223
Özal, Semra, 209–10
Özal, Turgut, 219

Pacific Princess, 231
Packard, Vance, 14–17, 19, 36–37, 137–38, 332
paint sludge, 43, 61
paint solvents, 95–96
Pakistan, shipbreaking in, 202, 204, 205, 214
Palau, 200–201
Palestinians, informal waste economy and, 152–53
palladium, 146
palm oil, 50, 51
Pan-African unity, 132–33
Panama, 186, 187
Panama Canal, 276
Paris Agreement, 144–45
Parkes, Alexander, 250
patents, copyrights compared to, 139–41
Paul, Jean-Claude, 78
Pearce, David, 87
Pearson, Wayne, 266
Peker, Sedat, 208
Pelicano, 79
Pentagon, 39
Pearl and Hermes Atoll, 258
Pericles, 238

Perks, Harry M., 77
Peru, 44
pesticide residues, 61
pesticide trade, 89
petrochemical industry
  on advanced chemical recycling, 330
  in China, 281–82
  in Chios, 238
  disposal of plastic and, 261, 274, 333
  global waste trade and, 248
  growth rate of, 256
  incineration of plastic and, 258
  plastic manufacturing and, 249–50, 253, 254, 256–57, 266, 283, 330, 332, 333–34
  profit model of, 336
  recycling of plastic and, 262, 263, 264, 265–66
  in Turkey, 13, 228
  in United States, 251–52
petrodollar recycling, 51
pharmaceutical waste, 84
Philadelphia, Pennsylvania, trash ash waste shipment from, 76–81, 91, 92, 289, 290
Philippines
  electronic waste shipped to, 151–52
  nickel mining in, 145
  plastic waste shipped to, 10, 286, 287
PIERS (Port Import/Export Reporting Service), 88
Piraeus, Greece, 238–39
planetary mine, 178, 179, 180, 182. *See also* scrap metal economy
Planty, Donald J., 72
plastic burning
  in Accra, 118
  burner boys and, 118, 119–21, 125
  in China, 279
  in Indonesia, 297
  near Adana, Turkey, 3–4, 6, 12
  recycling and, 98
  toxic effects of, 5
  for Turkish cement factories, 11
Plastic China (documentary), 280–81
plastic consumption, promotion of, 255, 259–60

385

INDEX

plastic manufacturing
   growth of, 256–58, 331, 335
   history of, 249–50, 252–55
   in Turkey, 13
plastic recycling
   economics of, 263
   feasibility of, 261–66, 267, 268, 282, 283, 331
   GPS chips for tracing of, 12
   promotion of, 264–65, 269, 330
   rate of, 180
   toxins and, 263–64, 294, 318–19, 325, 333
   in Turkey, 7, 11
   types of plastic and, 262, 272, 277, 284
Plastics Recycling Foundation, 266
plastic waste trade
   in Central America, 284
   in China, 8–9, 10, 182, 269, 271–73, 274, 275–82, 283, 286
   consequences of, 184, 333–34
   in Eastern Europe, 274
   economic opportunity and, 87
   in Global South, 53, 254, 283–84
   in Indonesia, 246–48
   in Java, 246, 291, 293–94
   in Netherlands, 9, 12, 274
   in Northern Ireland, 87–88
   in Philippines, 10, 286
   in Poland, 9, 12, 275
   in Russia, 275
   statistics on, 26
   toxicity of, 98, 99, 181, 334
   in Turkey, 8, 10, 11, 13, 181, 273, 285
   in United States, 9, 10, 016
plastiglomerates, 259
platinum, 146, 151
Poland
   plastic waste shipped to, 9, 12, 275, 284, 335
   scrap metal trade and, 220
   Western multinationals in, 86
Polluters Out, 331
polychlorinated biphenyls (PCBs), 43, 59, 85, 131, 187, 202, 331
polycyclic aromatic hydrocarbons, 187
polyethylene, 13, 262, 277, 285
polyethylene bags, 7, 253

polyethylene terephthalate, 24
polypropylene, 262
polystyrene products, 253, 262
polyvinyl chloride (PVC), 118, 119, 120, 251, 262, 272, 273
Porter, John Edward, 289
Portland, Oregon, 180
Port of Los Angeles, California, 203–4
Poux, Joseph, 26, 335
Procter & Gamble, 24–25
Proctor, Nathan, 140
promethium, 149
propane dehydrogenation plant, 13
Public Interest Network, 140
Puckett, Jim, 187, 280
Puerto Barrios, Guatemala, 29–35, 33, 49–50, 64–65
Puerto Rico, 77
Pure Earth, 106, 119
Puthucherril, Tony George, 199

Quartey, Henry, 156
Quaynor, Nii Narku, 133–34, 136

radioactive waste, in Benin, 49
Rainbow Warrior (Greenpeace vessel), 88
Ram, Bonnie, 81–82
rare earth elements
   mining of, 144, 145, 147, 148–49, 150, 151
   myths concerning, 147
Rawlings, Jerry, 131
Reagan, Ronald, 331
recyclable materials
   burning of, 98
   marketing of, 97
recycled waste
   China's processing of, 8–9, 276
   overproduction and, 22
   profits from, 25
Reich, Charles, 256
Reilly, William K., 79, 92
Resource Conservation and Recovery Act (RCRA), 43
Rey Rosa, Magalí, 72
rhodium, 151
Ritter, Donald, 92

INDEX

Roberts, Jonathan, 126
Robinson, Arthur, 53
Rogers, Kenny, 59
Romania, 274, 284–85
Roma scrap metal dealers, 177, 178, 225
Roosevelt, Theodore, 30
Russia
    as petrostate, 145
    plastic waste trade and, 275
    scrap metal trade in, 179–80
ruthenium, 151

Saba, 45
Sahel, the, 59, 130, 159
St. Damian, 230
St. Kitts, 200
Salas, Bernardo, 75
Salazar, Carlos, 73–74
Samsung, 24–25
Saudi Arabia, 145
Schoenberg, Germany, 53
Schwartz, Norman B., 69
scrap metal economy
    environmental advantages of, 181, 333
    in Greece, 177–78
    redistribution of resources and, 179, 180, 247
    scrap hoarding and, 181–82
    shipbreaking process and, 182, 183–86, 187–88, 189, 192–97, 198, 199, 202–4, 207–8
    in Turkey, 218, 234
    worker hazards and, 183, 189, 192, 193–94, 196–97, 199, 202, 204–6, 212, 213–14, 229, 234, 235, 241
Sears Island, Maine, plastic waste on shoreline of, 87–88
Seattle, Washington
    plastic waste trade in, 294, 320
    scrap metal trade in, 180
Serrano Elías, Jorge, 34–35, 72
Setyorini, Daru, 298
sewage sludge, 43, 53, 64, 71, 94, 247
Sey, Omar, 53, 85
Shipbreaking Platform, 188, 238
shipping industry
    automation of, 203–4
    climate change and, 240
    environmental regulation of, 185–86, 187, 189, 196, 200, 211–13, 214, 215, 232–36
    future of, 237–40
    history of, 186–87, 199
    manufacturing and, 184–85
    regulation of, 186–87, 188, 193, 195
    shipbreaking and, 182, 183–84, 185, 187–89, 192–97, 198, 199, 200–208, 210–14, 231–33, 235–36, 238, 239
Ship Recycling Industrialists Association of Turkey, 212
Ship Recycling Regulation, 193
silver, 115, 146
Singapore, 79, 200, 208, 298, 335
Sivas, Turkey, 189, 190–91, *191*, 208, 235
Sky News, 279
Slater, Ransford, 126
Slovakia, electronics recycling in, 114
Slovenia, 273
smartphones
    activation locks on, 141
    carbon footprint of, 139
    deliberate obsolescence and, 138–39, 140, 141
    overproduction of, 144
    as software-reliant devices, 139–40
Smith, Cornelius, 89–90
Society of Plastics Engineers, 257
Society of the Plastics Industry, 97, 251, 252, 255, 260, 265, 266
Sohm, Matthew, 54–55, 209–10
solid waste, designation of, 43
Somalia, 61–62
Somaliland, 58
South Africa, 59, 151, 182
Southeast Asia
    export markets of, 132
    palm oil exports in, 50
    plastic waste shipped to, 288, 289, 291, 299
South Korea, 199
Soviet Communism, 16, 17

INDEX

Soviet Union
    fall of, 20, 21, 220, 230, 268
    rare earth mineral mining in, 147
    toxic waste trade and, 48
    US aid to developing countries and, 81
    West Africa and, 127
steel
    recycling economy for, 180–81
    scrap metal trade and, 179
Stouffer, Lloyd, 255
Straub, Roland, 59–60
Styrofoam packaging / insulation
    burner boys and, 118, 120
    toxicity of, 98
Sudan, 45
Suez Canal, 239
Sujan, Ariful Islam, 205
sulfide waste, 95
Summers, Lawrence, 86
Sünmez, Can, 196
Surabaya, Java, 245, 248, 293, 298–302, 306, 309, 311, 314
sustainable development, as growth model, 86–87
Suyono, Teguh, 304–5, 308–14, 319–20
Synar, Mike, 80–81, 82

Taiwan, 199, 205
Tamale, Ghana, *105*, 122–23, 126, 128–30, 156, 169–70
tantalum, 151
Tanzania, 151, 182
Taşkın, Enes, 192, 197, 241
Taşkın, Oguz, 192–97, 204, 209, 213–14, 218, 232, 241
Taşkın family, 191–92, 195, 197, 241
Tāufaʻāhau Tupou IV (king of Tonga), 44
technology industry
    deliberate obsolescence and, 138–41
    electronic waste and, 143–44
    overproduction in, 137, 144
    profits of, 140–41
Tel Aviv, Israel, 152
Tema, Ghana, *105*, 113–15, 119, 135, 136
Temurtaş, Mehmet, 225
Temurtaş Metal Hurda, 225

terbium, 149
Tesla, Nikola, 310
Thailand
    industrialization of, 95
    plastic waste shipped to, 9, 287, 335
Thilafushi island, landfill of, 18–19
Thomas, Larry, 97, 260, 265
Thompson, J. Walter, 255
Thompson, Richard, 259
throwaway culture, 137
thulium, 149
Togo, 132
Tonga, 44, 89
toxic terrorism, 21–22
toxic waste dumping
    in Africa, 54, 85
    in Benin, 48
    in Global South, 49, 52, 53, 97
    in Guatemala, 32, 66–67, 70–71
    in Guinea-Bissau, 84
    hiding of, 61–62, 66
    in Ivory Coast, 95
    in Marshall Islands, 57
    on Native American reservations, 94–95
    renaming of waste, 94
    tracking of, 83–84
    uneven environmental regulations and, 49, 50, 52, 60–61, 62
    of United States, 80–82, 89, 91–93
Trafigura, 25
trash disposal
    global problem of, 19–23
    management of, 23
    toxicity of, 23–24
trash economies, in Global South, 22–23
trash elimination, "zero-waste" initiatives, 6–7
Trinidad and Tobago, 53
Trump, Donald, 331
Tunisia, 9, 79
Turkey
    carbon footprint reduction in, 13
    earthquake of, 11
    economy of, 216–19, 221–23

INDEX

electric arc furnaces in, 219–20
Europe's proximity and, 209, 210
garment industry of, 11
Greece and, 226, 227
hazardous waste shipped to, 61, 210
lack of environmental regulation in, 209
landfills in, 7
map of, *191*
modernization program in, 219
networks of public fountains in, 6–7
nuclear waste in, 228
plastic waste dumpsite in southeastern Turkey, 8, 10, 11, 13, 181, 273, 285, 287, 336–38
refugees and, 227
scrap metal trade and, 182, 220–25, 229
shipbreaking in, 202, 204, 207–8, 211–14, 224, 229–34
Zero Waste Project in, 6–8, 10, 11, 209, 285
Turkish Wealth Fund, 13
Turner, R. Kerry, 87

Uganda, 135, 182
Ukraine, 287
Unilever, 24–25
United Arab Emirates, 182
United Fruit Company, 30–32, 64–65
United Kingdom
  Ghana and, 127
  Gold Coast and, 126
  hazardous waste shipped from, 59
  plastic waste shipped from, 10–11, 12, 279
  shipbreaking in, 199
United Nations, 26, 186–87
UN Capital Development Fund, 209
UN Economic Commission for Africa, 131–32
UN Environment Programme, 55
United States
  Basel Convention and, 91, 93
  chemicals exported from, 65
  colonialism in waste trade and, 47
  consumer culture in, 17–18, 19, 22, 36, 246
  contraceptive devices exported from, 65–66
  economic growth of, 15–17, 21
  electronic waste trade and, 144
  environmental movement in, 37, 38–39, 43, 86, 93, 257–58, 333
  forced consumption in, 16–17
  foreign policy of, 81–82
  globalization of economy, 92
  manufacturing sector of, 82, 91, 93, 97, 139
  nuclear weapons testing in Marshall Islands, 55, 57
  pharmaceutical waste from, 84
  plastic manufacturing in, 251–53
  plastic recycling in, 296
  plastic waste shipped from, 9, 10, 246–47, 294–95
  rare earth mineral mining in, 147
  scrap metal trade in, 180, 218
  shipbreaking in, 199
  toxic waste exports of, 80–82, 89, 91–93
  trade talks with Kenya, 329–30
  waste management in, 80–82
  waste production in, 16–18
US Agency for International Development, 40
US Chamber of Commerce, 275
US Congress
  on global waste trade, 91–93
  House Committee on Government Operations, 80–82
US Navy, 39
US State Department, 39, 104, 128, 277
US Treasury Department, 39
Utomo, Didik, 309

Vallette, Jim, 87–90, 92–93, 96
Vieira, João Bernardo, 84
Vietnam
  ban on plastic waste, 291
  plastic waste shipped to, 287
Vinyl Institute, 263
Volta River, Ghana, 127
voodoo traditions, 169

INDEX

Ward, Simon, 212
Washington, 55–56
waste. *See also* global waste trade
   as commodity, 86, 87, 96, 317–18
   defining of, 92, 96
   as resource, 96
   utilization of, 86, 87
waste brokers
   Agbogbloshie and, 105, 113
   incentives for exporting garbage, 10, 44–45
   licensing of, 10
   manipulation of regulations, 96
   Tema and, 114, 136
waste diversion, profits from, 45, 58
waste management firms, transference of industrial waste, 44–45
waste production
   consumption patterns and, 18, 20, 22, 332
   in United States, 16–17
waste trade. *See* global waste trade
water bottles, 7
water pollution, 5
West Africa
   declaration against global waste trade, 90
   electronic recycling in, 114–15
   foreign waste flows in, 131
   gold in, 52
   hazardous waste shipped to, 54, 61
   Soviet Union and, 127
   structural adjustment in, 128
   toxic waste dumping in, 95
   urbanization of, 104
West Bank, secondhand electronics in, 152–53
Western imperialism
   colonialism and, 31–32, 47, 52, 54, 84–85
   poorer countries exploited by, 20–21
Western Samoa, legislation against global waste trade, 89

West Germany
   bribes for acceptance of waste from, 45–46
   East Germans entering, 85
   hazardous waste shipped to Eastern Bloc from, 53
   hazardous waste shipped to Turkey from, 61, 210
   industrial residue shipped from, 45
William & Mary university, 136
Wilmot, Patrick, 54
Wong, Sheng Yun, 269–70, 272
Wong, Steve, 268–73, 274
Woods Hole Oceanographic Institution, 258
World Bank, 51, 69, 86, 127–28, 134, 142, 209, 293, 328
World Health Organization, 40, 106
World Reuse, Repair and Recycling Association, 135
World Trade Organization, 113
World War II, 126, 250–51, 253

Xerxes I, 227
Xi, Jinping, 281
Xia, Ying, 280
Xing, Demao, 278–79

Yasan, Abah, 314–15
Yıldırım, Binali, 208
Yıldırım, Ercan, 196
Yin, Yeo Bee, 289
Yom Kippur War, 50
Yugoslavia, 20, 79

Zaire, 45
Zara, Turkey, 191–92, *191*, 194, 208–9, 235
Zeff, Robert, 61
Zimbabwe, 39–40, 54, 131

ABOUT THE AUTHOR

ALEXANDER CLAPP is a journalist and writer based in Greece. His reporting has appeared in publications including the *New York Times, The Economist,* the *London Review of Books,* and *The Guardian Long Read.* He is the recipient of numerous journalism awards, among them a Whiting Creative Nonfiction Grant, Matthew Power Literary Reporting Award, Robert B. Silvers Foundation Grant, and Pulitzer Center Breakthrough Journalism Award. He has also received the Alistair Horne Visiting Fellowship at Oxford and a Berggruen Fellowship in Los Angeles.